Intelligent Recognition and Detection for Unmanned Systems

Intelligent Recognition and Detection for Unmanned Systems

Editors

Bo Li
Chunwei Tian
Daqing Chen
Ming Yan

MDPI

Basel • Beijing • Wuhan • Barcelona • Belgrade • Novi Sad • Cluj • Manchester

Editors

Bo Li
Northwestern Polytechnical University
Xi'an
China

Chunwei Tian
Northwestern Polytechnical University
Xi'an
China

Daqing Chen
London South Bank University
London
United Kingdom

Ming Yan
Communication University of China
Beijing
China

Editorial Office
MDPI AG
Grosspeteranlage 5
4052 Basel, Switzerland

This is a reprint of articles from the Special Issue published online in the open access journal *Drones* (ISSN 2504-446X) (available at: https://www.mdpi.com/journal/drones/special_issues/6BHXE2FU7K).

For citation purposes, cite each article independently as indicated on the article page online and as indicated below:

Lastname, A.A.; Lastname, B.B. Article Title. *Journal Name* **Year**, *Volume Number*, Page Range.

ISBN 978-3-7258-2223-2 (Hbk)
ISBN 978-3-7258-2224-9 (PDF)
doi.org/10.3390/books978-3-7258-2224-9

Contents

About the Editors

Bo Li

Bo Li is a professor and doctoral supervisor at Northwestern Polytechnical University. Born in July 1978 in Tai'an, Shandong, he holds a PhD in Systems Engineering and a postdoctoral degree in Electronic Science and Technology. He previously was a visiting scholar at the London South Bank University in the UK. Li is a committee member of the Youth Work Committee of the Chinese Association for Artificial Intelligence and the Specialty Committee on Artificial Intelligence and Security, as well as a member of the Mixed Intelligence Specialty Committee of the Chinese Automation Society and the Swarm Intelligence and Cooperative Control Specialty Committee of the Chinese Society of Command and Control. Additionally, he is a member of the Aeronautical Society, Computer Society, Electronics Society, and System Simulation Society. Li serves as an Editorial Board Member for the international journal Drones and Journal of Cyber-Physical-Social Intelligence, is the deputy Editor-in-Chief of the Chinese core journal Journal of Armament and Equipment Engineering and is a youth Editorial Board Member for Tactical Missile Technology. His main research areas include intelligent decision making in air combat, the intelligent command and control of unmanned systems, and the processing of uncertain information. He has led more than 30 national and provincial-level scientific research projects, received one second prize for the National Defense Science and Technology Progress Award, and one second prize for the Science and Technology Progress Award from Shaanxi Provincial Higher Education Institutions. Li has published over 100 academic papers, including more than 40 indexed by SCI and EI, applied for over 30 invention patents (with more than 20 authorized), and published three monographs and two textbooks.

Chunwei Tian

Chunwei Tian is an Associate Professor at the School of Software, Northwestern Polytechnical University. He also holds a Ph.D. in Computer Application Technology from Harbin Institute of Technology. His achievements include being listed among the top 2% scientists in the world in 2022/2023/2024. He has published over 70 papers in prestigious international journals and participated in conferences such as IEEE TNNLS, IEEE TMM, IEEE TSMC, IEEE TGRS, IEEE TCSVT, IEEE TIV, Pattern Recognition, Neural Networks, Information Sciences, Information Fusion, and ICASSP. Among these, six are ESI highly cited papers, three are ESI hot papers, four are cover papers in top journals, and five are included in the international Benchmark List for the super-resolution field. Furthermore, he serves as an Editorial Board Member for several SCI journals, including CAAI Transaction on Intelligence Technology and Dense Technology.

Daqing Chen

Daqing Chen is a Senior Lecturer in Data Science, the Deputy Head of Division of Computer Science and Informatics, and the course director for MSc Data Science in the School of Engineering, London South Bank University. He mainly teaches Data Mining and Analytics, and Introduction to Data Science. His research interests include deep learning algorithms and systems with applications in lip-reading, high-dimensional data embedding and visualization, medical image diagnosis, high-volume data labelling, and business intelligence. In recent years, he has conducted several business-oriented projects across a variety of industry sectors, including Machine-based Learning for Automatic Construction Cost Indexing (the Royal Institution of Charted Surveyors), M&E cost life-cycle and analytics (The Royal Institution of Charted Surveyors), big data/fact-driven decision-making systems (London Borough of Lambeth), RESEARCHER LINK: Smart Technology for Fighting Zika Virus Epidemics (remotely and highly distributed sensed data, analytics and visual sensing of micro-macro environment health welfare, British Council and FACEPE/FAPITEC Brazil), student retention analysis and prediction (London South Bank University), donor segmentation for not-for-profit organizations (The Muscular Dystrophy Campaign, UK), and consumer-centric business intelligence for online retailers (The Rex International, Ltd, UK).

Ming Yan

Ming Yan is a Professor and Doctoral Supervisor, Head of the National Quality Course, Young Top Talent in Communication University of China. He received his Ph.D. in Communication and Information Systems from Communication University of China in 2011. From March 2014 to March 2015, he conducted visiting scholar research in the Department of Electrical and Electronic Engineering at the University of Melbourne. His main research interests include high-efficiency communication networks, intelligent media computing, and multi-access edge computing. He has presided over more than 20 projects, including the National Natural Science Foundation of China General Projects, sub-topics of National Key R&D Programs, International Cooperation Projects, projects from the National Radio and Television Administration, China Internet Society, China Mobile Research Institute, and entrusted projects from enterprises and institutions. He has participated in more than 20 national, provincial, and other scientific research projects. He has published more than 60 academic papers, including more than 10 SCI journal papers as the first author or corresponding author. As an inventor, he has been granted six national invention patents. He has published one textbook and one monograph. So far, he has hosted the United Nations Internet Governance Forum several times (IGF2020, IGF2021, IGF2022, IGF2023), served as a Guest Editor for several international academic journals, and organized multiple international academic conferences.

Preface

Unmanned systems (i.e., droned, robots and other intelligent systems) play an important role in many fields, i.e., disaster relief, intelligent transportation, intelligent medical service and space exploration. Furthermore, object recognition and detection have been employed extensively in these tasks. However, due to complex application environments, artificial intelligence techniques suffer from challenges in terms of robustness and flexibility. Thus, designing efficient and stable CNNs and other AI algorithms for object recognition and detection in unmanned systems is critical.

Inspired by this, this Special Issue aims to collect the research accomplishments of researchers from academia and industry. It also aims to present recent research results in the field of deep learning for object recognition and detection, and understand how governance strategies can influence it.

Bo Li, Chunwei Tian, Daqing Chen, and Ming Yan
Editors

drones

MDPI

Article

Weld Seam Identification and Tracking of Inspection Robot Based on Deep Learning Network

Jie Li [1], Beibei Li [2], Linjie Dong [1], Xingsong Wang [1,*] and Mengqian Tian [1]

1 School of Mechanical Engineering, Southeast University, Nanjing 211189, China
2 No. 703 Research Institute of CSSC, Harbin 150028, China
* Correspondence: xswang@seu.edu.cn

Abstract: The weld seams of large spherical tank equipment should be regularly inspected. Autonomous inspection robots can greatly enhance inspection efficiency and save costs. However, the accurate identification and tracking of weld seams by inspection robots remains a challenge. Based on the designed wall-climbing robot, an intelligent inspection robotic system based on deep learning is proposed to achieve the weld seam identification and tracking in this study. The inspection robot used mecanum wheels and permanent magnets to adsorb metal walls. In the weld seam identification, Mask R-CNN was used to segment the instance of weld seams. Through image processing combined with Hough transform, weld paths were extracted with a high accuracy. The robotic system efficiently completed the weld seam instance segmentation through training and learning with 2281 weld seam images. Experimental results indicated that the robotic system based on deep learning was faster and more accurate than previous methods, and the average time of identifying and calculating weld paths was about 180 ms, and the mask average precision (AP) was about 67.6%. The inspection robot could automatically track seam paths, and the maximum drift angle and offset distance were 3° and 10 mm, respectively. This intelligent weld seam identification system will greatly promote the application of inspection robots.

Keywords: inspection robot; weld seam identification; wall-climbing robot; deep learning

Citation: Li, J.; Li, B.; Dong, L.; Wang, X.; Tian, M. Weld Seam Identification and Tracking of Inspection Robot Based on Deep Learning Network. *Drones* **2022**, *6*, 216. https://doi.org/10.3390/drones6080216

Academic Editors: Bo Li, Daqing Chen, Ming Yan and Chunwei Tian

Received: 11 July 2022
Accepted: 18 August 2022
Published: 20 August 2022

1. Introduction

With the extensive use of welding technologies in petroleum, bridges, ships and other fields, the use of large-scale welding devices, such as natural gas pressurized spherical tanks, is greatly increasing. Weld seams at equipment joints need to be regularly tested to ensure the safe and stable operation of equipment. The traditional manual non-destructive testing (NDT) requires abundant experience and is dangerous for workers, and the NDT is time-consuming and labor-intensive. Automated wall-climbing robots can replace manual inspection, and the application of weld inspection robots has become the research focus.

In the research field of wall-climbing robots, many excellent wall-climbing robots have been developed, such as magnetic wheel climbing robots [1–3], negative pressure adhesion wall-climbing robots [4–6] and crawler wall-climbing robots [7–9]. Through permanent magnet adsorption or electromagnetic adsorption, these robots can be operated on metal walls to provide a platform for further work. However, most of the wall-climbing robots [10] only use camera devices to carry out the weld seam identification and positioning, and some robots [11] can identify the weld seam, which is obviously different from the surroundings. Low accuracy makes it difficult for robots to achieve industrial applications. It is difficult to distinguish the weld seam after surface painting from the surrounding environment in terms of color, so the identification by computer image processing is basically invalid. According to some methods combining laser scanning with image processing [12,13], the seam position can be determined by identifying the uneven surface of the weld seam. However, these methods have low efficiency and poor accuracy and are prone to interference from surrounding impurities.

Accurate environment recognition and localization in complex workspaces can improve robot inspection and operation automation. An autonomous navigation method in a 3D workspace was proposed to drive a non-holonomic under-actuated robot to the desired distance from the domain and then full scan of this surface within a given range of altitudes [14]. Novel central navigation and planning algorithms for the hovering autonomous underwater vehicle on ship hulls were developed and applied [15,16]. Acoustic and visual mapping processes were integrated to achieve closed-loop control relative to some features on the open hull. Meanwhile, large-scale planning routines were implemented to achieve full imaging coverage of all the structures on the complex area. An environment recognition and prediction method was developed for autonomous inspection robots on spherical tanks [17,18]. A group of 3D perception sources, including a laser rangefinder, light detection and a ranging and depth camera, were used to extract some environment characteristics to predict the storage tank dimensions and estimate robot position. Weld seams on spherical tank surfaces are prominent environmental features. Fast weld seam identification facilitates robot navigation and positioning on complex spherical surfaces.

It is difficult to achieve the complete and accurate identification of a weld seam path by general computer image processing. Weld seam image pre-processing (such as brightness adjustment, filtering and noise reduction) before identification is indispensable and cumbersome. Moreover, inaccurate and unstable recognition effects limit the application of wall-climbing robots. The rapid development of deep learning [19] in recent years has promoted the development of recognition and classification technologies for intelligent robots. From 2015, we have applied convolutional neural networks (CNN) and other algorithms in weld seam identification, such as sub-region BP neural networks [20], the AdaBoost algorithm [21] and Faster R-CNN [22]. The results are still exciting and deep learning can identify weld seams accurately. However, the training process of CNNs requires powerful hardware support, and early identification results of CNNs are not accurate enough. The identification process takes a long time, thus hindering the further application of inspection robots.

In the study, improved Mask R-CNN [23] was used in the inspection robot, which could flexibly climb on spherical tanks. Mask R-CNN can be regarded as a combined neural network structure of Faster R-CNN and a fully convolutional network (FCN) [24]. After training and learning, the inspection robot could identify and track weld seam with high precision. This paper introduces system design, weld seam identification and weld path tracking of the inspection robot in detail. Section 2 explains and analyzes the composition of the designed robotic system. Section 3 explains the deep learning method for weld seam identification. Weld path fitting and robot tracking movement are introduced in Section 4. In Section 5, the experimental results for weld seam identification and robot tracking are provided. Next, the conclusions and further works are given in Section 6.

2. Robotic System Implementation

2.1. Robot Mechanical Design

Wall-climbing robots for tank inspection should have stable adsorption and movement performance. The inspection robots can be used for NDT and maintenance of weld seams on tank surfaces, such as grinding, cleaning and painting. It is more difficult to climb on spherical tanks than ordinary metal walls since the robot needs to be adapted to the curved spherical surface and provide reliable adsorption force. A series of wall-climbing robot prototypes [25–27] have been designed and explored based on various performance indicators. Figure 1 shows the preliminary test of our developed wall-climbing robot prototype in different positions on a 3000 m^3 spherical tank.

Figure 1. Developed wall-climbing robot on a spherical tank. Note: Spherical tanks put forward higher requirements on wall-climbing robots and the robot must adapt to the curved surfaces to ensure stable adsorption.

According to the results of preliminary tests [25–27], an upgraded inspection robot for weld seam identification and tracking was designed. As shown in Figure 2, the inspection robot is composed of four mecanum wheels, four elastic suspensions, four dampers, four permanent magnets and an adjustable robot frame. A camera is installed at the front of the robot for weld seam identification and two detection probes are installed at its bottom for defect detection in the weld seams.

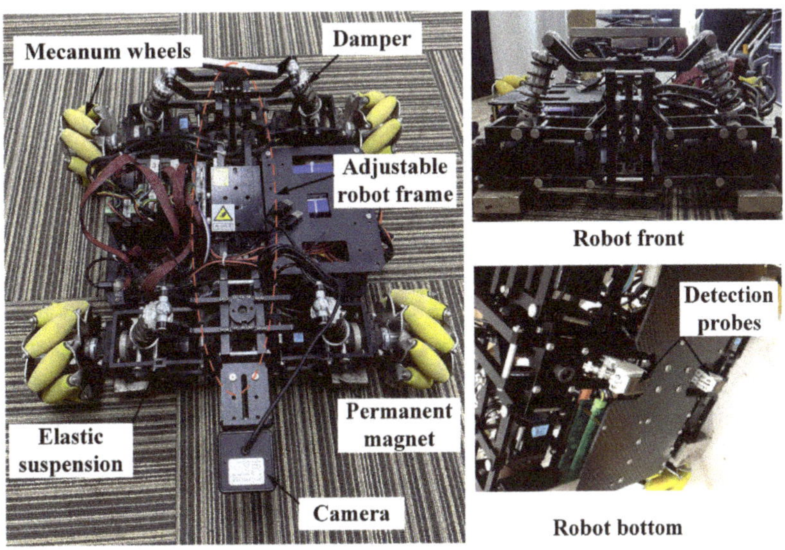

Figure 2. Mechanical structure of the inspection robot. Note: An industrial camera is installed at the front of the robot for weld seam identification; two detection probes are installed at its bottom for defect detection of weld seams.

The inspection robot should provide a sufficient adsorption force at any position of the tank, especially on the vertical and bottom surfaces. On the tank surfaces of 0° and 90°, the adsorption force F_N should satisfy the following formulas, respectively: $\mu F_N > G_t$ and $\mu F_N > G_t$, where G_t is the gravity of the robot and μ is the frictional coefficient of wheels. Compared to electromagnetic adsorption or other means, permanent magnet adsorption can provide greater adsorption force, save energy and avoid the risk of falling.

The robot with mecanum wheels has omnidirectional movement ability, which reduces the risk and energy consumption of swerving and turning around when the robot climbs on the spherical tank.

The inspection robot adopts four-wheel independent elastic suspensions and the adjustable robot frame, which can adjust the inclination of mecanum wheels to be adapted to working surfaces with different curvatures. Each elastic suspension includes an independent damping mechanism, which reduces the instability of the robot and absorbs excess vibration energy. Meanwhile, four permanent magnets can provide sufficient adsorption force to ensure the smooth climbing of the robot on spherical tank surfaces.

When climbing on spherical tanks, the inspection robot has two states: the normal climbing state and the obstacle-surmounting state. The obstacle-surmounting capacity is an important performance of the robot. On the working surface of spherical tanks, the height of weld seams is about 3–4 mm. The robot needs to surmount weld seams during the running process. Dampers installed on independent suspensions can automatically adjust the robot's posture for smoothly surmounting weld seams. The mechanical structural design of the robot can meet the operational requirements on spherical tanks.

The relevant parameters of the inspection robot with mecanum wheels are shown in Table 1. The self-weight of the robot is reduced from 20 kg to 13.75 kg, and the robot payload capacity is increased from 5 kg to 10 kg. To benefit from improved elastic suspensions, the adsorption force of the robot is increased from 180 N to 204 N, and the obstacle-surmounting height is also increased to 5 mm. The maximum climbing velocity is set as 0.2 m/s, and the maximum continuous working time about is 120 min. The improvement in the performance of the inspection robot could increase the application scope and reduce operational risks.

Table 1. Performance and parameters of the inspection robot.

Symbols	Meanings	Unit	Quantitative Values
M_G	Self-weight	kg	13.75
M_l	Maximum payload	kg	10
$V_{\max}$	Maximum velocity	m/s	0.2
F_N	Adsorption force	N	204
h_m	Obstacle-surmounting height	mm	5
T	Working time	min	120

2.2. Robotic System Composition

The weld seam identification and tracking system of the inspection robot is a composite system and its main functions include weld seam identification, path calculation, motion control and data transmission. As shown in Figure 3, the robotic system includes the following subsystems: robot motion control system, weld seam identification system, detection system and remote computer. The robot motion control system mainly includes an industrial personal computer (IPC), motion controller, motor drivers, DC motors, a gyroscope and a remote control unit. It realizes the movement function of the robot, position adjustment and remote control. The weld seam identification system is used to identify weld seams and calculate seam path, including an identification computer (GPU RTX 2060) and an industrial camera. The detection system is an additional device of the robot, and the defect detection of weld seams is performed by ultrasonic probes. The remote computer is connected to the robot through a wireless router for data analysis and storage.

Problems that can be solved by the robotic system include:

(1) Accurate identification of weld seams by machine vision;
(2) Weld path extraction and fitting;
(3) Weld seam tracking by the inspection robot.

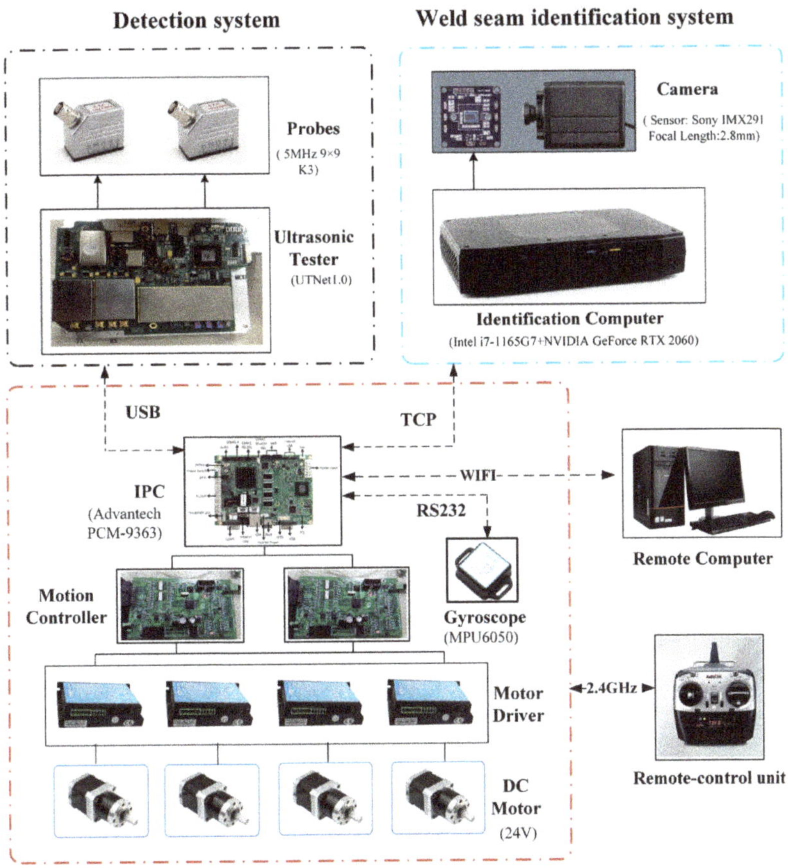

Figure 3. Weld seam identification and tracking system. Note: The robotic system includes robot motion control system, weld seam identification system, detection system and remote computer.

3. Weld Seam Identification

3.1. Weld Seam Images

In the field of computer vision, image processing is a general and fast method. Image processing can extract feature information in a certain image or identify lines that are distinguished in colors. However, in some complex environments, where the acquired images have similar colors (such as painted weld seams) or too much interference, it is difficult to obtain useful information through image processing. In the case of weld seams without obvious distinction, weld seams cannot be accurately identified and extracted by image processing. Even if some features are acquired, the path information is largely discontinuous, unclear and distorted [28].

Real-time weld path tracking can make the inspection robot more automatic and intelligent. However, it is difficult to realize the weld seam identification and path line extraction with high-precision. After long-term work, the weld seams on spherical tank surfaces are covered by dirt and rust, which seriously affect identification accuracy. Weld seam images under the daytime and nighttime lighting conditions acquired by the inspection robot are shown in Figure 4. According to distribution characteristics, weld seams on tank surfaces include transverse weld seams, longitudinal weld seams, diagonal weld seams, crossed weld seams, etc. Different weld seam categories increase the difficulty in fitting weld paths. Due to the small distinction between weld seams and the surrounding feature identification by image processing is usually incomplete and it is difficult to extract the seam path information. Some image processing techniques [28], such as edge detection algorithms and digital morphology, have been tested to extract weld seam path lines, but the extraction results are not satisfactory.

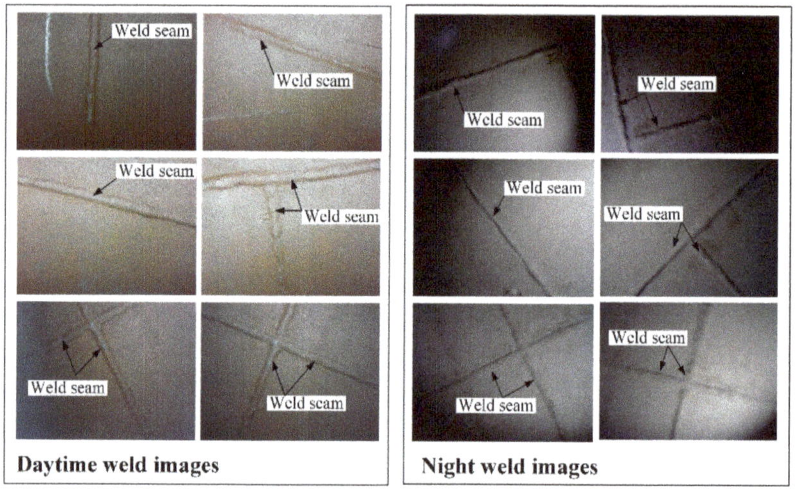

Figure 4. Weld seam images acquired by the inspection robot. Note: Weld seams include transverse weld seams, longitudinal weld seams, diagonal weld seams and crossed weld seams.

3.2. Weld Identification Workflow

Figure 5 shows the workflow of weld seam identification and tracking, which is divided into four steps. Firstly, the weld images are captured by the camera; secondly, weld seams in images are identified by deep learning networks; thirdly, seam paths are extracted and fitted; finally, the robot is controlled to track weld paths based on path information. The purpose of the deep learning networks we used is to accurately segment weld seams. The ultimate goal is to extract weld seams and output path information.

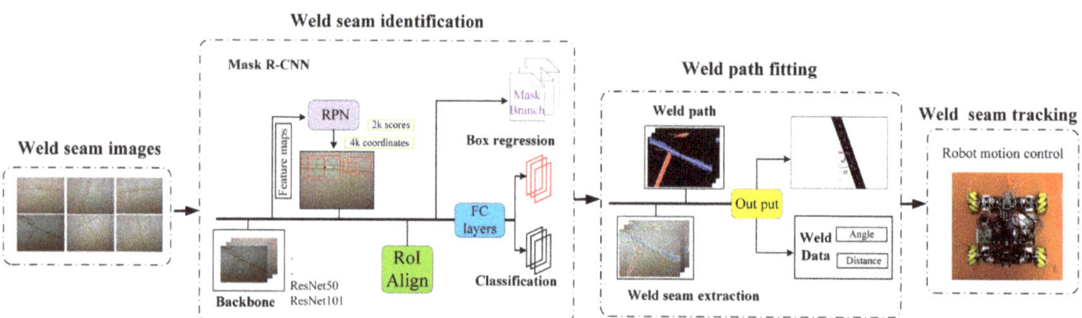

Figure 5. Weld seam identification and tracking workflow.

In the initial phase, the camera of the robot acquires the images of the local environment. Subsequently, a Mask R-CNN model is used to identify weld seams from images and perform instance segmentation. After identifying weld seams, weld path lines are extracted and fitted through Hough transform and image processing. Here, the position parameters of weld paths are estimated, including drift angle α and offset distance d. α is the inclination angle of the robot relative to the chosen weld seam and d is the shortest distance from the path center line to the image center. These two parameters directly reflect the direction angle and climbing velocity of the robot. Finally, the inspection robot continuously changes its movement position to automatically track the chosen weld seam.

3.3. Networks Model

Mask R-CNN is a deep learning mode based on CNNs [23]. It can accomplish the segmentation of weld seams in images. Mask R-CNN performs well in weld seam identification in terms of accuracy and time, which can meet the requirements of the inspection robot. The process of Mask R-CNN for weld identification mainly includes: backbone networks, regional proposal networks (RPN), RoIAlign layers, classification, bounding-box regression and mask generation.

In the beginning, a series of CNN layers (such as VGG19, GoogLeNet [29], ResNet50 [30] and ResNet101) are used to extract feature maps. Deeper CNNs can extract deeper features of weld images, but they ignore the features of small-sized objects. At this stage, feature pyramid networks (FPN) [31] are used to fuse the feature maps from the bottom layer to the top layer to fully utilize the features at different depths. For example, in ResNet50-FPN, the feature map used is C2–C5.

The feature map processed by the backbone networks will be passed into RPN. The purpose of RPN is to recommend a region of interest (RoI) and it is a fully convolutional network. RPN takes weld images as the input and uses nine anchors of different sizes to extract the features of original images. RPN outputs a set of rectangular object proposals, and each region proposal has a suggested score. The anchor sizes have three categories: 128×128, 256×256 and 512×512. There are three kinds of proportional relationships: 2:1, 1:2 and 1:1.

Each sliding window (9 anchors) is mapped to a lower-dimensional feature, which is fed into two sibling fully-connected layers: a box-regression layer and a box-classification layer. At each sliding-window location, RPN simultaneously predicts multiple region proposals and the number of maximum possible proposals for each location is denoted as k. In the stage of RPN, by encoding the coordinates of k boxes, the regression layer gives $4k$ outputs, and the classification layer outputs $2k$ scores (objective scores or non-objective scores) to estimate the probability for k proposals.

Due to the different sizes of the region proposals obtained by RPN, these obtained regional proposals are sent to the non-quantization layer for processing, which is called RoIAlign. RoIAlign uses bilinear interpolation instead of quantization operations to extract fixed-size feature maps from each RoI (for example, 7×7). There is no quantization operation in the whole process. In other words, the pixels in the original image are completely aligned with the pixels in the feature map, and there is no deviation. In this way, the detection accuracy is improved, and the instance segmentation process is simplified.

Mask R-CNN finally outputs three branches: classification, bounding-box regression and mask prediction. After the RoIAlign layer, on the one hand, the RoI is fed into two fully connected layers for image classification and bounding box regression. The classification layer determines the category of weld seams. The bounding-box regression layer refines the location and size of the bounding box. On the other hand, pixel-level segmentation of weld seams is acquired through FCN. FCN uses convolutional layers instead of fully connected layers for pixel-to-pixel object mask prediction. The mask prediction branch uses FCN to segment objects in the image by pixels and has a Km^2-dimensional output for each RoI (K is the number of categories). In the dataset for the identification of weld seams, the number of categories was 2 (including background and weld seams), the network depth of the classification layer and regression layer was 2, and output network size of the mask prediction branch was $28 \times 28 \times 2$ pixels.

3.4. Loss Function

Weld seam images are segmented by Mask R-CNN. There are three output layers: the classification layer, bounding-box regression layer and mask branch. During the training process, the relevant variables and their meanings are shown in Table 2. The total loss function is defined as:

$$L = L_{cls+box} + L_{mask-branch} \tag{1}$$

$$L_{cls+box} = L_{cls} + L_{box} = \frac{1}{N_{cls}}\sum_i L_{cls}(\rho_i, \rho_i^*) + \lambda_1 \frac{1}{N_{reg}}\sum_i \rho_i^* L_{reg}(\tau_i, \tau_i^*) \tag{2}$$

Table 2. Variables and meanings in the loss function.

Symbols	Meanings
L	Total loss
$L_{cls+box}$	Sum of classification loss and bounding-box loss
$L_{mask-branch}$	Mask branch loss
L_{cls}	Classification loss
L_{box}	Bounding-box loss
L_{mask}	Average binary cross-entropy loss
N_{cls}	Number of corresponding anchors
N_{reg}	Number bounding boxes
i	Anchor index
ρ_i	Predicted classification probability of anchor i
ρ_i^*	Ground-truth label probability of anchor $i; \rho_i^*$ is 1 for the positive anchor and 0 for the negative anchor
$\tau_i = (\tau_x, \tau_y, \tau_w, \tau_h)$	Difference between the prediction bounding box and the ground-truth label box in four parameter vectors (horizontal coordinate, vertical coordinate, width and height)
$\tau_i^* = \left(\tau_x^*, \tau_y^*, \tau_w^*, \tau_h^*\right)$	Differences between the ground-truth label box and the positive anchor in four parameter vectors
λ_i, γ_i	Hyper-parameters to balance the training losses of the regression and mask branch

Classification loss L_{cls} can be defined as:

$$L_{cls}(\rho_i, \rho_i^*) = -\log \rho_i^* \rho_i \tag{3}$$

Bounding-box loss L_{box} is defined over a tuple of true bounding-box regression targets:

$$L_{reg}\ (\tau_i, \tau_i^*) = \text{smooth}_{L1}(\tau_i^* - \tau_i) \tag{4}$$

$$\text{smooth}_{L1}(x) = \begin{cases} 0.5x^2 & ,\text{ if } |x| < 1 \\ |x| - 0.5 & ,\text{otherwise} \end{cases} \tag{5}$$

Mask branch loss $L_{mask-branch}$ is defined as:

$$\begin{aligned} L_{mask-branch} &= L_{mask-branch}(\rho_i, \rho_i^*, \tau_i, \tau_i^*, \sigma_i, \sigma_i^*) \\ &= \frac{1}{N_{cls}}\sum_i L_{cls}(\rho_i, \rho_i^*) + \lambda_2 \frac{1}{N_{reg}}\sum_i \rho_i^* L_{reg}(\tau_i, \tau_i^*) + \gamma_2 \frac{1}{N_{\text{mask}}}\sum_i L_{\text{mask}}\ (\sigma_i, \sigma_i^*) \end{aligned} \tag{6}$$

L_{mask} is defined as the average binary cross-entropy loss by used a per-pixel sigmoid. The definition of L_{mask} allows the network to generate masks for every category without competition.

3.5. Training

During the training phase of Mask R-CNN, 2281 weld seam images at different angles were collected (more images might have the better effect). The image size was fixed at 320 × 240. In the collected images, 1500 images were used as the training dataset; 500 images were used as the verification set; and 281 images were used as the testing set.

These dataset images were labelled and processed in advance. Some data augmentation methods, such as random flip, random crop, color jitter and noise addition, were used to extend the dataset. Image transformations and noise addition were beneficial to avoid overfitting. After data augmentation (24 types), the number of weld seam images in the training set, validation set and test set were 37,500, 12,500 and 7025, respectively. Figure 6 visualizes results of the weld seam dataset. Regarding the label number, 1 (or 2) represents weld seams and 0 represents the background.

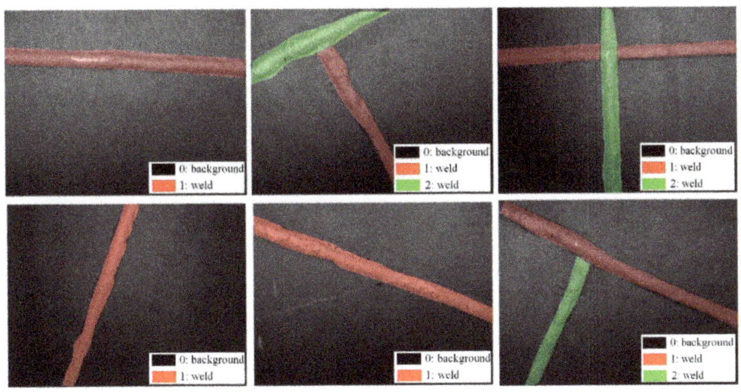

Figure 6. Visualization results of the weld seam dataset.

The pre-trained Mask R-CNN weight was inherited from the COCO dataset and the backbone network was ResNet101 + FPN. The number of categories was 2 (background and weld). The model output the category scores, bounding boxes and masks of weld seams for each input image after training. The training parameters were set as follows: the initial learning rate of the network was 0.001; the weight attenuation coefficient was 0.0005; the momentum coefficient was 0.85. During the deep learning training process of the weld seam dataset, the loss function and accuracy curve changes as shown in Figure 7. After 10,000 iterations, the training loss function curve stabilizes at 0.15–0.2, and the accuracy of the network model is greater than 0.97. Comparing the learning effect of 3000, 5000 and 8000 iterations, due to smaller weld images and the number of categories is only 2, the effect after 3000 iterations is already good and after 5000 times is basically stable.

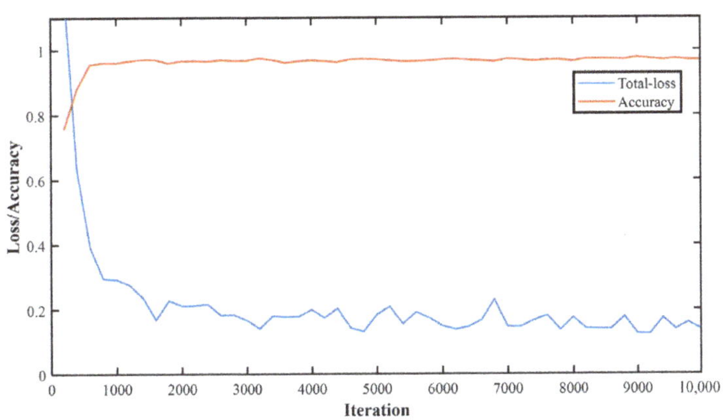

Figure 7. Loss function and accuracy during training process.

4. Weld Path Tracking

4.1. Path Selection Problem

Since weld images not only include a single weld seam (such as transverse and longitudinal weld seams) but sometimes include multiple weld seams (such as T-shape and crossed weld seams). When there are multiple weld seams in an image, the following possibilities exist in the weld path extraction stage:

(1) Extract the path of weld seam 1
(2) Extract the path of weld seam 2
(3) Extract one wrong path

As shown in Figure 8, in order to avoid the interference of multiple paths, the acquired weld seams are identified and distinguished by different colors, and the centerline of each weld seam is fitted separately. If there are T-shape or crossed weld seams in the weld images, multiple weld seams are segmented.

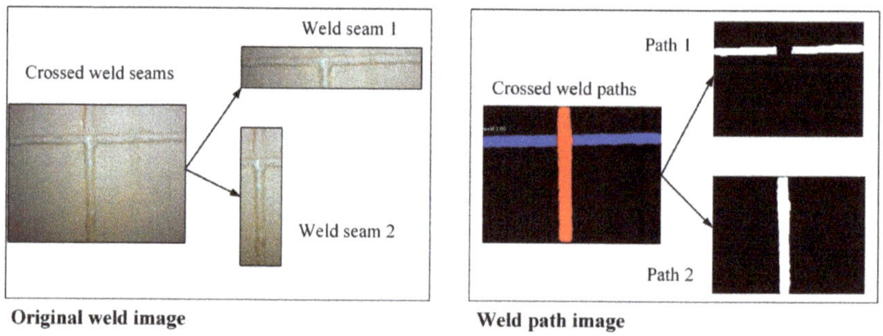

Figure 8. Path selection for crossed weld seams.

In general, the crossed weld seam is considered to be a special welding joint, so the crossed weld seam is defined as a combination of two weld seams. Therefore, the optimal weld path can be efficiently processed. In the weld identification phase, Mask R-CNN has a more obvious advantage in segmenting multiple weld seams; these weld seams are distinguished by different color pixels.

4.2. Weld Path Fitting

Binarized images of weld paths are generated through image processing, and image erosion and filtering are also executed. Hough transform is used to fit the path line of a single weld seam. In the image coordinate system (x, y), the path line of the weld seam can be supposed as:

$$y = a_0 x + b_0 \tag{7}$$

After mapping the line to Hough space (a, b), the line equation can be expressed as:

$$b_0 = -x a_0 + y \tag{8}$$

Each line in the image space can be described as a point in the Hough space. As shown in Figure 9, point (x_i, y_i) and point (x_j, y_j) correspond to two straight lines in Hough space, respectively:

$$\begin{cases} b = -x_i a + y_i \\ b = -x_j a + y_j \end{cases} \tag{9}$$

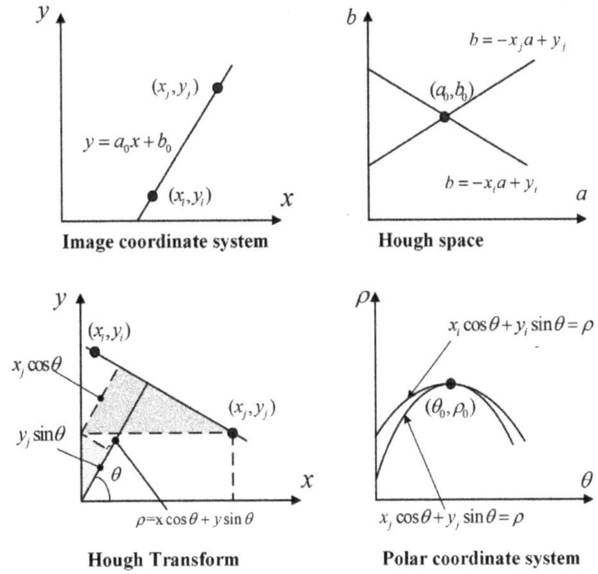

Figure 9. Weld path line fitted by Hough transform.

The straight line equation can be converted into polar coordinate system (θ, ρ):

$$x\cos\theta + y\sin\theta = \rho \tag{10}$$

So, the straight line equation of the weld path can be expressed as:

$$y = \frac{-\cos(\theta_0)}{\sin(\theta_0)}x + \frac{\rho_0}{\sin(\theta_0)} \tag{11}$$

a_0 and b_0 can be expressed as:

$$\begin{cases} a_0 = -\cos(\theta_0)/\sin(\theta_0) \\ b_0 = \rho_0/\sin(\theta_0) \end{cases} \tag{12}$$

In the polar coordinate system, point (θ_0, ρ_0) represents a straight line in the image coordinate system. The number of curves passing through point (θ_0, ρ_0) can represent the quantity of points in a straight line in the image coordinate system. The straight lines in the image coordinate system are converted into the points in the polar coordinate system, and the straight line with the most points in the image coordinate system can be determined as the weld seam path line.

Through Hough transform, multiple straight line functions in the weld image can be fitted. The line with the maximum linear length is selected as the weld path line. There is a slight deviation between the solved path line and the actual path centerline, but it can meet the requirement of robot tracking.

According to the weld path line parameters a_0 and b_0, the drift angle α and offset distance d of weld paths can be expressed as:

$$\begin{cases} \alpha = \frac{\arctan(-a_0)\cdot 180}{\pi} \\ d = \frac{x_c - b_0}{k} - y_c/2 \end{cases} \tag{13}$$

where (x_c, y_c) is the center coordinate of weld line images.

4.3. Robot Kinematics

The inspection robot with mecanum wheels can achieve omnidirectional movement. Based on elastic suspensions and the adjustable robot frame, the robot can move flexibly on curved surfaces, so the weld seam tracking on spherical tanks can be simplified as a plane adjustment motion. In this process, the robot continuously corrects drift angle α and offset distance d to climb forward along welding seams.

As shown in Figure 10, assuming that the robot forward velocity is v_0, the rotation angular velocity and lateral velocity to be adjusted are $\Delta\alpha$ and Δd, respectively. Then the inverse kinematics equation of tracking movement is:

$$\begin{bmatrix} \dot{\theta}_1 \\ \dot{\theta}_2 \\ \dot{\theta}_3 \\ \dot{\theta}_4 \end{bmatrix} = \frac{1}{R_c}\begin{bmatrix} 1 & -\cot\beta_c & -(W_c + \cot\beta_c \cdot L_c) \\ 1 & \cot\beta_c & W_c + \cot\beta_c \cdot L_c \\ 1 & -\cot\beta_c & W_c + \cot\beta_c \cdot L_c \\ 1 & \cot\beta_c & -(W_c + \cot\beta_c \cdot L_c) \end{bmatrix}\begin{bmatrix} v_0 \\ \Delta d \\ \Delta\alpha \end{bmatrix} \tag{14}$$

$$V_i = R\dot{\theta}_i (i = 1, 2, 3, 4) \tag{15}$$

where V_i and $\dot{\theta}i$ are the velocities and angular velocities of four mecanum wheels, respectively. W_c and L_c are the half-width and half-length of the robot frame. R is the radius of the mecanum wheels and β_c represents the angle between the wheel roller and the wheel axis. Substituting with robot design parameters, the Jacobian matrix (defined as K_c) of inverse kinematics of the robot is:

$$K_c = \frac{1}{0.0635}\begin{bmatrix} 1 & -1 & -0.788 \\ 1 & 1 & 0.788 \\ 1 & -1 & 0.788 \\ 1 & 1 & -0.788 \end{bmatrix} \tag{16}$$

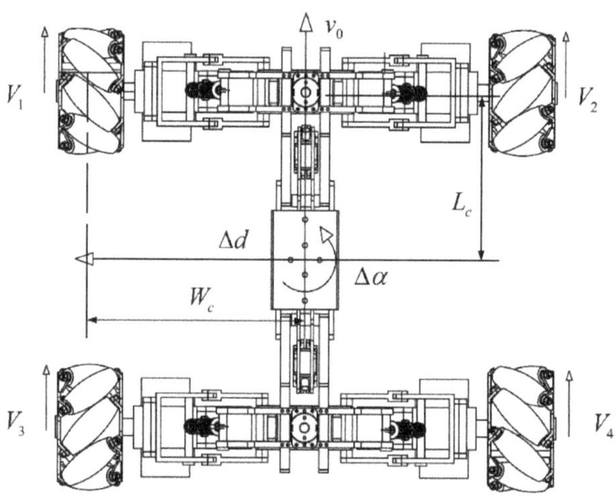

Figure 10. Robot dimensions and motion parameters.

5. Experiments and Results

As shown in Figure 11, the experimental platform is a cylindrical tank with a diameter of 4000 mm, a height of 2800 mm and a thickness of 10 mm. Weld seams are distributed on the surface of the experimental tank for identification and tracking by the robot. The laser tracker was used to record the running position of the inspection robot.

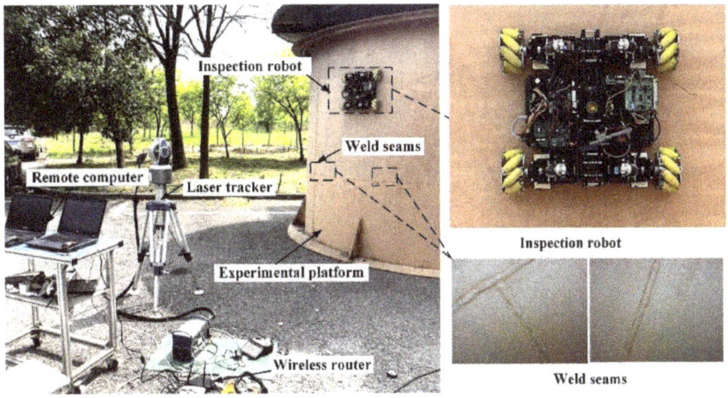

Figure 11. Experimental platform and the inspection robot.

5.1. Weld Seam Identification Experiments

Figure 12 shows weld seam identification results by the robot. From the perspective of identification effect, accurate outline descriptions of weld seams have been generated with over 98% identification probability. Generated pixel-level masks can cover weld seams, which is beneficial for extracting and fitting weld paths. Table 3 shows the statistical data of weld seam box AP (bounding box AP) and mask AP. The average precision calculation formula is as:

$$AP = \int_0^1 P(r)dr \tag{17}$$

where P and r represent precision and recall of each image, respectively. Mask AP and box AP are 67.5% and 78.9%, respectively. Both AP_{50} and AP_{75} are greater than 90%. Results of AP_s, AP_m and AP_l show that the size of weld seams is mainly concentrated in the medium pixel size (32^2–96^2).

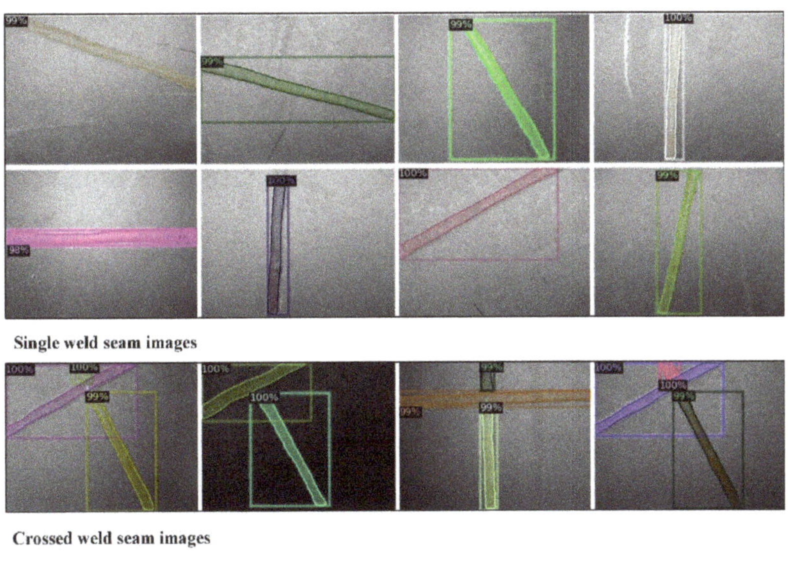

Figure 12. Identification results of weld seam images.

Table 3. Box AP and mask AP of weld seam identification.

(%)	AP	AP_{50}	AP_{75}	AP_s	AP_m	AP_L
Box	78.947	96.193	90.802	nan	86.079	3.791
Mask	67.596	90.802	90.802	nan	71.252	0.000

5.2. Weld Path Fitting Experiments

As shown in Figure 13, multiple weld paths are extracted using different colors. When the weld images are crossed weld seams, two weld path lines are fitted by dividing them into red and blue. Figure 14 shows image binarization and fitting results of weld paths. The robot can accurately fit path lines to estimate their positional parameters (drift angle and offset distance).

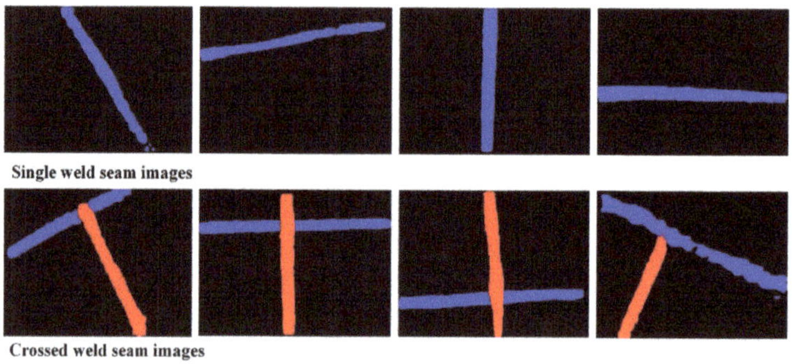

Figure 13. Weld path extraction results with different colors. Note: Two weld paths are extracted by dividing them into red and blue.

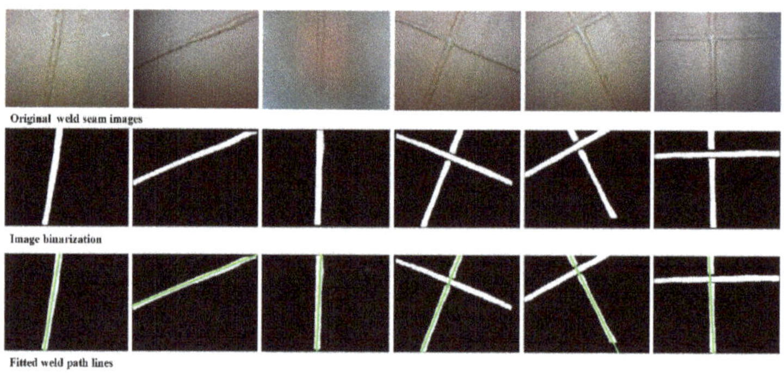

Figure 14. Image binarization and fitting results of weld paths by the robot. Note: The binarized seam path images are convenient for processing and path line fitting.

Figure 15 shows the deviations between the calculated values and the actual measured value of drift angle α and offset distance d. In deviation values of single weld seams, the average deviation of drift angle α between calculated paths and actual paths is 1.01°, and the maximum deviation is 2.78° (Figure 15a). The average deviation of offset distance d is 2.21 pixels, and the maximum value is 5.57 pixels (Figure 15b). In terms of the deviation of crossed weld seams, the average and maximum deviations of drift angle α are about 0.87° and 2.37° (Figure 15c), and the average and maximum deviations of offset distance d are 2.62 pixels and 7.83 pixels (Figure 15d). These deviations are generally kept at small values which have little effect on the weld path tracking by the robot.

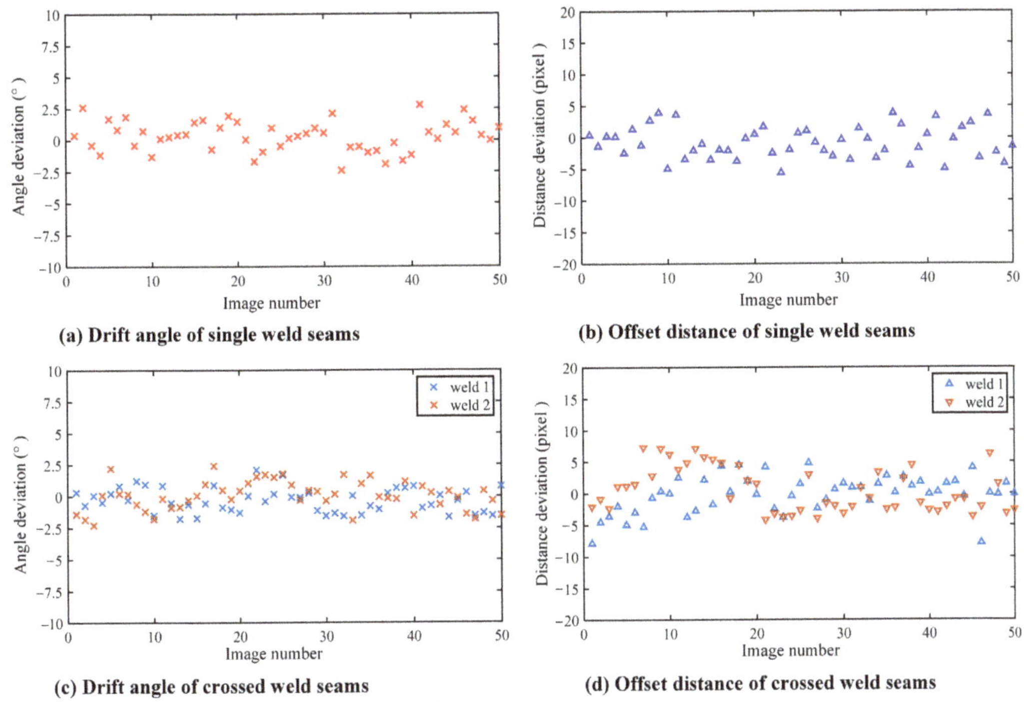

Figure 15. Deviations between the calculated values and the actual measured value of drift angle and offset distance.

Table 4 shows the time consumption of weld seam real-time identification and processing, which mainly includes image loading loss time, deep learning identification time, image processing time and total time. In the process of continuous weld seam identification, the total processing time of each image is kept between 0.15 s and 0.18 s, the average time of the deep learning network of each image is 0.137 s, the average image loading loss time is about 0.006 s and the average image processing time is about 0.018 s.

Table 4. Time consumption of weld seam identification and processing.

Image Number	1	2	3	4	5	...	196	197	198	199	200	Mean Time (s)
$Time_0$ (s)	0.172	0.236	0.226	0.176	0.142	...	0.155	0.163	0.161	0.156	0.156	0.162
$Time_1$ (s)	0.007	0.007	0.003	0.007	0.003	...	0.006	0.007	0.006	0.006	0.007	0.006
$Time_2$ (s)	0.147	0.181	0.202	0.144	0.130	...	0.129	0.133	0.131	0.131	0.130	0.137
$Time_3$ (s)	0.019	0.049	0.021	0.023	0.009	...	0.019	0.023	0.023	0.019	0.019	0.018

Note: $Time_0$ = total time; $Time_1$ = loading time; $Time_2$ = identification time; $Time_3$ = image processing time.

In the real-time weld seam identification, the robotic system can output five sets of path data (drift angle and offset distance) in one second. Due to the low running velocity of the robot, it can meet the automated inspection work. Weld seam identification speed can be accelerated by optimizing deep learning identification time and clearing the loss time of loading images.

5.3. Robot Tracking Experiments

According to the imaging ratio of the camera, the actual offset distances between weld paths and the inspection robot were calculated. Based on the robot kinematics, the rotation velocities of four motors were adjusted, and the attitude and position of the robot were continuously changed to track the identified weld seam. Due to the drive mode of mecanum wheels, the robot could achieve omnidirectional movement without frequently changing its heading angle. In robot tracking experiments, when the drift angle was over $\pm 3^{\circ}$, the climbing robot would rotate to correct its heading angle.

Figure 16 shows the drift angles and offset distances of robot while tracking the identified weld seams. The robot can successfully correct its angle and velocities to track weld seams. The maximum drift angle and offset distance were about 3° and 10 mm, respectively. The accuracy of tracking weld seams could meet the operational requirements of the inspection robot.

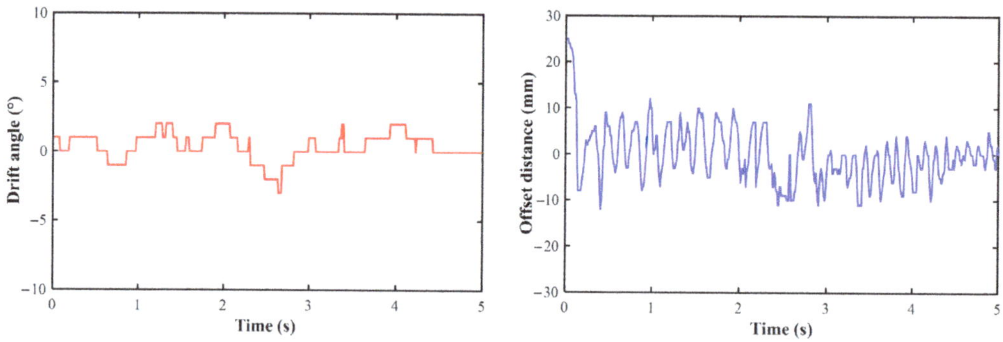

(a) Drift angle of the robot (b) Offset distance of the robot

Figure 16. Drift angle and offset distance when robot tracking weld paths.

6. Conclusions

In this study, an intelligent inspection robotic system based on deep learning was developed to achieve the weld seam identification and tracking. The inspection could complete the segmentation of weld images and output the masks of weld paths using Mask R-CNN. Deep learning weld seam identification makes up for the low accuracy of image processing. By using specific colors to distinguish multiple weld seams, possible errors in the fitting weld path can be avoided. The real-time test results indicated the deep learning model had higher accuracy in weld seam identification, and the average processing time of each image was about 180 ms. Weld path fitting experiments were carried out to test weld path extraction deviation. The maximum deviations of drift angle α and offset distance d were within 3° and 8 pixels, respectively. The results of robot tracking experiments demonstrated the inspection robot could accurately track weld seams, with a maximum tracking deviation of 10 mm.

The further training and adjusting of the network structure will be explored to speed up processing. In the construction process of the entire robotic system, due to the combination of multiple methods and algorithms, optimizing the system composition and connection is also one of our priorities in the future. In terms of engineering applications, this robotic system can meet the basic operational and inspection requirements and the robot performance will be promoted in further research.

Author Contributions: J.L. contributed significantly to robot design, analysis and manuscript preparation. X.W. contributed to supervision, reviewing and validation. B.L. and M.T. contributed to robot experiments and result analysis. L.D. performed software debugging and image data collection. All authors have read and agreed to the published version of the manuscript.

Funding: This research was funded by the Science and Technology Project of Quality and Technical Supervision Bureau of Jiangsu Province grant number KJ175933 and the National Prevention Key Technology Project for Serious and Major Accidents in Work Safety of China grant number jiangsu-0002-2017AQ.

Institutional Review Board Statement: Not applicable.

Informed Consent Statement: Not applicable.

Data Availability Statement: All data included in this study are available upon request by contact with the corresponding author.

Conflicts of Interest: The authors have no conflict of interest to disclose.

References

1. Wu, M.; Gao, X.; Yan, W.X.; Fu, Z.; Zhao, Y.; Chen, S. New Mechanism to Pass Obstacles for Magnetic Climbing Robots with High Payload, Using Only One Motor for Force-Changing and Wheel-Lifting. *Ind. Rob.* **2011**, *38*, 372–380. [CrossRef]
2. La, H.M.; Dinh, T.H.; Pham, N.H.; Ha, Q.P.; Pham, A.Q. Automated Robotic Monitoring and Inspection of Steel Structures and Bridges. *Robotica* **2019**, *37*, 947–967. [CrossRef]
3. Lee, G.; Wu, G.; Kim, J.; Seo, T. High-Payload Climbing and Transitioning by Compliant Locomotion with Magnetic Adhesion. *Rob. Auton. Syst.* **2012**, *60*, 1308–1316. [CrossRef]
4. Zhou, Q.; Li, X. Experimental Investigation on Climbing Robot Using Rotation-Flow Adsorption Unit. *Rob. Auton. Syst.* **2018**, *105*, 112–120. [CrossRef]
5. Zhu, H.; Guan, Y.; Wu, W.; Zhang, L.; Zhou, X.; Zhang, H. Autonomous Pose Detection and Alignment of Suction Modules of a Biped Wall-Climbing Robot. *IEEE/ASME Trans. Mechatron.* **2015**, *20*, 653–662. [CrossRef]
6. Sakagami, N.; Yumoto, Y.; Takebayashi, T.; Kawamura, S. Development of Dam Inspection Robot with Negative Pressure Effect Plate. *J. Field Robot.* **2019**, *36*, 1422–1435. [CrossRef]
7. Schmidt, D.; Berns, K. Climbing Robots for Maintenance and Inspections of Vertical Structures—A Survey of Design Aspects and Technologies. *Rob. Auton. Syst.* **2013**, *61*, 1288–1305. [CrossRef]
8. Park, C.; Bae, J.; Ryu, S.; Lee, J.; Seo, T. R-Track: Separable Modular Climbing Robot Design for Wall-to-Wall Transition. *IEEE Robot. Autom. Lett.* **2021**, *6*, 1036–1042. [CrossRef]
9. Gao, X.; Shao, J.; Dai, F.; Zong, C.; Guo, W.; Bai, Y. Strong Magnetic Units for a Wind Power Tower Inspection and Maintenance Robot. *Int. J. Adv. Robot. Syst.* **2012**, *9*, 189. [CrossRef]
10. Wang, Z.; Zhang, K.; Chen, Y.; Luo, Z.; Zheng, J. A Real-Time Weld Line Detection for Derusting Wall-Climbing Robot Using Dual Cameras. *J. Manuf. Process.* **2017**, *27*, 76–86. [CrossRef]
11. Maglietta, R.; Milella, A.; Caccia, M.; Bruzzone, G. A Vision-Based System for Robotic Inspection of Marine Vessels. *Signal Image Video Process.* **2018**, *12*, 471–478. [CrossRef]
12. Zhang, L.; Ye, Q.; Yang, W.; Jiao, J. Weld Line Detection and Tracking via Spatial-Temporal Cascaded Hidden Markov Models and Cross Structured Light. *IEEE Trans. Instrum. Meas.* **2014**, *63*, 742–753. [CrossRef]
13. Zhang, L.; Ke, W.; Ye, Q.; Jiao, J. A Novel Laser Vision Sensor for Weld Line Detection on Wall-Climbing Robot. *Opt. Laser Technol.* **2014**, *60*, 69–79. [CrossRef]
14. Matveev, A.S.; Ovchinnikov, K.S.; Savkin, A.V. A Method of Reactive 3D Navigation for a Tight Surface Scan by a Nonholonomic Mobile Robot. *Automatica* **2017**, *75*, 119–126. [CrossRef]
15. Hover, F.S.; Eustice, R.M.; Kim, A.; Englot, B.; Johannsson, H.; Kaess, M.; Leonard, J.J. Advanced Perception, Navigation and Planning for Autonomous in-Water Ship Hull Inspection. *Int. J. Robot. Res.* **2012**, *31*, 1445–1464. [CrossRef]
16. Englot, B.; Hover, F.S. Sampling-Based Coverage Path Planning for Inspection of Complex Structures. In Proceedings of the ICAPS 2012—22nd International Conference on Automated Planning and Scheduling, Sao Paulo, Brazil, 24–28 June 2012; pp. 29–37.
17. Teixeira, M.A.S.; Santos, H.B.; De Oliveira, A.S.; De Arruda, L.V.R.; Neves, F. Environment Identification and Path Planning for Autonomous NDT Inspection of Spherical Storage Tanks. In Proceedings of the 2016 XIII Latin American Robotics Symposium and IV Brazilian Robotics Symposium LARS/SBR 2016, Recife, Brazil, 8–12 October 2016; pp. 193–198. [CrossRef]
18. Teixeira, M.A.S.; Santos, H.B.; Dalmedico, N.; de Arruda, L.V.R.; Neves-Jr, F.; de Oliveira, A.S. Intelligent Environment Recognition and Prediction for NDT Inspection through Autonomous Climbing Robot. *J. Intell. Robot. Syst. Theory Appl.* **2018**, *92*, 323–342. [CrossRef]
19. Lecun, Y.; Bengio, Y.; Hinton, G. Deep Learning. *Nature* **2015**, *521*, 436–444. [CrossRef] [PubMed]
20. Liang, G.; Wang, S.; Tu, C.; Wang, X. Existing Weld Seam Recognition and Tracking Based on Sub Region Neutral Network. In Proceedings of the M2VIP 2016—Proceedings of 23rd International Conference on Mechatronics and Machine Vision in Practice, Nanjing, China, 28–30 November 2016.
21. Wang, S.; Wang, X. Existing Weld Seam Recognition Based on Sub-Region BP-Adaboost Algorithm. In Proceedings of the M2VIP 2016—Proceedings of 23rd International Conference on Mechatronics and Machine Vision in Practice, Nanjing, China, 28–30 November 2016.
22. Ren, S.; He, K.; Girshick, R.; Sun, J. Faster R-CNN: Towards Real-Time Object Detection with Region Proposal Networks. *IEEE Trans. Pattern Anal. Mach. Intell.* **2017**, *39*. [CrossRef] [PubMed]
23. He, K.; Gkioxari, G.; Dollár, P.; Girshick, R. Mask R-CNN. In Proceedings of the 2017 IEEE International Conference on Computer Vision (ICCV), Venice, Italy, 22–29 October 2017; pp. 2980–2988.

24. Shelhamer, E.; Long, J.; Darrell, T. Fully Convolutional Networks for Semantic Segmentation. *IEEE Trans. Pattern Anal. Mach. Intell.* **2017**, *39*. [CrossRef] [PubMed]
25. Li, J.; Wang, X.S. Novel Omnidirectional Climbing Robot with Adjustable Magnetic Adsorption Mechanism. In Proceedings of the 2016 23rd International Conference on Mechatronics and Machine Vision in Practice (M2VIP), Nanjing, China, 28–30 November 2016; pp. 1–5. [CrossRef]
26. Zheng, K.; Li, J.; Tu, C.L.; Wang, X.S. Two Opposite Sides Synchronous Tracking X-Ray Based Robotic System for Welding Inspection. In Proceedings of the M2VIP 2016—Proceedings of 23rd International Conference on Mechatronics and Machine Vision in Practice, Nanjing, China, 28–30 November 2016.
27. Tu, C.I.; Li, X.D.; Li, J.; Wang, X.S.; Sun, S.C. Bilateral Laser Vision Tracking Synchronous Inspection Robotic System. In Proceedings of the 2017 Far East NDT New Technology & Application Forum (FENDT), Xi'an, China, 22–24 June 2017; pp. 207–215. [CrossRef]
28. Liang, G.A.; Zheng, K.; Tu, C.L.; Wang, S.S.; Wang, X.S. Existing Weld Seam Recognition Based on Image Processing. In Proceedings of the 2017 IEEE Far East NDT New Technology and Application Forum, FENDT 2017, Xi'an, China, 22–24 June 2017.
29. Szegedy, C.; Liu, W.; Jia, Y.; Sermanet, P.; Reed, S.; Anguelov, D.; Erhan, D.; Vanhoucke, V.; Rabinovich, A. Going Deeper with Convolutions. In Proceedings of the IEEE Computer Society Conference on Computer Vision and Pattern Recognition, Boston, MA, USA, 7–12 June 2015.
30. Xie, S.; Girshick, R.; Dollár, P.; Tu, Z.; He, K. Aggregated Residual Transformations for Deep Neural Networks. In Proceedings of the Proceedings—30th IEEE Conference on Computer Vision and Pattern Recognition, CVPR 2017, Honolulu, HI, USA, 21–26 July 2017.
31. Lin, T.Y.; Dollár, P.; Girshick, R.; He, K.; Hariharan, B.; Belongie, S. Feature Pyramid Networks for Object Detection. In Proceedings of the Proceedings—30th IEEE Conference on Computer Vision and Pattern Recognition, CVPR 2017, Honolulu, HI, USA, 21–26 July 2017.

drones

MDPI

Article

FEC: Fast Euclidean Clustering for Point Cloud Segmentation

Yu Cao [1,†], Yancheng Wang [2,†], Yifei Xue [3], Huiqing Zhang [1] and Yizhen Lao [1,*]

1 College of Computer Science and Electronic Engineering, Hunan University, Lushan Road (S), Yuelu District, Changsha 410012, China
2 Alibaba Group, Hangzhou 311121, China
3 Jiangxi Provincial Natural Resources Cause Development Center, Nanchang 330000, China
* Correspondence: yizhenlao@hnu.edu.cn
† These authors contributed equally to this work.

Abstract: Segmentation from point cloud data is essential in many applications, such as remote sensing, mobile robots, or autonomous cars. However, the point clouds captured by the 3D range sensor are commonly sparse and unstructured, challenging efficient segmentation. A fast solution for point cloud instance segmentation with small computational demands is lacking. To this end, we propose a novel fast Euclidean clustering (FEC) algorithm which applies a point-wise scheme over the cluster-wise scheme used in existing works. The proposed method avoids traversing every point constantly in each nested loop, which is time and memory-consuming. Our approach is conceptually simple, easy to implement (40 lines in C++), and achieves two orders of magnitudes faster against the classical segmentation methods while producing high-quality results.

Keywords: point cloud; instance segmentation; fast clustering

Citation: Cao, Y.; Wang, Y.; Xue, Y.; Zhang, H.; Lao, Y. To be added by editorial staff during production. FEC: Fast Euclidean Clustering for Point Cloud Segmentation. *Drones* **2022**, *6*, 325. https://doi.org/10.3390/drones6110325

Academic Editor: Joaquín Martínez-Sánchez

Received: 17 October 2022
Accepted: 21 October 2022
Published: 27 October 2022

1. Introduction

The point cloud is a typical data structure deemed natural and intuitive to represent geometric space information in specific scenarios, ranging from a pen on a table to illustrating the world. People acquire this kind of data mainly by reconstruction from pairs of 2D images, or directly utilizing LiDAR devices (often mounted on backpacks, vehicles, drones, or aircraft). However, raw point cloud data are not always ready for use due to redundant information, uneven distribution, and object occlusion [1]. As shown in Figure 1 that point cloud segmentation is a common approach to tackle this issue by classifying point clouds with similar properties. Hence, point cloud segmentation becomes essential in 3D perception (computer vision, remote sensing) and 3D reconstruction (autonomous driving, virtual reality). For example, a robot should identify obstacles nearby to interact with and move around the scene [2]. Achieving this goal requires distinguishing between distinct semantic labels and various instances with the same semantic label. Thus, it is crucial to investigate the problem of point cloud segmentation.

1.1. Related Works

There is a substantial amount of work that targets acquiring a global point cloud and segmenting it off-line which can be classified into four main categories:

1.1.1. Edge-Based Method

The most important procedure for Edge-based approaches is finding boundaries of each object section. Zucker and Hummel [3] extends edged detection methods from two-dimension (2D) to three-dimension (3D) level in geometrical way. This notion hypothesizes that the 3D scene can be divided by planes, and finding the edges is an intrinsic optimal way to get units. Then the segmentation problem simply to edge detection according to gradient, thresholds, morphological transform, filtering, template-matching [4–8]. Öztireli et al. [9]

introduce a parameter-free edge detection method with assistance of kernel regression. Rabbani et al. [10] summarise that this kind of method could be divided into two stages, namely border outlining and inner boundary grouping. However, the main drawback of edge-based methods is under-effective in noisy data [8].

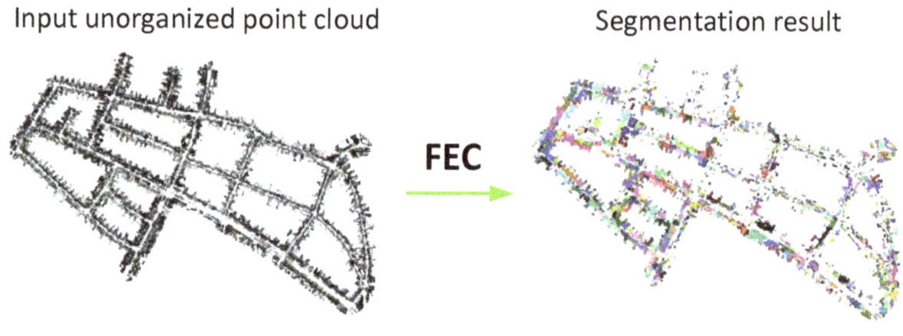

Figure 1. Unorganized point cloud instance segmentation results of proposed method FEC on one KITTI sequence [11] with 27 million points. FEC outperforms the existing segmentation approaches [10,12] with over 100× speedup rates.

1.1.2. Region Growing Based Method

Region growing (RG)-based methods focus more on extracting geometrical homogeneous surfaces than detecting boundaries. The assumption is that different parts of objects/scenes are visually distinguishable or separable. Generally speaking, these methods make point groups by checking each point (or derivatives, like voxels/supervoxels [13,14]) around one or more seeds by specific criteria. So researchers advocating RG-based method developed several criteria considering orientation [10,15], surface normal [10,16,17] , curvatures [18], et al. Li et al. [19] explore the possibility to apply RG algorithm to multiple conclusion cases (i.e., leaf phenotyping) and proves to be feasible. Typically, the conventional regional growth algorithm employs the RANSAC algorithm to generate seed spots. Considering internal features of point clouds, Habib and Lin [20] randomly select seeds and segment multi-class at once. Dimitrov and Golparvar-Fard [21] propose an RG-based method, which filters seed candidates by roughness measurement, for robust context-free segmentation of unordered point clouds based on geometrical continuities. However, the selection strategy of seed points is crucial cause if means to the computing speed and over- or under-segmentation [13]. In addition, these approaches are highly responsive to the imprecise calculation of normals and curvatures close to region borders [22].

1.1.3. Clustering Based Method

The clustering algorithms segment or simplify point cloud elements into categories based on their similarities or euclidean/non-euclidean distances. As a result, k-means [23], mean shift [24], DBSCAN [25], and euclidean cluster (EC) extraction [12] were employed on this task. The k-means clustering aims at grouping point data into k divisions constrained by average distances [26]. Since point cloud data is often acquired unevenly, k-means is an ideal tool to remove redundant dense points [27]. Kong et al. [28] introduces a k-plane approach to categorize laser footprints that cannot be accurately classified using the standard k-means algorithm. DBSCAN assumes density distribution is the critical hint for deciding sample categories as dense regions, making it noise-resistant [25]. Even so, the DBSCAN method suffers from high memory capacity requirements for storing distance data between points. Chehata et al. [29] improve normal clustering methods to a hierarchical one, which is shown to be suitable for microrelief terrains. In order to compress point clouds without distortion and losing information, Sun et al. [30] demonstrate a 3D range-image oriented clustering scheme. Although the clustering-based methods are simple, the high iterate rate of each point in the point cloud leads to a high computation burden and defeats efficiency.

1.1.4. Learning Based Method

While deep learning-based methods often provide interesting results, the understanding of the type of coding solutions is essential to improve their design in order to be used effectively. To alleviate the cost of collecting and annotating large-scale point cloud data, Zhang and Zhu [31] propose an unsupervised learning approach to learn features from an unlabeled point cloud dataset by using part contrasting and object clustering with deep graph convolutional neural networks (GCNNs). Xu et al. [32] incoprate clustering method with the proposed FPCC-Net specially for industial bin-picking. *PointNet* [33] is the first machine learning neural network to deal with 3D raw points, and senstive to each points' immediate structure and geometry. Other current methods use deep learning directly on point clouds [34–37] or projections into a camera image [38] to segment instances in point clouds. Learning-based methods provide in an indoor scene but commonly suffer from long runtime and process large-scale point clouds.

1.2. Motivations

Although classical segmentation approaches mentioned above achieve promising segmentation results, one main drawback is huge computation-consuming, restricting their application to real-world, large-scale point cloud processing. To overcome this deficiency, we summarized two main strategies existing to accelerate the point cloud segmentation:

1.2.1. GPU vs. CPU

Conventional segmentation methods depend on the CPU's processing speed to run computations in a sequential fashion. Buys and Rusu [39] provide GPU-based version of EC [12] in the PCL [40] library and further extended by Nguyen et al. [41] who achieve 10 times speedup than the CPU-based EC.

Despite the fact that GPU enables faster segmentation, it is not practical for hardware devices with limited memory capacity and computation resources, such as mobile phones and small robotic systems (e.g., UAV), not to mention the steep price.

1.2.2. Pre-Knowledge vs. Unorganized

Taking advantage of pre-knowledge about the point cloud such as layer-based organized indexing in LiDAR data [42], relative pose between the LiDAR sensor, or point cloud in each 3D scan(frame) [43] accelerate segmentation speed. These assumptions provided by the structured point cloud hold in specific scenarios such as autonomous vehicle driving.

However, these premises are not available in many applications since not all the point cloud data are generated from the vehicle-installed LiDAR. For example, airborne laser scanning, RGB-D sensors, and image-based 3D reconstruction supply general organized data instead, making the pre-knowledge approaches [42,43] fail.

From the discussion above, we found out that an *efficient* and *low-cost* solution to *general* point cloud segmentation is vital for real-world applications but absent from research literature.

1.3. Contributions

As shown in Table 1, we attack the general point cloud segmentation problem and place an emphasis on computational speed as compared to the works that are considered state-of-the-art. The process of segmentation is proposed to be completed in two parts by our approach: (i) ground points removal and (ii) the clustering of the remaining points into meaningful sets. The proposed solution underwent extensively rigorous testing on both synthetic and real data. The results provided conclusive evidence that our method is superior to the other existing approaches. A fast segmentation redirects precious hardware resources to more computationally demanding processes in many application pipelines. The following is a condensed summary of the contributions that this work makes:

- We present a new Euclidean clustering algorithm to the point could instance segmentation problem by using point-wise against the cluster-wise scheme applied in existing works.
- The proposed fast Euclidean clustering (FEC) algorithm can handle the general point cloud as input without relying on the pre-knowledge or GPU assistance.
- Extensive synthetic and real data experiments demonstrate that the proposed solution achieves similar instance segmentation quality but 100× speedup over state-of-the-art approaches. The source code (implemented in C++, Matlab, and Python) is publicly available on https://github.com/feihuzhang/LiDARSeg.

Table 1. Summary of the related works.

Ref	[37]	[44]	[42]	[43]	[39]	[35]	[34]	[12]	[10]	[16]	[18]	[18]	FEC (Ours)
GPU-free			✓	✓				✓	✓	✓	✓	✓	✓
Unorganized point cloud					✓		✓	✓	✓	✓	✓	✓	✓

2. Materials and Methods

Our method concludes two steps: (i) ground points removal and (ii) the clustering of the remaining points.

2.1. Ground Surface Removal

Cloud points on the ground constitute the majority of input data and reduce the computation speed. Besides, the ground surface affects segmentation quality since it changes input connectivity. Therefore, it is essential to remove the ground surface as a pre-processing. Many ground points extraction methods such as grid-based [45] and plane fitting [44] have been used in existing works. The cloth simulation filter (CSF) [46], which is robust on complicated terrain surfaces, was the one that we decided to utilize to extract and eliminate ground points for this research.

2.2. Fast Euclidean Clustering

Similar to EC [12], we employ Euclidean (L2) distance metrics to measure the proximity of unorganized points and aggregate commonalities into the same cluster, which can describe as:

$$\min\|\mathbf{P}_i - \mathbf{P}_{i'}\|_2 \geqslant d_{\mathbf{th}} \quad (1)$$

where $\mathbf{C}_i = \{\mathbf{P}_i \in \mathbf{P}\}$ is a distinct cluster from $\mathbf{C}_i' = \{\mathbf{P}_i' \in \mathbf{P}\}$, and $d_{\mathbf{th}}$ is a maximum distance threshold.

Algorithm 1 describes the algorithmic processes and illustrates them with an example displayed in Figure 2. Note that the proposed algorithm uses **point-wise scheme**, which loop points with the input numbering order against the **cluster-wise scheme** used in EC and RG. The deployment of the proposed FEC is simple, requiring only 40 lines of code written in C++.

Time complexity. The complexity of constructing the kd-tree in the Big-O notation format is $O(3N \log N)$, where N is the input size of the total point number in $\{\mathbf{P}\}$. The cost of main loop is $O(N^2 \nu)$, where ν is an constant number determined by the 3D point density ρ and the neighbor radius threshold $d_{\mathbf{th}}$ as $\nu = \frac{4}{3}\pi {d_{\mathrm{th}}}^3 \rho$. Since $N^2 \nu > 3n \log N$, thus the overall cost is $O(N^2 \nu)$.

Algorithm 1: PROPOSED FEC

```
Input:
 • p_i ∈ P: unorganized point cloud
 • d_th: the neighbor radius threshold
 • Th_max: the max number of neighbor points
Output:
 • P.lab: point cloud data with labels
// Initialization:
P.lab ← 0: initialize all point labels as 0;
segLab ← 1: Segment label ;
Constructing kd-tree for P;
// Main loop:
foreach p_i.lab in P do
    if p_i.lable = 0 then
        P_NN = FindNeighbor(p_i,d_th);
        if ∃Nonzero(P_NN.lab) then
            minSegLab = min(Nonzero( P_NN.lab),SegLab);
        else
            minSegLab = SegLab;
        end
        // Segment merge loop:
        foreach p_j in P_NN do
            if p_j.lab > minSegLab then
                foreach p_k.lab in P do
                    if p_k.lab = p_j.lab then
                        p_k.lab = minSegLab;
                    end
                end
            end
        end
        SegLab + +;
    end
end
return point labels
```

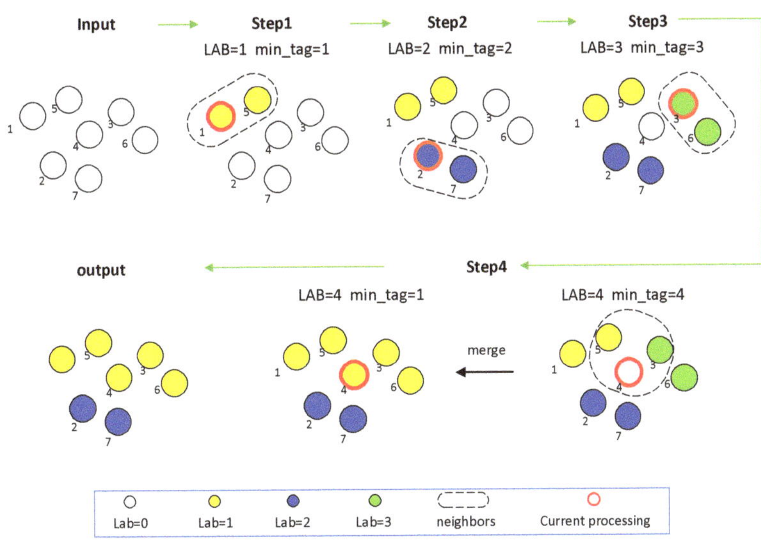

Figure 2. An example of FEC to point cloud segmentation. Note that FEC utilizes point-wise scheme with point index order.

2.3. Efficiency Analysis

The running time of the proposed FEC algorithm is analyzed in comparison to two different segmentation methodologies (i.e., RG [10] and EC [12]) on several typical examples in order to gain an understanding of the algorithm's effectiveness. This result intuitively explains the observations in our experiments that the FEC has significant advantages in terms of efficiency against others.

Setting. We assume a point cloud $\{\mathbf{P}\}$ with n clusters $\{\mathbf{C}_l \mid l \in [1, n]\}$. The number of points in $\{\mathbf{C}_i\}$ is m_i. Without losing the generality to investigate the limits of running time of EC, RG, and FEC, we draw three extreme distribution cases as shown in Figure 3. k_n represents the number of operations on all points in the inner loop of the algorithm, which is affected by the size of the point cloud. k_{sc} represents the number of searches for the kd-tree, which is affected by the number of clusters. Without loss of generality, we verify the value ranges of k in three extreme situations: Case 1, all points are merged into one cluster; Case 2, each point belongs to a cluster; Case 3, all points are distributed on a line. In all three cases, the value of k_{sc} is significantly less than k_n. While according to the Equation (3), the running time of RG and EC is positively correlated with k_n, and the running time of FEC is only affected by k_{sc}. Note that all three algorithms use a triple nested loop. Thus, we denote the single kd-tree search $\Omega(.)$ takes t_1 time while the other $O(1)$ cost processes take t_2, such as value assignment. The function for counting numbers is denoted by $\pi(.)$. Since FEC generates clusters by merging sub-clusters, we assume $\Theta(\mathbf{C}_l)$ is the function to get the sub-cluster sets.

Time consuming. With such a setting, the processing time of EC, RG, and FEC on $\{\mathbf{P}\}$ are:

$$\begin{aligned} t^{\mathrm{EC}} &= Nt_1 + k_{\mathrm{n}}t_2 \\ t^{\mathrm{RG}} &= Nt_1 + 2k_{\mathrm{n}}t_2 \\ t^{\mathrm{FEC}} &= k_{\mathrm{sc}}t_1 + k_{\mathrm{sc}}Nt_2 \end{aligned} \quad \text{with,} \quad \begin{aligned} 2N - 2n \leqslant k_{\mathrm{n}} < N^2 \\ 1 \leqslant k_{\mathrm{sc}} \leq N - n \end{aligned} \tag{2}$$

where k_{n} and k_{sc} defined as:

$$k_{\mathrm{n}} = \sum_{l=1}^{n} \pi(\Omega(\sum_{\in\{\mathbf{C}_l\}} \mathbf{P}_i)), \quad k_{\mathrm{sc}} = \sum_{l=1}^{n} \pi(\Theta(\mathbf{C}_l)) \tag{3}$$

Note that EC and RG loop every point in $\mathbf{C}_i$ and thus constantly consume Nt_1. Alternatively, FEC calls the number of sub-clusters times $\Omega(.)$, depending on the numbering order.

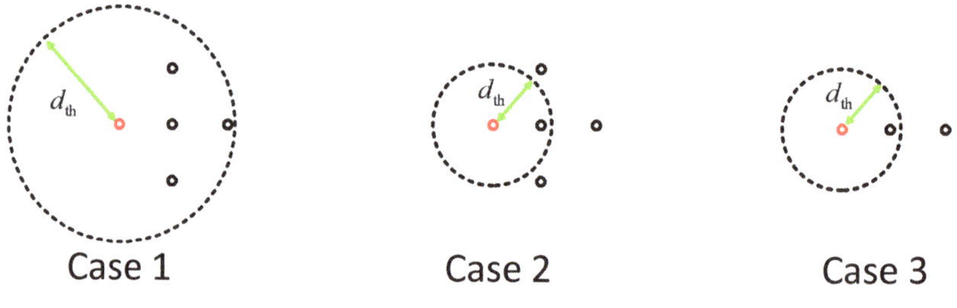

Figure 3. Values of k_{n} and k_{sc} in three extreme cluster distribution cases. Case 1 (maximus d_{th}): $k_{\mathrm{sc}} = 1$, $k_{\mathrm{n}} = N^2$; Case 2 (minimus d_{th}): $k_{\mathrm{sc}} = N - n$, $k_{\mathrm{n}} = 2N - 2n$; Case 3 (all points aligned in a line): $k_{\mathrm{sc}} = \frac{1}{2}(N - n)$, $k_{\mathrm{n}} = 2N - 2n$.

In the inner loops, EC and RG check if the neighbourhood points meets the criterion (minimum distance for EC, normal and curvature for RG) to be added into $\mathbf{C}_l$, and leads to $k_{\mathrm{n}}t_2$ and $2k_{\mathrm{n}}t_2$ respectively. While FEC needs to loop whole $\{\mathbf{P}\}$ with Nt_2 consuming in every sub-cluster merging.

To analyze the efficacy of the aforementioned three algorithms, we investigate the ratio between the total running time of the suggested FEC and that of the others:

$$\frac{t^{\text{FEC}}}{t^{\text{RG}}} < \frac{t^{\text{FEC}}}{t^{\text{EC}}} = \frac{k_{\text{sc}}(t_1 + Nt_2)}{Nt_1 + k_n t_2} < \frac{k_{\text{sc}}(t_1 + Nt_2)}{Nt_1 + Nt_2} < 1 \quad (4)$$

- In general setting, $k_{\text{sc}} \ll N$ which makes $\frac{t^{\text{FEC}}}{t^{\text{RG}}} < \frac{t^{\text{FEC}}}{t^{\text{EC}}} \approx \frac{k_{\text{sc}}}{N} \ll 1$ and lifting efficiency of FEC over EC and RG.
- By substituting the lower bound of k_n and upper bound of k_{sc} in Equation (2) into t^{EC}, t^{RG} and t^{FEC}, FEC is faster than EC and RG even with the extreme setting.

We interpolate the improvement of FEC raised by the point-wise scheme against the cluster-wise scheme used in EC and RG. This difference leads to significantly fewer calls to kd-tree search $\Omega(.)$ in the first loop.

3. Experiments and Results

3.1. Method Comparison

In our experiments, the proposed method *FEC* was compared to five state-of-the-art point cloud segmentation solutions:

- ***EC***: Classical Euclidean clustering algorithm [12] implemented in PCL library [40] (*EuclideanClusterExtraction* function).
- ***RG***: Classical region growing based point cloud segmentation solution [10] implemented in PCL library [40] (*RegionGrowing* function).
- ***SPGN***: Recent learning-based method [34] which is designed for small-scale indoor scenes (https://github.com/laughtervv/SGPN, Salt Lake City, USA).
- ***VoxelNet***: Recent learning-based method [35] which learns sparse point-wise features for voxels and use 3D convolutions for feature propagation (https://github.com/steph1793/Voxelnet, Salt Lake City, USA).
- ***LiDARSeg***: State-of-the-art instance segmentation [37] for large-scale outdoor LiDAR point clouds (https://github.com/feihuzhang/LiDARSeg, Paris, France).

Note that the value of d_{th} and Th_{max} were set as the same to **EC**, and **RG**. For a fair comparison, we remove the ground points in the point cloud (Section 2.1) and then use them as the input for **EC**, **RG**, and **FEC**.

3.2. Metrics Evaluation

We provide three metrics (average prevision, time complexity, space complexity) to demonstrate that the proposed **FEC** outperforms baselines on efficiency without penalty to effectiveness.

- ***Average precision:*** The **average precision (AP)** is a widely accepted point-based metric to evaluate the segmentation quality [37], as well as similar to criterion for COCO instance segmentation challenges [47]. The equation for AP can be presented as:

$$AP = \frac{TP}{TP + FP} \quad (5)$$

where TP is true positive, and FP is false positive. We use 0.75 as the point-wise threshold for true positives in the following experiments.
- ***Complexity:*** Both running time and memory consumption of real data experiments are designed to evaluate the time complexity and the space complexity. Our method was executed on a 32GB RAM, Intel Core I9 computer and compared to two other classical geometry-based methods: **EC** and **RG**. The NVIDIA 3090 GPU was utilized to evaluate both the learning-based methods **SPGN** and **VoxelNet**.

3.3. Synthetic Data

In this experiment, we evaluate the performances of **EC**, **RG**, and **FEC** respectively on synthetic point cloud data with increasing scale. Note that we generate unorganized points where the state-of-the-art learning-based methods **SPGN**, **VoxelNet**, and **LiDARSeg** which use single LiDAR scan as input, are not feasible in such setting.

3.3.1. Setting

We first divide 3D space into multiple voxels, and then generate clusters (segments) by filling m 3D points evenly inside the randomly selected n voxels. Under such a setting, we can control the cluster number by varying the value of n. Similarly, the cluster density can be determined by varying the value of m. Besides, we can also simulate clusters with different uniformities by changing the strategy of filling 3D points into voxel from even to random (even filling followed by random shift with variance σ). Note that we randomly set the 3D point index to make the synthetic an unorganized point cloud.

3.3.2. Varying Density of Cluster

In this experiment, we increase the density of synthetic data from 10 to 500 under varying total cluster numbers from 100 to 2200. As illustrated in Figure 4, the running time of both **RG** and **EC** grew significantly with increasing cluster densities under varying cluster numbers. An interesting observation is that the growth curve of **RG** is linear while **EC** provides exponential growth and overtakes the running time with a density of cluster larger than 500. In contrast, the proposed method **FEC** provides stable running times which are at least 100× faster performance over **EC** and **RG** with increasing density (number of points in unit volume) of each cluster.

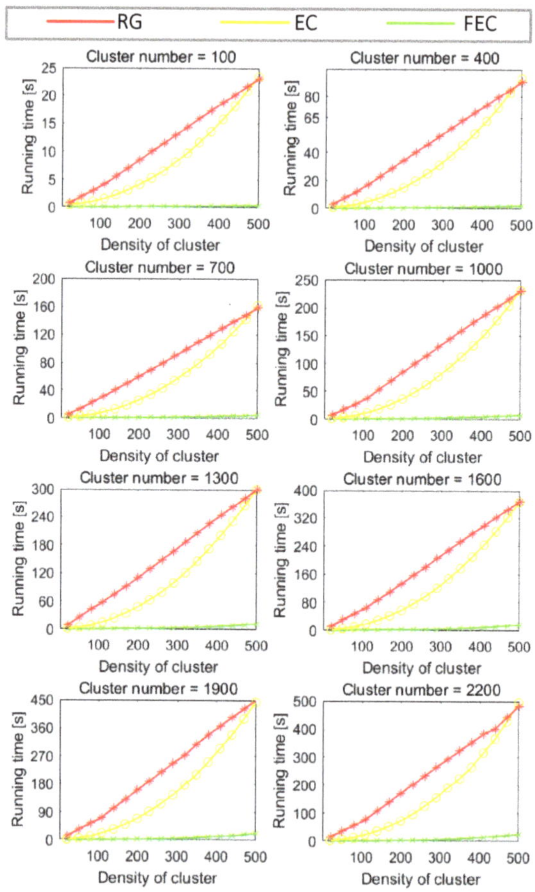

Figure 4. Running time for **EC** (yellow), **RG** (red) and **FEC** (green) with increasing density of different number of cluster.

3.3.3. Varying Number of Clusters

In this experiment, we fix the density of clusters to 200, and increase the number of clusters from 100 to 1900. Figure 5a demonstrates that the running time of **EC** and **RG** grow dramatically with an increasing number of clusters from milliseconds level to hundred seconds level. In contrast, **FEC** runs significantly faster with < 1 s performance under all configurations.

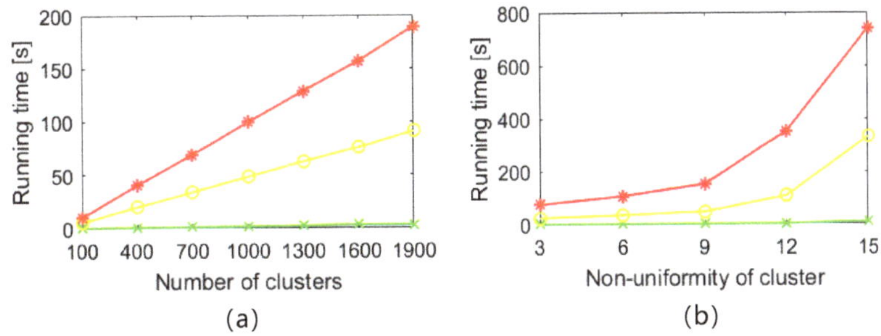

Figure 5. Running time for **EC** (yellow), **RG** (red) and **FEC** (green) with increasing number of clusters (**a**), and non-uniformity (**b**) of each cluster.

3.3.4. Varying Uniformity of Cluster

We alter the uniformity of each cluster in this experiment by simulating an increasing number of normal distribution sub-clusters. The results in Figure 5b reveal that the uniformity of the cluster obviously drags the running time of **EC** and **RG** with a hundredfold increase. In contrast, the uniformity of the cluster has slightly affected **FEC** without significant growth in running time. Besides, as shown in Figure 6 that the proposed method achieve fastest and stable performance with the increasing number of cluster and cluster density jointly.

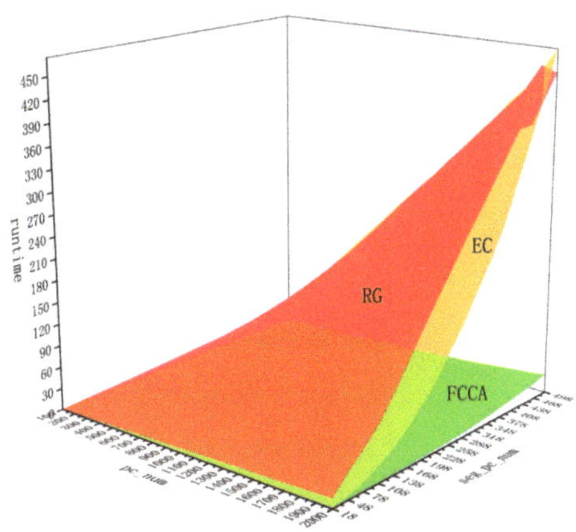

Figure 6. Running time for **EC** (yellow), **RG** (red) and **FEC** (green) with increasing dual-mixing of cluster number and density.

3.3.5. Segmentation Quality

In the synthetic data experiment, we observe that all three geometry-based methods **EC**, **RG**, and **FEC** provide segmentation precision approximate to 1 without significant difference.

3.4. Real Data

In this experiment, we evaluate the performance of all methods on a publicly available dataset, namely, KITTI odometry task [11] with point-wise instance label from semanticKITTI [48].

Following the instructions of [34,35,37], we trained the **SPGN**, **VoxelNet** and **LiDARSeg** using semanticKITTI. Thus, we compare the proposed method **FEC** against to classical geometry-based approaches **EC**, **RG** and learning based solutions **SPGN**, **VoxelNet**, **LiDARSeg** in 12 sequences of semanticKITTI.

3.4.1. FEC vs. Geometry-Based Methods on KITTI

We tested the **EC**, **RG** and the proposed **FEC** on real point cloud datasets from KITTI odometry point cloud sequence [11] with two common segmentation styles in practice, namely:

- **Inter-class segmentation:** the input for inter-class segmentation is a point cloud representing a single class, such as car, building, or tree, for example. Following the completion of the classification step, instance segmentation is carried out in a manner that is distinct for each of the classes.
- **Intra-class segmentation:** as input, intra-class segmentation utilizes multiple-class point clouds. In such a mode, the original LiDAR point cloud is utilized as input without classification.

Efficiency. As shown in Table 2 that quantitatively **FEC** achieves an average of 50× speedup against existing method **EC** and 100× **RG** under intra-class segmentation mode on 12 sequences form #03 to #11. In inter-class mode, **FEC** achieves average 30× and 40× speedup for car and building segmentation, 10× and 20× speedup for tree segmentation against to **EC** and **RG**. Besides, an interesting observation is that the running time of the three methods on #03 and #11 is nearly 2–3 times longer against the rest sequences under the intra-class mode. This is raised because the total number of instances in #03 and #11 is more extensive than the others, especially the tiny objects such as bicycles, trunks, fences, poles, and traffic signs. Since the geometry-based approaches will call more cluster processes in the loop with many small instances, thus, **RG**, **EC**, and **FEC** take much more running time on #03 and #11 sequences over the others.

Table 2. Experimental results on point clouds from KITTI vision benchmark [11]. Running time (in seconds s) of **EC**, **RG** and **FEC** on 11 sequences (#00-11 from odometry task) based on inter-class and intra-class styles are reported [in seconds]. Best results are shown in green.

Dataset (Point Number)	Mode	Class (Point Number)	EC [12] (s)	RG [10] (s)	FEC [Ours] (s)
#00 (1,498,667)	intra-class	-	414	780	140
	inter-class	car (55,048)	23.0	26.9	0.5
		building (131,351)	13.0	58.1	1.3
		tree (351,878)	51.9	167.0	9.9
#01 (1,249,590)	intra-class	-	289.4	651.0	103.3
	inter-class	car (62,469)	21.7	30.0	0.5
		building (204,911)	26.3	95.9	2.2
		tree (455,519)	150.3	232.7	6.3
#02 (1,406,316)	intra-class	-	323.2	721.3	105.9
	inter-class	car (85,528)	37.7	42.0	0.87
		building (209,704)	21.1	98.3	2.4
		tree (476,077)	118.1	242.2	23.3
#03 (1,714,700)	intra-class	-	931.1	908.1	90.4
	inter-class	car (89,337)	31.9	42.4	2.8
		building (210,300)	21.4	97.8	2.39
		tree (582,929)	167.1	298.6	17.2

Table 2. *Cont.*

Dataset (Point Number)	Mode	Class (Point Number)	EC [12] (s)	RG [10] (s)	FEC [Ours] (s)
#04 (1,324,991)	intra-class	-	313.6	682.6	82.1
	inter-class	car (89,646)	36.1	43.9	0.9
		building (212,195)	98.2	99.3	2.33
		tree (653,342)	292.4	335.9	18.0
#05 (1,239,293)	intra-class	-	319.3	631.3	44.0
	inter-class	car (96,472)	45.8	48.2	0.9
		building (241,977)	173.6	122.7	2.6
		tree (658,529)	127.3	335.1	21.0
#06 (1,276,859)	intra-class	-	319.3	631.3	44.0
	inter-class	car (99,480)	39.5	48.3	0.9
		building (241,977)	36.0	124.0	3.1
		tree (679,665)	158.9	345.8	25.5
#07 (1,556,851)	intra-class	-	413.1	801.3	123.5
	inter-class	car (102,205)	38.8	51.0	1.0
		building (274,741)	195.1	134.6	3.4
		tree (751,354)	156.6	383.3	32.2
#08 (1,828,399)	intra-class	-	265.7	417.3	49.1
	inter-class	car (113,404)	33.5	54.2	1.8
		building (307,844)	55.4	156.8	3.9
		tree (754,161)	189	395.9	26.3
#09 (1,116,058)	intra-class	-	225.7	563.5	62.7
	inter-class	car (115,489)	36.0	55.8	1.3
		building (296,751)	55.9	153.53	3.8
		tree (828,874)	229.3	428.6	35.1
#10 (1,172,405)	intra-class	-	280.4	598.5	59.9
	inter-class	car (115,617)	39.4	55.8	1.0
		building (311,047)	43.0	152.9	3.9
		tree (841,600)	230.4	431.2	24.8
#11 (2,002,769)	intra-class	-	948.2	1062.9	120.9
	inter-class	car (127,417)	51.7	61.6	1.3
		building (337,911)	48.6	166.6	4.3
		tree (870,785)	274.7	447.1	23.2

Memory-consuming. As shown in Table 3 that the proposed **FEC** consume only one third and half of memory against to **EC** and **RG**.

Table 3. Running time (in seconds), memory-consuming (in MB), and AP (average precision defined by [37]) over 11 sequences in KITTI [11] odometry task are reported. Best and second best results are shown in green and blue.

Method	Time				Memory (MB)	AP (%)
	Intra-Class	Inter-Class				
		Car	Building	Tree		
EC	387.8	36.3	65.7	195.6	211	65.8
RG	706.0	46.7	121.7	337.0	145	61.1
FEC	87.8	1.2	3.0	21.9	89	65.5

Effectiveness. As shown in Table 3 quantitatively, all three geometry-based methods provide similar instance segmentation quality with AP larger than 60%. Specifically, **EC** provides the best segmentation accuracy at 65.8% while the proposed **FEC** achieves a slightly lower score at 65.5%. The qualitative segmentation results are shown in Figure 7 with sequence #00 and #11 as two examples. It verifies our statement that our method and baseline methods provide similar segmentation quality globally. Interestingly, we found out that the proposed **FEC** provides better segmentation quality in handling details segment.

As shown in Figure 8, for the points on the tree, which are sparer and nonstructural than the other classes, EC and RG often suffer from over-segmentation and under-segmentation problems. For example, in the second row of Figure 8, the three independent trees are clustered into the building by EC and RG while FEC successfully detects them. In summary, as shown in Table 3 demonstrates that our method achieves significant improvement in efficiency while ***without penalty to the performance (quality)***.

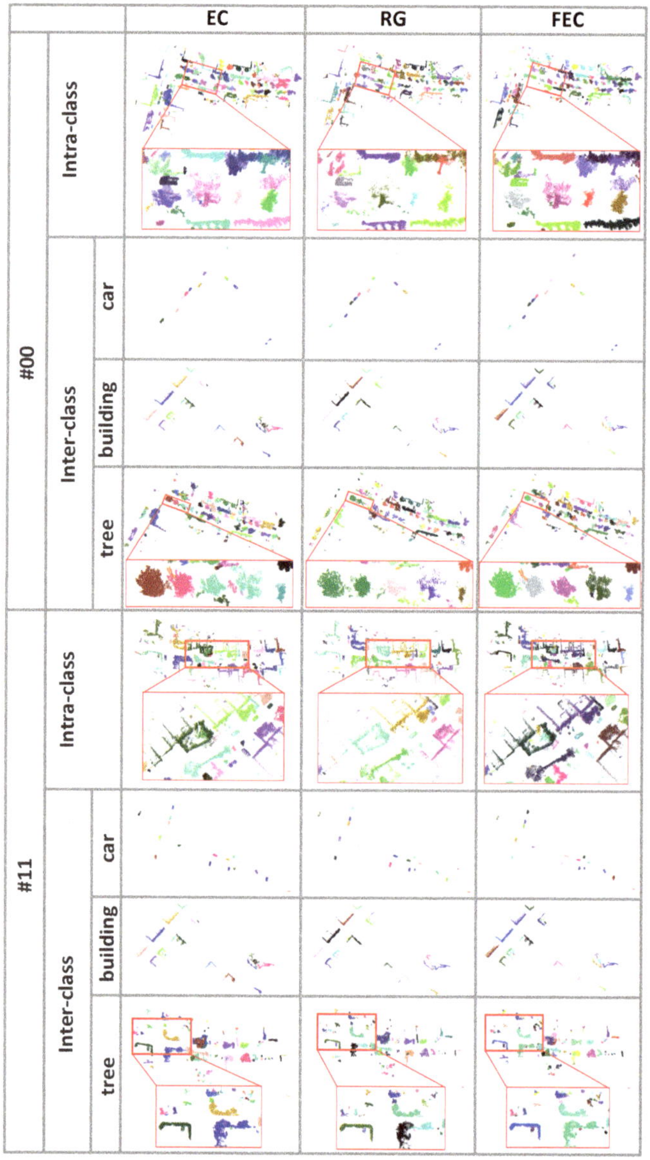

Figure 7. Qualitative segmentation results of *EC*, *RG* and *FEC* on KITTI odometry dataset [11] sequence #00 and #11.

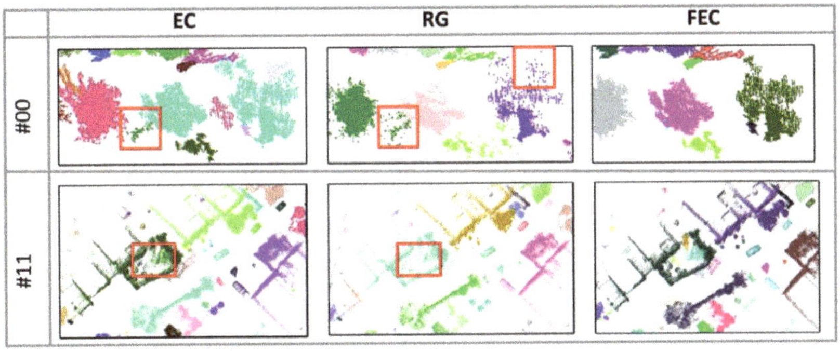

Figure 8. Details and highlights of segmentation results of *EC*, *RG* and *FEC* on KITTI odometry dataset [11] sequence #00 and #11.

3.4.2. FEC vs. Learning-Based Methods on KITTI

In this experiment, we compare the proposed method **FEC** to state-of-the-art learning-based solutions on sequence #00-#11 of the KITTI odometry dataset [48]. We trained the learning approaches **SPGN** [34], **VoxelNet** [35] and **LiDARSeg** [37] semanticKITTI labeling [48], and evaluated the running time and average precision (AP) defined in [37]. Note that all the learning-based methods were tested with GPU while the geometry solutions **EC** [12], **RG** [10], and **FEC** run with CPU only.

Data input. Note that in the Section 3.4.1 we use the point cloud of the whole sequence as input to three geometry-based methods. However, such large scale input is not feasible to the state-of-the-art deep learning based approaches for instance segmentation or 3D detection, namely, **SGPN**, **VoxelNet**, and **LiDARSeg**. Thus, in this experiment, we alternatively spilt the 12 sequences in the KITTI odometry dataset into a single scan as input. Besides, we also report the performances of **EC** and **RG**.

Training and hardware. Particularly, we use the ground-truth point-wise instance label from semanticKITTI panoptic segmentation dataset [48] to train textbfSGPN, **VoxelNet**, and **LiDARSeg** with a Nvidia 3090 GPU while forcing the CPU-only mode for **EC**, **RG** and proposed **FEC**.

Efficiency. Please note that since **LiDARSeg** [37] is designed for segmenting the single LiDAR scan, thus the running time we recorded below is the average process time of each scene instead of the whole sequence. As shown in Table 4 that **FEC** achieves 5× speed up against to **EC**, **VoxelNet** and **LiDARSeg**, 10× to **RG**, and 20× to **SPGN**. Since all the geometry-based methods require ground surface removal as pre-process, thus for a fair comparison, the time-consumings of ground surface detection and removal have been considered in the total running time. It is important to notice that **FEC** relies on CPU calculation only while all the learning-based approaches are accelerated with GPU in inference without mentioning the huge time-consuming in the training stage.

Effectiveness. As shown in Table 4 that quantitatively all three geometry-based methods **EC**, **RG** and **FEC** achieve similar instance segmentation quality with AP around 62%. Note that the segmentation accuracy of geometry-based approaches is even slightly better than the leaning-based ones with AP at 59% of LiDARSeg, 55.7% of **VoxelNet** and 46.3% of **SPGN**. The qualitative comparisons are shown in Figure 9 with 2 scans in sequence #00 as examples. We can observe that the proposed solution **FEC** method provides similar segmentation quality as the state-of-the-art learning-based method **LiDARSeg**.

Table 4. Comparisons to state-of-the-art learning based solutions sequence #00-#11 of KITTI odometry dataset [48]. Note that since **LiDARSeg** [37] is designed for segmenting the single LiDAR scan, thus the running time we reported is the average process time of each scene instead of the whole sequence. Best results are shown in green.

Sequence	GPU: Nvidia 3090X1			CPU: Intel I9		
	SPGN [34] (ms)	**VoxelNet [35] (ms)**	**LiDARSeg [37] (ms)**	**FEC [Ours] (ms)**	**RG [10] (ms)**	**EC [12] (ms)**
#00	1510.9	442.7	470.2	140.8	780.8	414.6
#01	1705.4	454.1	471.9	103.3	651.0	289.4
#02	1605.6	427.4	504.8	105.9	721.3	323.2
#03	1875.3	380.6	386.0	90.4	908.1	931.1
#04	1523.2	469.0	506.2	82.2	682.7	313.7
#05	1675.0	453.4	383.1	44.0	631.4	319.3
#06	1749.4	606.8	481.1	71.7	653.7	318.5
#07	1496.2	460.1	439.9	53.6	801.4	123.6
#08	1816.3	644.2	452.7	49.1	417.3	165.8
#09	1509.2	472.7	465.9	62.8	563.5	225.8
#10	1558.0	396.4	485.6	60.0	598.5	280.4
#11	1485.6	362.9	422.0	121.0	1062.9	948.3
Average time (ms)	1625.8	464.2	455.8	87.9	706.0	387.8
AP (%)	46.3	55.7	59.0	63.2	61.0	63.1

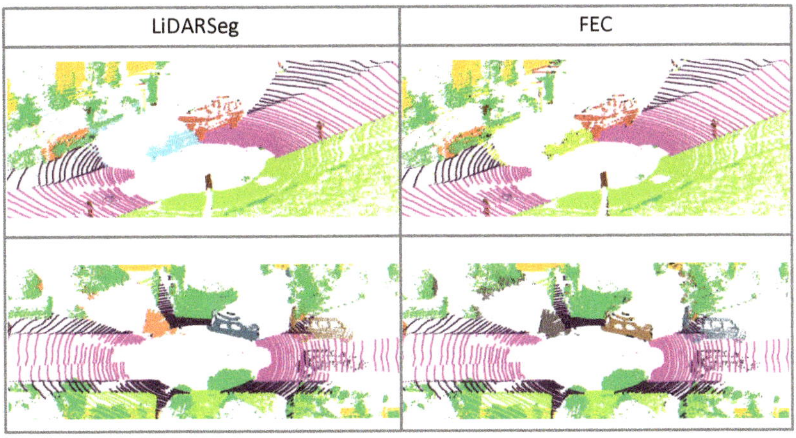

Figure 9. Comparisons to state-of-the-art learning based solution **LiDARSeg** [37] on KITTI odometry dataset [48] with semanticKITTI labeling [48]. Note that **LiDARSeg** [37] only applies to a single LiDAR scan as input instead of a long sequence.

4. Discussion

If FEC is faster?

Both the synthetic (Section 3.3) and real data (Section 3.4) experiments demonstrate that the proposed **FEC** outperform the state-of-the-art geometry-based and learning-based methods with a significant margin. Specifically, our method is an order of magnitude faster than **RG** [10] and nearly 5 times faster than **EC** [12]. Besides, without GPU acceleration, **FEC** is nearly 5 times faster **LiDARSeg** [37], 6 times faster than **VoxelNet** [35], and nearly 20 times faster than **SPGN** [34].

If FEC losses accuracy?

One may have concern that if **FEC** will lose segmentation accuracy in order to accelerate the processing speed. Both the synthetic (Section 3.3) and real data (Section 3.4) experiments verify that **FEC** provides similar segmentation quality as the conventional geometry-based methods **RG** [10], **EC** [12], and even slightly better than the learning-based solutions **LiDARSeg** [37], **VoxelNet** [35] and **SPGN** [34]. Thus, we point out that the proposed solution **FEC** achieves significant improvement in efficiency while ***without penalty to the performance (quality)***.

Why FEC is faster?

Based on the intuitive analysis in the Section 2.3, we interpret the **FEC** brings significant improvement , n efficiency mainly due to the point-wise scheme over the cluster-wise scheme used in existing works **RG** [10] and **EC** [12]. Such a novel point-wise scheme leads to significantly fewer calls to kd-tree search in the loop, which is the key to reducing the running time significantly.

Where we can use FEC?

The proposed solution **FEC** is a GPU-free faster point cloud instance segmentation solution. It can handle general point cloud data as input without relying on scene scale (e.g., single LiDAR scan) or structure pre-knowledge (e.g., scan line id). Thus, we can apply **FEC** to large-scale point cloud instance segmentation in various 3D perception (computer vision, remote sensing) and 3D reconstruction (autonomous driving, virtual reality) tasks.

5. Conclusions

This paper introduces an efficient solution to a general point cloud segmentation task based on a novel algorithm named faster Euclidean clustering. Our experiments have shown that our methods provide similar segmentation results but with 100× higher speed than the existing approaches. We interpret this improved efficiency as using the point-wise scheme against the cluster-wise scheme in existing works.

Future work. The current implementation of **FEC** is based on a serial computation strategy. Since the point-wise scheme contains multiple associations between the outer and inner loops, thus the parallel computation strategy could be applied to **FEC** for a potential acceleration.

Author Contributions: Conceptualization, Y.C., Y.X. and Y.L.; methodology, Y.C., Y.W. and Y.L.; validation, Y.C. and H.Z.; formal analysis, Y.C. and Y.L.; investigation, Y.W. and H.Z.; resources, Y.L.; data curation, Y.C. and H.Z.; writing—original draft preparation, Y.C., Y.X. and Y.L.; writing—review and editing, Y.X. and Y.L.; visualization, Y.C., Y.L.; supervision, Y.X. and Y.L.; project administration, Y.L.; funding acquisition, Y.X. and Y.L. All authors have read and agreed to the published version of the manuscript.

Funding: This research was funded by (1) Nature Science Foundation of China (Grant No. 62102145), (2) Jiangxi Provincial 03 Specific Projects and 5G Program (Grant No. 20212ABC03A09), (3) Scientific and Technological Innovation Project of Jiangxi Provincial Department of Natural Resources (Grant No. ZRKJ20222312).

Institutional Review Board Statement: Not applicable.

Informed Consent Statement: Not applicable.

Data Availability Statement: Not applicable.

Conflicts of Interest: The authors declare no conflict of interest.

References

1. Vo, A.V.; Truong-Hong, L.; Laefer, D.F.; Bertolotto, M. Octree-based region growing for point cloud segmentation. *ISPRS J. Photogramm. Remote. Sens.* **2015**, *104*, 88–100. [CrossRef]
2. Dewan, A.; Caselitz, T.; Tipaldi, G.D.; Burgard, W. Motion-based detection and tracking in 3D LiDAR scans. In Proceedings of the 2016 IEEE International Conference on Robotics and Automation (ICRA), Stockholm, Sweden, 16–21 May 2016; pp. 4508–4513.
3. Zucker, S.W.; Hummel, R.A. A Three-Dimensional Edge Operator. *IEEE Trans. Pattern Anal. Mach. Intell.* **1981**, *PAMI-3*, 324–331. [CrossRef]
4. Monga, O.; Deriche, R. 3D edge detection using recursive filtering: Application to scanner images. In *1989 IEEE Computer Society Conference on Computer Vision and Pattern Recognition*; IEEE Computer Society: Washington, DC, USA, 1989; pp. 28–35.
5. Monga, O.; Deriche, R.; Rocchisani, J.M. 3D edge detection using recursive filtering: Application to scanner images. *CVGIP Image Underst.* **1991**, *53*, 76–87. [CrossRef]
6. Wani, M.; Batchelor, B. Edge-region-based segmentation of range images. *IEEE Trans. Pattern Anal. Mach. Intell.* **1994**, *16*, 314–319. [CrossRef]
7. Sappa, A.D.; Devy, M. Fast range image segmentation by an edge detection strategy. In Proceedings of the International Conference on 3-D Digital Imaging and Modeling, Quebec City, QC, Canada, 28 May–1 June 2001.
8. Wani, M.A.; Arabnia, H.R. Parallel Edge-Region-Based Segmentation Algorithm Targeted at Reconfigurable MultiRing Network. *J. Supercomput.* **2003**, *25*, 43–62. [CrossRef]
9. Öztireli, A.C.; Guennebaud, G.; Gross, M. Feature Preserving Point Set Surfaces based on Non-Linear Kernel Regression. *Comput. Graph. Forum* **2009**, *28*, 493–501. [CrossRef]
10. Rabbani, T.; Van Den Heuvel, F.; Vosselmann, G. Segmentation of point clouds using smoothness constraint. *Int. Arch. Photogramm. Remote. Sens. Spat. Inf. Sci.* **2006**, *36*, 248–253.
11. Geiger, A.; Lenz, P.; Urtasun, R. Are we ready for autonomous driving? the KITTI vision benchmark suite. In Proceedings of the 2012 IEEE Conference on Computer Vision and Pattern Recognition, Providence, RI, USA, 16–21 June 2012; pp. 3354–3361.
12. Rusu, R.B. Semantic 3D object maps for everyday manipulation in human living environments. *KI-Künstliche Intelligenz* **2010**, *24*, 345–348. [CrossRef]
13. Xu, Y.; Hoegner, L.; Tuttas, S.; Stilla, U. Voxel- and graph-based point cloud segmentation of 3D scenes using perceptual grouping laws. *ISPRS Ann. Photogramm. Remote. Sens. Spat. Inf. Sci.* **2017**, *4* , 43–50. [CrossRef]
14. Huang, M.; Wei, P.; Liu, X. An Efficient Encoding Voxel-Based Segmentation (EVBS) Algorithm Based on Fast Adjacent Voxel Search for Point Cloud Plane Segmentation. *Remote Sens.* **2019**, *11*, 2727. [CrossRef]

15. Jagannathan, A.; Miller, E.L. Three-dimensional surface mesh segmentation using curvedness-based region growing approach. *IEEE Trans. Pattern Anal. Mach. Intell.* **2007**, *29*, 2195–2204. [CrossRef] [PubMed]
16. Klasing, K.; Althoff, D.; Wollherr, D.; Buss, M. Comparison of surface normal estimation methods for range sensing applications. In Proceedings of the 2009 IEEE International Conference on Robotics and Automation, Kobe, Japan, 12–17 May 2009; pp. 3206–3211.
17. Che, E.; Olsen, M.J. Multi-scan segmentation of terrestrial laser scanning data based on normal variation analysis. *ISPRS J. Photogramm. Remote. Sens.* **2018**, *143*, 233–248. [CrossRef]
18. Belton, D.; Lichti, D.D. Classification and segmentation of terrestrial laser scanner point clouds using local variance information. *Int. Arch. Photogramm. Remote Sens. Spat. Inf. Sci* **2006**, *36*, 44–49.
19. Li, D.; Cao, Y.; Tang, X.s.; Yan, S.; Cai, X. Leaf Segmentation on Dense Plant Point Clouds with Facet Region Growing. *Sensors* **2018**, *18*, 3625. [CrossRef]
20. Habib, A.; Lin, Y.J. Multi-Class Simultaneous Adaptive Segmentation and Quality Control of Point Cloud Data. *Remote Sens.* **2016**, *8*, 104. [CrossRef]
21. Dimitrov, A.; Golparvar-Fard, M. Segmentation of building point cloud models including detailed architectural/structural features and MEP systems. *Autom. Constr.* **2015**, *51*, 32–45. [CrossRef]
22. Grilli, E.; Menna, F.; Remondino, F. A review of point clouds segmentation and classification algorithms. *Int. Arch. Photogramm. Remote. Sens. Spat. Inf. Sci.* **2017**, *42*, 339. [CrossRef]
23. Lavoué, G.; Dupont, F.; Baskurt, A. A new CAD mesh segmentation method, based on curvature tensor analysis. *Comput.-Aided Des.* **2005**, *37*, 975–987. [CrossRef]
24. Yamauchi, H.; Gumhold, S.; Zayer, R.; Seidel, H.P. Mesh segmentation driven by Gaussian curvature. *Vis. Comput.* **2005**, *21*, 659–668. [CrossRef]
25. Ester, M.; Kriegel, H.P.; Sander, J.; Xu, X. A density-based algorithm for discovering clusters in large spatial databases with noise. *KDD-96 Proc.* **1996**, *96*, 226–231.
26. Xu, R.; Xu, J.; Wunsch, D.C. Clustering with differential evolution particle swarm optimization. In Proceedings of the IEEE Congress on Evolutionary Computation, Barcelona, Spain, 18–23 July 2010; pp. 1–8. [CrossRef]
27. Shi, B.Q.; Liang, J.; Liu, Q. Adaptive simplification of point cloud using k-means clustering. *Comput.-Aided Des.* **2011**, *43*, 910–922. [CrossRef]
28. Kong, D.; Xu, L.; Li, X.; Li, S. K-Plane-Based Classification of Airborne LiDAR Data for Accurate Building Roof Measurement. *IEEE Trans. Instrum. Meas.* **2014**, *63*, 1200–1214. [CrossRef]
29. Chehata, N.; David, N.; Bretar, F. LIDAR data classification using hierarchical K-means clustering. In Proceedings of the ISPRS Congress Beijing 2008, Beijing, China, 2–10 September 2008; Volume 37, pp. 325–330.
30. Sun, X.; Ma, H.; Sun, Y.; Liu, M. A Novel Point Cloud Compression Algorithm Based on Clustering. *IEEE Robot. Autom. Lett.* **2019**, *4*, 2132–2139. [CrossRef]
31. Zhang, L.; Zhu, Z. Unsupervised Feature Learning for Point Cloud Understanding by Contrasting and Clustering Using Graph Convolutional Neural Networks. In Proceedings of the 2019 International Conference on 3D Vision (3DV), Quebec City, QC, Canada, 16–19 September 2019; pp. 395–404. [CrossRef]
32. Xu, Y.; Arai, S.; Liu, D.; Lin, F.; Kosuge, K. FPCC: Fast point cloud clustering-based instance segmentation for industrial bin-picking. *Neurocomputing* **2022**, *494*, 255–268. [CrossRef]
33. Charles, R.Q.; Su, H.; Kaichun, M.; Guibas, L.J. PointNet: Deep Learning on Point Sets for 3D Classification and Segmentation. In Proceedings of the 2017 IEEE Conference on Computer Vision and Pattern Recognition (CVPR), Honolulu, HI, USA, 21–26 July 2017; pp. 77–85. [CrossRef]
34. Wang, W.; Yu, R.; Huang, Q.; Neumann, U. Sgpn: Similarity group proposal network for 3d point cloud instance segmentation. In Proceedings of the IEEE Conference on Computer Vision and Pattern Recognition, Salt Lake City, UT, USA, 18–23 June 2018; pp. 2569–2578.
35. Zhou, Y.; Tuzel, O. Voxelnet: End-to-end learning for point cloud based 3d object detection. In Proceedings of the IEEE Conference on Computer Vision and Pattern Recognition, Salt Lake City, UT, USA, 18–23 June 2018; pp. 4490–4499.
36. Lahoud, J.; Ghanem, B.; Pollefeys, M.; Oswald, M.R. 3D instance segmentation via multi-task metric learning. In Proceedings of the IEEE/CVF International Conference on Computer Vision, Seoul, Korea, 27 October–2 November 2019; pp. 9256–9266.
37. Zhang, F.; Guan, C.; Fang, J.; Bai, S.; Yang, R.; Torr, P.H.; Prisacariu, V. Instance segmentation of lidar point clouds. In Proceedings of the 2020 IEEE International Conference on Robotics and Automation (ICRA), Paris, France, 31 May–31 August 2020; pp. 9448–9455.
38. Wang, B.H.; Chao, W.L.; Wang, Y.; Hariharan, B.; Weinberger, K.Q.; Campbell, M. LDLS: 3-D object segmentation through label diffusion from 2-D images. *IEEE Robot. Autom. Lett.* **2019**, *4*, 2902–2909. [CrossRef]
39. Rusu, R.B.; Blodow, N.; Marton, Z.C.; Beetz, M. Close-range scene segmentation and reconstruction of 3D point cloud maps for mobile manipulation in domestic environments. In Proceedings of the 2009 IEEE/RSJ International Conference on Intelligent Robots and Systems, St. Louis, MI, USA, 10–15 October 2009; pp. 1–6.
40. Rusu, R.B.; Cousins, S. 3D is here: Point cloud library (pcl). In Proceedings of the 2011 IEEE iNternational Conference on Robotics and Automation, Shanghai, China, 9–13 May 2011; pp. 1–4.

41. Nguyen, A.; Cano, A.M.; Edahiro, M.; Kato, S. Fast Euclidean Cluster Extraction Using GPUs. *J. Robot. Mechatron.* **2020**, *32*, 548–560. [CrossRef]
42. Zermas, D.; Izzat, I.; Papanikolopoulos, N. Fast segmentation of 3D point clouds: A paradigm on lidar data for autonomous vehicle applications. In Proceedings of the 2017 IEEE International Conference on Robotics and Automation (ICRA), Singapore, 29 May–3 June 2017; pp. 5067–5073.
43. Bogoslavskyi, I.; Stachniss, C. Fast range image-based segmentation of sparse 3D laser scans for online operation. In Proceedings of the 2016 IEEE/RSJ International Conference on Intelligent Robots and Systems (IROS), Daejeon, Korea, 9–14 October 2016; pp. 163–169.
44. Himmelsbach, M.; Hundelshausen, F.V.; Wuensche, H.J. Fast segmentation of 3D point clouds for ground vehicles. In Proceedings of the 2010 IEEE Intelligent Vehicles Symposium, La Jolla, CA, USA, 21–24 June 2010; pp. 560–565.
45. Thrun, S.; Montemerlo, M.; Dahlkamp, H.; Stavens, D.; Aron, A.; Diebel, J.; Fong, P.; Gale, J.; Halpenny, M.; Hoffmann, G.; et al. Stanley: The robot that won the DARPA Grand Challenge. *J. Field Robot.* **2006**, *23*, 661–692. [CrossRef]
46. Zhang, W.; Qi, J.; Wan, P.; Wang, H.; Xie, D.; Wang, X.; Yan, G. An easy-to-use airborne LiDAR data filtering method based on cloth simulation. *Remote Sens.* **2016**, *8*, 501. [CrossRef]
47. Lin, T.Y.; Maire, M.; Belongie, S.; Hays, J.; Perona, P.; Ramanan, D.; Dollár, P.; Zitnick, C.L. Microsoft coco: Common objects in context. In *European Conference on Computer Vision*; Springer: Cham, Switzerland, 2014; pp. 740–755.
48. Behley, J.; Garbade, M.; Milioto, A.; Quenzel, J.; Behnke, S.; Stachniss, C.; Gall, J. Semantickitti: A dataset for semantic scene understanding of lidar sequences. In Proceedings of the IEEE/CVF International Conference on Computer Vision, Seoul, Korea, 27 October–2 November 2019; pp. 9297–9307.

drones

MDPI

Article

Multidomain Joint Learning of Pedestrian Detection for Application to Quadrotors

Yuan-Kai Wang [1], Jonathan Guo [1] and Tung-Ming Pan [2,*]

[1] Department of Electrical Engineering, Fu Jen Catholic University, New Taipei 242, Taiwan
[2] Holistic Education Center, Department of Information Innovation and Digital Life, Fu Jen Catholic University, New Taipei 242, Taiwan
* Correspondence: macgyver@fju.edu.tw

Abstract: Pedestrian detection and tracking are critical functions in the application of computer vision for autonomous driving in terms of accident avoidance and safety. Extending the application to drones expands the monitoring space from 2D to 3D but complicates the task. Images captured from various angles pose a great challenge for pedestrian detection, because image features from different angles tremendously vary and the detection performance of deep neural networks deteriorates. In this paper, this multiple-angle issue is treated as a multiple-domain problem, and a novel multidomain joint learning (MDJL) method is proposed to train a deep neural network using drone data from multiple domains. Domain-guided dropout, a critical mechanism in MDJL, is developed to self-organize domain-specific features according to neuron impact scores. After training and fine-tuning the network, the accuracy of the obtained model improved in all the domains. In addition, we also combined the MDJL with Markov decision-process trackers to create a multiobject tracking system for flying drones. Experiments are conducted on many benchmarks, and the proposed method is compared with several state-of-the-art methods. Experimental results show that the MDJL effectively tackles many scenarios and significantly improves tracking performance.

Keywords: pedestrian detection; multiobject tracking; multitask learning; multidomain joint learning; MDJL; drone application

Citation: Wang, Y.-K.; Guo, J.; Pan, T.-M. Multidomain Joint Learning of Pedestrian Detection for Application to Quadrotors. *Drones* **2022**, *6*, 430. https://doi.org/10.3390/drones6120430

Academic Editor: Anastasios Dimou

Received: 17 November 2022
Accepted: 16 December 2022
Published: 19 December 2022

1. Introduction

Pedestrian detection and tracking applications involving intelligent robots or flying drones [1–3] facilitate the determination of the position of a human to avoid collisions and the appropriate direction of a camera when walking or acquiring images. Both applications, which are used in autonomous cars, can assist computers in determining whether braking should be performed to slow down or avoid pedestrians [4,5].

Pedestrian detection and tracking in flying drones are used as examples because these miniature aircraft can move arbitrarily in the sky; the acquired images will be different because of the flight height and angle of the mounted camera. Thus, pedestrian images obtained with a downward orientation are different from those acquired with a forward-pointed camera. This causes significant challenges when pedestrian detection and tracking are applied to drones.

However, to train a network that covers multiple domains in an application scenario, the collection of videos and the preparation of annotations to create a dataset that meets our needs is time-consuming and labor intensive. Such multiple domains can then be integrated to use the resulting multidomain datasets [6]. However, after referring to the relevant literature regarding this approach, we found that such datasets will influence each other in each domain, making it difficult for the network to learn how to extract key features, reducing performance. This paper proposes a training method based on domain-guided dropout [7] that uses network domain information to enhance the performance of the network when learning. In addition, our detection method was applied to a pedestrian tracking system that is suitable for drones.

In practice, it is time-consuming to collect video data and generate the corresponding annotations. The most intuitive method is to combine multiple ready-made datasets to train the model. However, in the experiments, the classification accuracy of the model trained with this method was slightly reduced when applied to the test data of each dataset. We believe that this phenomenon occurred because the different video datasets have different scenes, angles, shooting lengths, and equipment. Therefore, the features of the video data in each dataset varied. The neural network is easily affected by these differences during training, and learning a method for efficiently extracting pedestrian features is impossible, resulting in performance degradation.

In computer vision, a domain can be broadly referred to as a collection of data groups with the same features. For example, Figure 1 shows three pedestrian detection datasets that belong to different domains. The objective is to train the pedestrian detection network by combining the datasets of multiple domains so that the learned model can be more generalized to cope with the challenges of different domains.

(a) (b) (c)

Figure 1. Pedestrian detection datasets belonging to different domains: (**a**) videos recorded using the driving recorder, (**b**) videos of campus surveillance, and (**c**) aerial videos obtained using flying drones.

The emergence of superlarge datasets, such as ImageNet, has led to the flourishing of neural networks [8]. Starting from the pretrained model using a superlarge dataset to perform further fine-tuned training, also known as transfer learning, has a better effect than using a randomly generated initial network. However, it is difficult to obtain this large dataset in various domains. Therefore, research units in different domains often create smaller datasets, but the simultaneous use of multiple datasets to train a model is challenging. In this paper, we call this dataset, composed of multiple datasets of different domains, a multidomain dataset.

Domain-guided dropout (DGD) [7] is a dropout method wherein neurons that respond strongly to a specific domain may not respond as strongly to other domains. By analyzing the degree of response of each neuron in the network to each domain, the probability of neuron dropout during the training can be adjusted. This improves the specificity and sensitivity of the neuron to features, thereby improving accuracy.

Our proposed pedestrian detection method, multidomain joint learning (MDJL), is derived from multitask learning. It uses the DGD characteristics to allow the network to effectively learn the features during training with the multidomain dataset to improve accuracy and generalizability across the domains. It was applied to the multi-pedestrian tracking algorithm for use in the changeable environment of a flying drone.

Figure 2 shows the flowchart for the proposed method. First, initialization training is performed so that the network can initially learn the features of each domain. The neuron impact score (NIS) of each domain is then calculated. The dropout in the network is replaced with the DGD, and training is continued. Finally, domain-specific fine-tuning is performed for each domain-specific neuron.

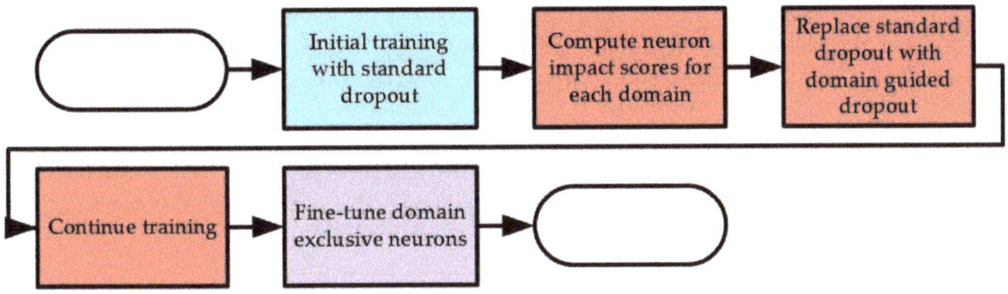

Figure 2. MDJL flow chart.

This process divides the network into several subnets, as shown in Figure 3, and each subnet is specialized to a specific domain. Given the characteristics of the DGD, subnets may overlap, which means that the domains have common data features.

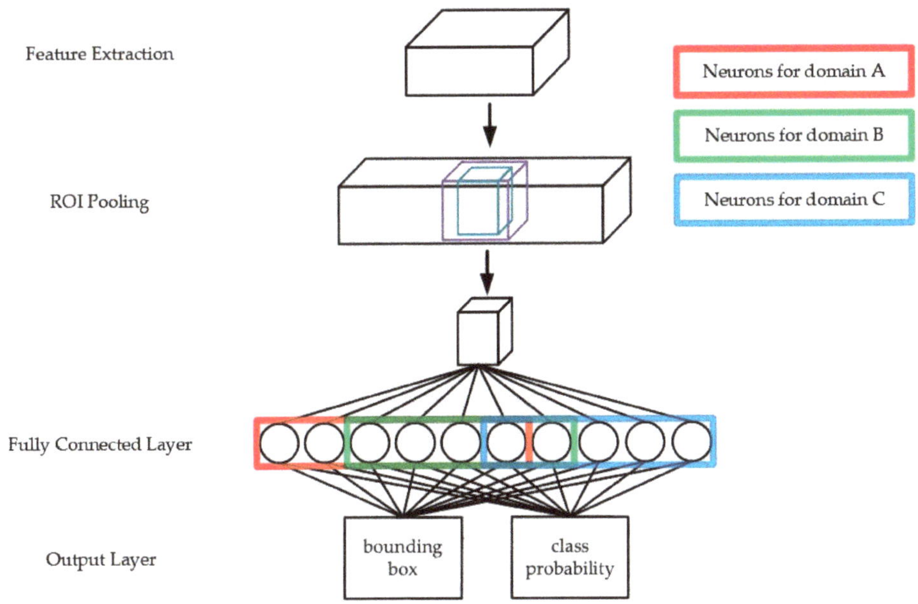

Figure 3. Schematic of MDJL subnet segmentation.

The remainder of this paper is organized into different sections. Section 2 reviews related deep learning methods and tracking schemes. The MDJL method is presented in Section 3. Experimental results and comparisons of the proposed technique with state-of-the-art methods are presented in Section 4. Finally, the conclusions and suggestions for future work are summarized in Section 5.

2. Related Works

Pedestrian detection and human pose estimation are typical surveillance applications [9–11]. Pedestrian detection algorithms can use traditional machine-learning methods on the basis of feature extraction and a learnable classifier. The architecture of this method extracts the specified features of the image (e.g., Haar feature [12], histogram of oriented gradients (HOG) [13], and deformable part models (DPM) [14]) and then uses sliding windows of different sizes to sample the feature images. These samples are finally sent to a trained classifier (e.g., support vector machine (SVM) [15]) for classification to achieve pedestrian detection.

Traditional object detection methods typically use image pyramid and sliding-window methods to address a problem in which the size of objects in an image is not fixed. In addition, to reduce the number of samples for a single image, some methods analyze additional information extracted from the image, such as the dominant ground plane, to limit the sliding-window range [16]. Methods such as Viola–Jones [12], HOG [13], convolutional channel filter [17], and checkerboard [18] are the representative methods of traditional machine learning.

Another pedestrian detection algorithm is the neural-network method [19]. However, the developed multilayer perceptron (MLP) cannot efficiently process multidimensional video data. LeCun et al. [20] proposed the convolutional neural network (CNN) as network architecture. CNN can be divided into two layers: feature extraction and classification networks. Its function is equivalent to that of the HOG or DPM feature extraction methods. The features extracted by the CNN are not manually designed but are generated by the network via a learning stage. The classification network is equivalent to an SVM classifier, which is a set of classifications that learns how to use the information provided by a previous network. This neural network does not rely on manually designed feature extraction methods.

The advantage of CNN is that it can effectively solve manually designed feature extraction problems using traditional methods. Provided that the network parameters are adjusted via backpropagation using end-to-end training, the CNN can learn a filter to effectively extract the key features of the object and categorize the feature classifier extracted by this filter.

Some object detection methods using neural networks also use the sliding-window method [21] when the size of objects in an image is not fixed but the computation time is generally long. Therefore, in recent years, the mainstream method has been to adopt the strategy of object proposal by analyzing the bounding box that is likely to appear as the target object in the image. These possible bounding boxes are then inputted into the neural network for further analysis to obtain the final result. Moreover, some methods use the bounding box regression algorithm to correct the error of the bounding box for improved accuracy. In this study, algorithms such as faster R-CNN [22], Deepparts [23], UDN [24], and MS-CNN [25] were used as the baseline methods of deep learning.

MS-CNN uses different depths of CNN features with varying sizes of receptive fields to increase the detection range of object sizes. The implemented MS-CNN follows the feature-sharing concept proposed in [26] and uses a CNN trunk to detect long-distance object features. The difference is that the image features used by MS-CNN are extracted from different depths of the CNN trunk and sent to the region proposal network for analysis. This approach can address the problem in which the features of objects that are too small disappear in the deep CNN layer because of max pooling.

To enable the application of MS-CNN to drones, DGD, which is used in traditional neural-network training, is used to replace dropout [27]. Different datasets are regarded as independent domains, and the dropout probability is adjusted by analyzing the degree of response of each neuron to each domain. Consequently, the network can learn how to extract the most valuable features in each domain to improve its overall performance.

Recently, multiobject tracking methods have adopted the tracking-by-detection approach, which primarily aims to obtain the best connection between the bounding boxes in each frame. Therefore, most multiobject tracking methods aim to maximize a posteriori probability (MAP). The solutions to MAP problems can be divided into two categories. The first type of deterministic optimization problem is amenable to offline tracking methods. A commonly used method is based on a graphical solution, which uses the detection information to define a directed graph and obtain the best association tracklet via cost minimization [28,29]. Alternatively, an energy function can be utilized to detect the scene's fixed action mode and the constraint rules. Energy minimization is then used to obtain the best solution [30].

The second method uses probabilistic inference and a probability inference method based on current detection information to estimate the target state distribution. Given that only present and past information is used, it is more suitable for online tracking methods. It uses single-object tracking to predict the location, surface feature, and motion to model the target. These two models are used to predict the location of a target object. The data association algorithm is then used to associate the relevant bounding box. Finally, the target model is updated, and the aforementioned steps are performed in a loop to achieve online multiobject tracking. The Kalman filter combined with the Hungarian algorithm for data association has been used to achieve multiobject tracking [31,32]. In addition to using the Kalman filter as the motion model, mean-shift tracking has been used as an appearance model, using the Hungarian algorithm to perform data association. The Markov decision-process (MDP) tracker [33] introduces the concept of learning to automatically obtain the similarity function for data association and uses the MDP model to determine the object's state in the adopted tracking strategy.

3. MDJL Method

Multitask learning (MTL) is a training method that has been commonly used in deep learning networks in recent years. The application of MTL includes natural language processing, speech recognition, and computer vision.

Figure 4 shows the differences between single- and multitask learning. Single-task learning is optimized for a specific task or even a particular domain during training and learning. However, MTL combines training information from multiple different but related tasks, such that the backpropagation results between different domains can compete to achieve a more generalized network model.

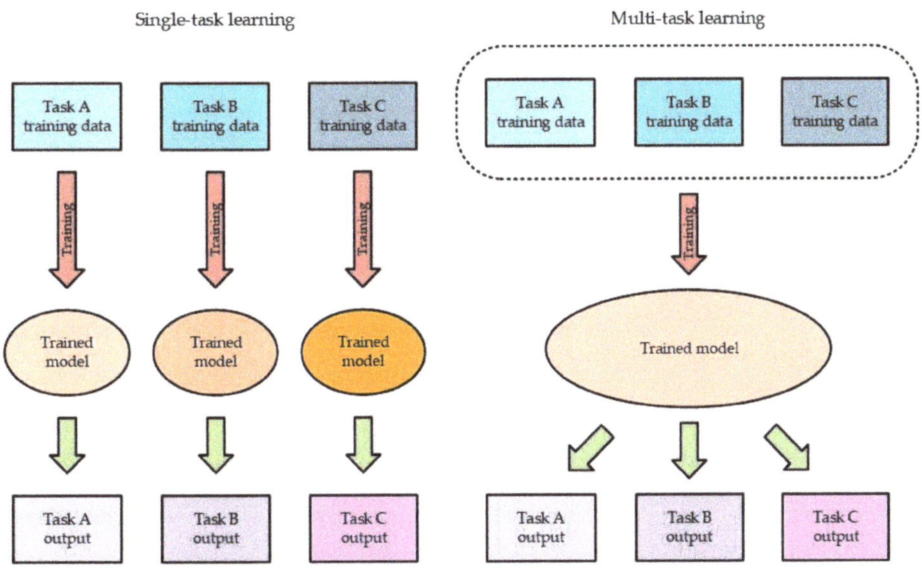

Figure 4. Single-task learning vs. multitask learning.

Labeled training data are available for a set of k tasks $\mathcal{T} = \{\mathcal{T}_1, \mathcal{T}_2, \ldots, \mathcal{T}_k\}$, where each task is associated with a different domain, $\mathcal{D} = \{\mathcal{D}_1, \mathcal{D}_2, \ldots, \mathcal{D}_k\}$ [34]. It is nearly impossible to reliably estimate the empirical joint distribution $\hat{P}_k(X, Y)$ by using data from the kth domain only at the kth task, where $X = \{x_1, x_2, \ldots, x_n\}$ is the set of samples from the feature space and $Y = \{y_1, y_2, \ldots, y_n\}$ is the label of the sample. A better approximation for $\hat{P}_k(X, Y)$ is learned by exploiting the training data from all domains and learning all tasks simultaneously. Thus, in the training process, the exchange of information between tasks and the competition between domains can improve the performance of each task.

In [35], the MTL of deep learning networks was divided into two categories: hard and soft parameter sharing. Hard parameter sharing primarily uses shared hidden layers in a deep learning network with multiple task outputs, as shown in Figure 5a. This approach can force the network to find a more generalized model that can satisfy all tasks during training and is easy to overfit. Soft parameter sharing means that different tasks have independent models and parameters; however, each model mutually restricts the distance between its parameters during training, as shown in Figure 5b. This method can encourage each model to improve each other's similarities, but they retain the ability to complete the specified task. There are many ways to calculate the distance, such as the Euclidean distance or trace norm.

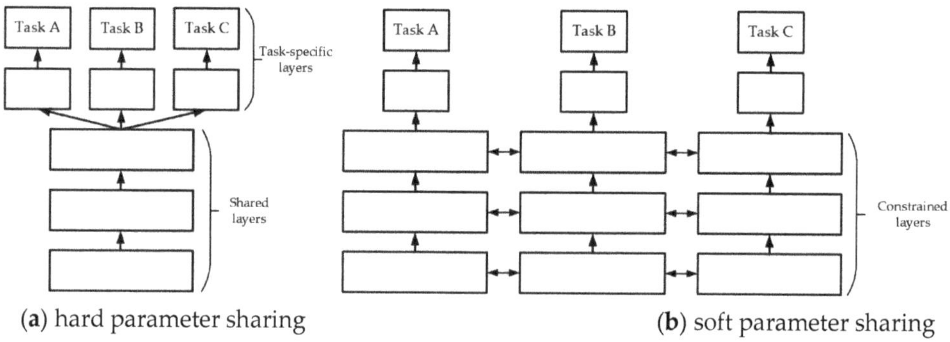

Figure 5. Parameter-sharing diagram.

An inductive bias obtained via MTL seems intuitively plausible. However, the following features highlight the advantages of this approach. First, MTL can indirectly increase the amount of training information. The second feature is that this method allows the deep learning network to focus on more-valuable features during training. The third feature is that MTL causes the network to exhibit a representative bias when learning. In addition, the MTL can import inductive bias during in-depth learning network training to make the network model less easy to overfit. Finally, it can also add an eavesdropping training mechanism to make it easier for the network to learn different tasks.

3.1. Multidomain Joint Learning

The MDJL approach proposed in this paper addresses the following problem: when a multidomain dataset is used to train a deep learning network, the final model performs worse on individual domains than on separate training on single domains. As defined in [34] for MTL, if we treat pedestrian detection in different domains as a different task, we can treat the aforementioned problem as an MTL problem.

The utilized MTL method has the hard parameter-sharing architecture described in [35]. It includes a shared feature extraction network, a task-specific hidden layer, and the output of each task. However, given that all the outputs have the same class probability and bounding box, we combine and share the outputs of various tasks. Figure 6 shows a simple architectural diagram. The colored blocks in the figure represent task-specific hidden layers. The objective is that the network will independently construct each task block during the training process. Therefore, we used DGD to facilitate the self-allocation of hidden layer resources in the network, according to the response of neurons to different tasks in the hidden layer.

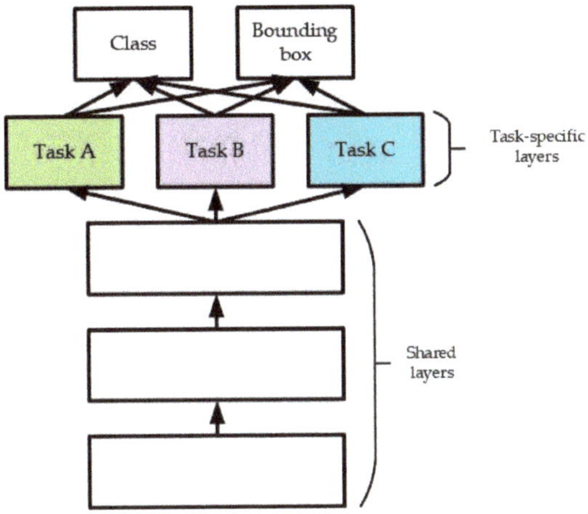

Figure 6. MDJL hard parameter-sharing architecture diagram.

MDJL divides the classification network into the same number of subnetworks according to the number of domains in the multidomain dataset. The division standard was determined on the basis of the response strength of each neuron in the classification network of different domains. Thus, each subnet can be specialized in a specific domain to more efficiently learn the information of various domains contained in the multidomain dataset and improve the network's overall performance in each domain.

To calculate the degree of response of a specific neuron to a specific domain, we used the NIS in the DGD, which is defined in Equation (1) as follows:

$$s_i = L\left(g(x)_{\setminus i}\right) - L(g(x)), \quad (1)$$

where $g(x)_{\setminus i}$ is the output of the entire network when the ith neuron is ignored, $g(x)$ is the original output, and $L(x)$ is a loss function. Because our network has two outputs, a bounding box and category probability distribution, we use the multitask loss to evaluate the network's performance.

A DGD has two modes: deterministic and stochastic. Equation (2), which expresses the deterministic mode, is used to drop out all neurons with a nonpositive impact score (NIS) of less than zero.

$$\begin{cases} keep, \ if \ s_i > 0 \\ dropout, \ if \ s_i \leq 0 \end{cases} \quad (2)$$

The stochastic mode uses the sigmoid function to determine the dropout probability of a neuron, which is defined in Equation (3) as follows:

$$p(s_i) = \frac{1}{1 + e^{-s_i/T}}, \quad (3)$$

where T is an adjustable parameter, and its value determines the degree of the NIS influence probability. If $T \to 0$, then it becomes deterministic. If $T \to \infty$, it returns to the traditional dropout. In this paper, MDJL adopts the deterministic mode in the preliminary training stage, whereas the stochastic mode is used for fine-tuning.

Given that MDJL has two types of outputs (class and bounding box), the bounding box's output is meaningful only when the object is classified as nonbackground to facilitate end-to-end training in the network and obtain the best parameters W^*. Therefore, multitask loss was used as the evaluation basis for training the network. The sample set used in training was $S = \{(X_i, Y_i)\}_{i=1}^{K}$, where X_i is the sample image, $Y_i = (y_i, b_i)$ is the category $y_i \in \{1, 2, 3, \ldots, K\}$ to which the sample belongs, and the bounding box position of the actual object is $b_i = \left(b_i^x, b_i^y, b_i^w, b_i^h\right)$. Using this information, we can write the multitask loss function as follows, in Equation (4):

$$(\mathbf{W}) = \sum_{m=1}^{M} \sum_{i \in S^m} \alpha_m l^m (X_i, Y_i | \mathbf{W}), \tag{4}$$

where M is the number of branches, α_m is the weight of l^m, and S^m in $S = \left\{S^1, S^2, \ldots, S^M\right\}$ is the collection of objects in charge of this branch. Thus, l^m calculates the loss from two aspects: the correctness of classification and the accuracy of the bounding box. It takes the form of Equation (5):

$$l(X, Y|\mathbf{W}) = L_{cls}(p(X), y) + \lambda[y \geq 1] L_{loc}\left(b, \hat{b}\right) \tag{5}$$

$p(X) = \{p_0(X), p_1(X), \ldots, p_K(X)\}$ is the probability distribution of each category, and L_{cls} is the classified cross-entropy loss, which is defined in Equation (6) as follows:

$$L_{cls}(p(X), y) = -\log p_y(X) \tag{6}$$

$\hat{b} = \left\{b_x, b_y, b_w, b_h\right\}$ are the calculated bounding boxes. MS-CNN uses Equation (7), defined in [26], to calculate the bounding box loss.

$$L_{loc} = \frac{1}{4} \sum_{j \in \{x,y,w,h\}} smooth\left(b_j, \hat{b}_j\right) \tag{7}$$

From Equation (5), L_{loc} is valid only for nonbackground categories, and λ is the weight parameter. After the error has been obtained, the best parameter $\mathbf{W}^* = argmin_{\mathbf{W}}\ \mathcal{L}(\mathbf{W})$ can be obtained using the stochastic gradient descent method.

To equitably calculate the NIS of each domain, it is necessary to calculate the NIS of all validation samples in the domain, followed by the calculation of the average to obtain the final score. Figure 7 shows a schematic of the MDJL network. To replace the initially used dropout with DGD in the MS-CNN and reduce the error caused by the difference in the implementation, we adopted the DGD three-stage training method. The detailed training process is explained in Section 3.3.

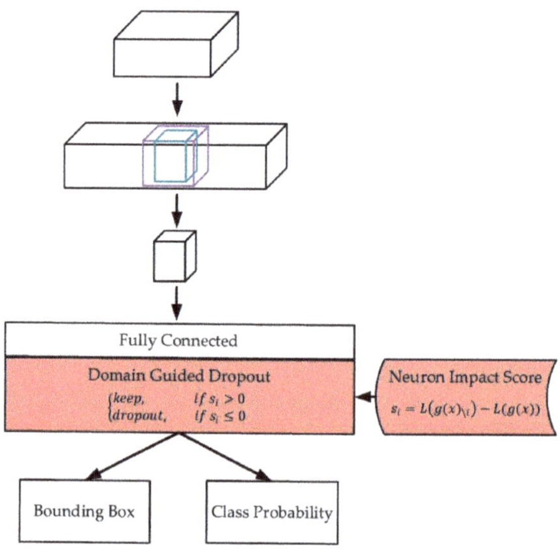

Figure 7. MDJL schematic.

3.2. Domain-Oriented Fine-Tuning

To further improve the network's performance, the consensus in the research community is to fine-tune a preliminarily trained network. In this section, domain-oriented fine-tuning is presented as a method that uses domain information to effectively fine-tune the MDJL.

We use Equation (1) to calculate the NIS of each neuron for each domain and use the rules of Equation (2) to allocate the domain for which the neuron is responsible. Because the same neuron may be responsible for more than one domain, we can represent the distribution of the neurons that are responsible for different domains by using a Venn diagram, as shown in Figure 8. This figure shows that each domain has unique neurons (red, yellow, and blue blocks). These neurons are known as domain-exclusive neurons (DENs). DENs are responsible for only a single domain; therefore, it is feasible that fine-tuning these neurons can further improve the network's performance.

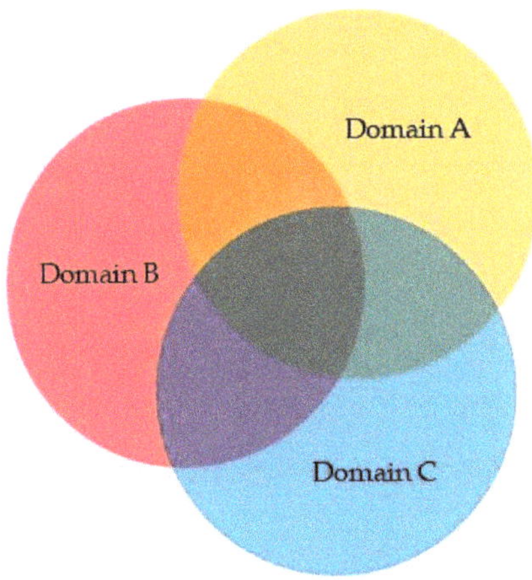

Figure 8. Schematic of neuron distribution in MDJL according to NIS classification.

By using the characteristics of stochastic DGD, the drop rate can be significantly increased during the training of neurons that do not belong to DENs by manipulating the NIS of each neuron for different domains. Thus, the fine-tuned DENs in the network can be further improved. To obtain the DENs ε that belongs to each domain, we used the previously calculated NIS score $\boldsymbol{\mathcal{S}}_j = \{s_{1j}, s_{2j}, s_{3j}, \ldots, s_{ij}, \ldots, s_{nk}\}$ for all n neurons for all k domains $\boldsymbol{\mathcal{D}} = \{\mathcal{D}_1, \mathcal{D}_2, \mathcal{D}_3, \ldots, \mathcal{D}_j, \ldots, \mathcal{D}_k\}$ and Equation (2) to determine the neuron set $\mathcal{N}_j$ responsible for domain $\mathcal{D}_j$. Then, if $I = \{1, 2, 3, \ldots, k\}$, we obtain the DENs $\varepsilon_j = \mathcal{N}_j \backslash (\cup_{i \in I \,\&\, i \neq j} \mathcal{N}_i)$ that belong to domain $\mathcal{D}_j$. Finally, we used Equation (8) to obtain the adjusted NIS of each neuron n_i, called the exclusive neuron impact score (eNIS). R, known as the reserve constant, is a parameter less than 0 that can be adjusted to modify the reserve rate of neurons that do not belong to the DENs so that it is smaller than the reserve rate of neuron DENs. The reserve rate used in fine-tuning is the result calculated using Equation (3) and using eNIS, as shown in Figure 9.

$$\hat{s}_{ij} = \begin{cases} s_{ij}, \ if\ n_i \in \varepsilon_j \\ R, \ otherwise \end{cases} \tag{8}$$

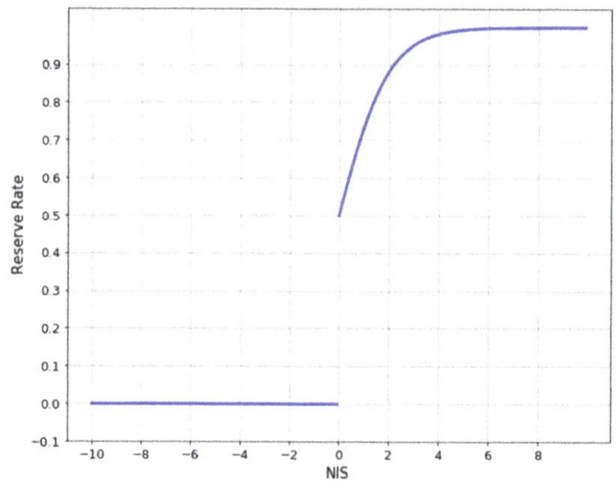

Figure 9. Adjusted reserve rate.

3.3. Multidomain Joint Learning Training

Backpropagation was used as the training algorithm. Training of the MDJL was divided into three stages. The first stage synthesized the initial training of all the datasets. We used VGG16 as our initial model to obtain a basic network that had learned all the domains. The second stage analyzed the NIS of the basic network obtained in the first stage, replaced the dropout in the network with DGD, and performed subsequent training actions. In this stage, the DGD dropped the specified neuron according to the NIS of the domain to which the input sample belonged during the training process, thereby strengthening the network's concept of the domain to improve its performance. In the final stage, we propose domain-oriented fine-tuning to optimize the network model further.

The network architecture used in this paper was MS-CNN; therefore, the first stage of initialization training was conducted according to the MS-CNN training method. This method also has two training stages. The main difference between the two stages is the negative sampling mode. The first stage adopts the random sampling mode, and the second stage adopts the bootstrapping sampling mode. Bootstrapping is a type of hard-negative mining-sampling method. It can be used to determine whether a sample can be used for training according to the objectness score of the sample. Its goal is to facilitate easier misjudging of the collected sample, thereby reducing the network chance of misjudgment.

During training, the training sample set of the branch detection layer m is defined as $S^m = \{S^m_+, S^m_-\}$, where S^m_+ represents positive samples and S^m_- represents negative samples. Assume that a single sample is defined as $S = (X_i, Y_i)$, where X_i is the sample image and $Y_i = (y_i, b_i)$ is the category to which the sample belongs and the location of the bounding box. Here, o^*, as defined in Equation (9), determines whether the sample is positive or negative.

$$o^* = \max_{i \in S_{gt}} IoU(b, b_i) \tag{9}$$

S_{gt} is the ground truth, and IoU is the intersection between the two bounding boxes. Thus, when o^* is greater than or equal to 0.5, the sample is added to the positive sample set; when o^* is less than 0.2, the sample is added to the negative sample set, and all other samples are discarded. Typically, the distribution of objects and nonobjects in an image is asymmetric. Therefore, to achieve a balance between the two, we set $|S^m_-| = \gamma |S^m_+|$. Random sampling and bootstrapping sampling are then used to obtain the final negative sample set.

To ensure that each detection layer detects objects only within a fixed scale range, the training set of each layer is composed of a subset of S, and the samples of this subset correspond to the scale range of the layer detection. However, this method may cause a specific layer to have no available positive samples and indirectly cause learning instability.

Therefore, it is necessary to modify the cross-entropy, as shown in Equation (10), to solve this problem.

$$L_{cls} = \frac{1}{1+\gamma}\frac{1}{|S_+|}\sum_{i\in S_+} -logp_{y_i}(X_i) + \frac{1}{1+\gamma}\frac{1}{|S_-|}\sum_{i\in S_-} -logp_0(X_i) \quad (10)$$

3.4. Multiobject Tracking System for Flying Drones

The operation process of the proposed online multiobject tracking system is shown in Figure 10. It can be divided into two parts: the object detector and the tracker. The MDJL algorithm proposed in this paper is used for the former, whereas the MDP tracker algorithm is used for the latter. The MDP tracker is an online tracking algorithm that can instantaneously calculate and output results by using new data input. These algorithms are more suitable for applications that require real-time feedback, applications such as flying drones.

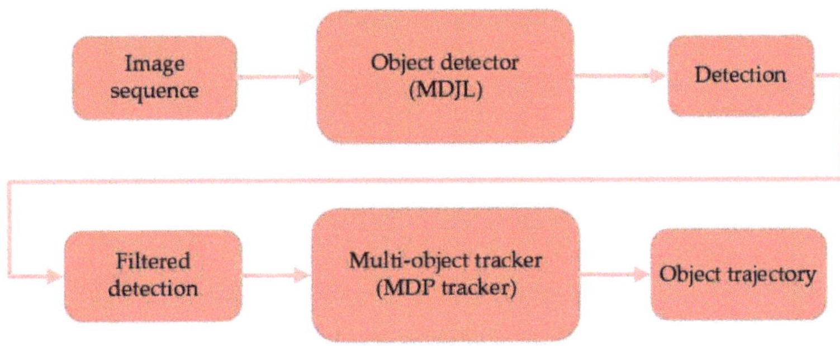

Figure 10. Flow chart of multiobject tracking system.

Our online tracking system utilized MDJL as the pedestrian detector, and the MDP was used as the tracker. By combining the pedestrian detection methods proposed in this paper, it is expected that the multiobject tracking system will be more effective in multidomain situations. In an MDP tracker, each object builds an MDP model. This model is composed of four elements (), where S represents the state of the object; A and $T(\cdot)$ represent the action and transfer equations executed when the object is transferred, respectively; and $R(\cdot)$ represents the reward function after the A action has been performed.

In the MDP model, S has four states: active, tracked, lost, and inactive. Figure 11 is a flowchart of the status of the MDP tracker. First, when the object detector detects an object, it is initialized and enters the active state. In this state, it determines whether subsequent tracking should be performed and changes the status to either tracked or inactive. When the status is tracked, the target is continuously tracked and maintained in the tracked status. However, when tracking is not possible, such as when the target is obscured or exits the field of view of the camera, the MDP changes the status to lost. If the target appears again and is successfully associated, the state is changed to tracked; otherwise, it remains lost. When the status is lost for more than a period, the MDP changes to an inactive status. Moreover, the MDP aborts all tracking actions and does not change the status.

During tracking, each target has an appearance model consisting of several appearance templates that were created during the tracking process. These templates record the appearance of the target during tracking. To maintain the validity of the appearance model, they must be updated. Therefore, these templates adopt a lazy-update rule. It is updated for the current appearance only when the tracked target status changes from lost to tracked.

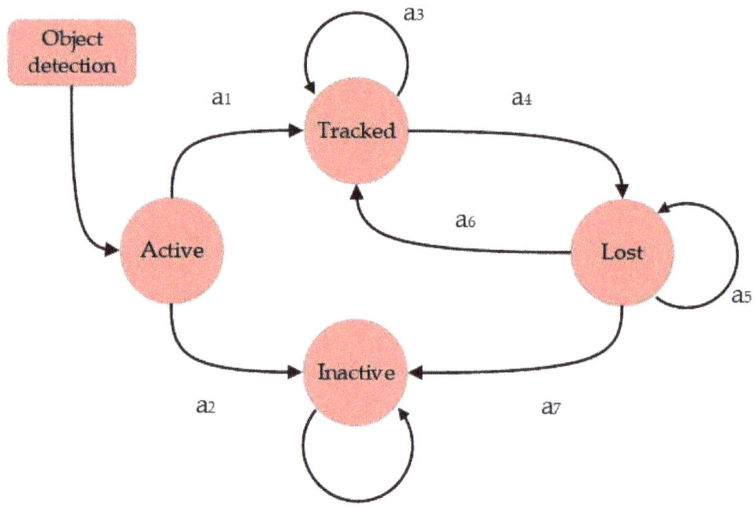

Figure 11. Flowchart of MDP tracker status.

The decision on the lost state can be considered as a data association problem. Therefore, the MDP tracker continues to be associated with the new bounding box in the new frame in this state. If there is an association with the bounding box, it changes the state to track; otherwise, it continues to maintain the lost state. When the lost state is maintained for more than a specific frame number, the target is considered to have left the camera's field of view, and the state is changed to inactive, to terminate tracking.

During the training process, the MDP adopts reinforcement learning. First, given multiple video sequences as training data, it can track all objects in these videos, and the training goal is to facilitate the successful tracking of all objects. The MDP tracks the target according to this strategy for each state, uses the reward equation to evaluate the performance of the tracker, and maximizes the result of the final tracker's reward equation by adjusting the parameters of the reward equation for the lost state. The training algorithm processes all the training videos and tracks the training targets in a loop, and the training does not end until all the targets have been successfully tracked. During training, data association is updated only when an error occurs during this process. There are two possible errors: the first case is associated with a nontarget bounding box, and the second case occurs when a target bounding box appears but is not associated. When these errors occur, an incorrect association is added to the negative training sample. Conversely, a correct association was added to the positive training sample for training.

4. Experimental Results

The experiment was conducted in two parts. In the first part, we evaluated the performance of our proposed MDJL pedestrian detection method. In the second part, we adjusted the pedestrian tracking parameters on the basis of our MDJL for comparison with other advanced tracking algorithms.

The computer hardware specifications used in the experiment included an Intel Core i7-930 CPU with 24 GB RAM and an Nvidia GTX 1080Ti GPU with 11 GB RAM. The operating system was Ubuntu 14.04 LTS, and CUDA 8.0, CUDNN 5.1, and MATLAB R2014b were the primary software programs.

In the experiment, we used datasets primarily from three sources. The Caltech pedestrian detection dataset [36] (Caltech dataset) is a driving video. The format of the video is 640×480 pixels at 30 Hz. This dataset includes images with approximately 250,000 frames, 350,000 bounding boxes, and 2300 unique pedestrians. Among them, set00–set05 was defined as the training dataset, and set06–set10 was defined as the testing dataset. To ensure data integrity, set05 in the training dataset was excluded from the validation dataset that was used during training. The shooting angle used in this dataset was of a forward orientation, as shown in Figure 12a.

Figure 12. Different shooting angles of each dataset: (**a**) Caltech dataset—forward orientation angle, (**b**) PETS09 dataset—top view angle, and (**c**) DroneFJU—45-degree downward and (**d**) 90-degree downward orientations.

The primary purpose of the PETS09 dataset [37] was to track an individual within a crowd and to detect flow and crowd events. Eight cameras and eight angles were used to capture a designated area at the same time on the campus. The video format used was 768 × 576 pixels at 7 Hz. In this paper, to prevent the PETS09 and Caltech datasets from using videos with repeated shooting angles, only the downward-oriented (top view angle) videos in PETS09 were used, as shown in Figure 12b. We divided these seven videos into three sets: training (S1L2-1, S2L1, S2L3, and S3MF1); validation (S1L1-1 and S1L1-2); and testing (S2L2).

The DroneFJU dataset was obtained on the Fu Jen Catholic University campus using a quadrotor. The equipment included a DJI Matrice 100 + Zenmuse-integrated PTZ camera. The shooting format was 1080 × 1920 pixels at 60 Hz, and the shooting angle was divided into a 45-degree downward-facing orientation (Figure 12c) and a 90-degree downward orientation (Figure 12d). The latter shooting angle was the challenging aspect of this dataset. In addition, because a wide-angle lens was used for video acquisition, the pedestrian's face and overall outline changed depending on their position (as shown in Figure 12d, red box) and various body postures.

In Section 4.1, the performance of the proposed MDJL pedestrian detection method is analyzed by using an ablation study; moreover, the network characteristics of the MDJL are analyzed in Section 4.2. In Section 4.3, domain-oriented fine-tuning is implemented, and its ability to improve network performance is evaluated. In Section 4.4, we discuss and optimize the parameters of domain-oriented fine-tuning and compare it with the current state-of-the-art pedestrian detection method.

Finally, we used the PET09 and the TUD datasets with pedestrians as tracking objects and compared them with the original MDP and several more advanced tracking methods. In Section 4.5, we introduce the evaluation methods and present the adjustments we performed to facilitate pedestrian detection, in addition to the influence of the pedestrian tracking parameters on the final results. We present the optimization results of our method using an experimental video and a comparative analysis with other tracking methods.

4.1. Ablation Study

MDJL is a network training method designed for a multidomain dataset. Our objective is to use this method to address the reduction in network accuracy that occurs when using a multidomain dataset for training. In this experiment, we discuss the aforementioned issues and evaluate the effectiveness of our method.

In the following experiments, we used the Caltech dataset (referred to as c), PETS09 (referred to as p), and DroneFJU (referred to as d) for different sizes and different shooting methods, to combine them into several multidomain datasets and use these datasets to train the baseline network (MS-CNN) and our proposed method (MDJL). The method of evaluating the network uses an appropriate setting of the Caltech pedestrian detection benchmark to assess the network's average miss rate and the receiver-operating characteristic (ROC) curve in the test set of the aforementioned three datasets.

Table 1 summarizes the number of training, validation, and testing samples for each dataset. The number of training samples of PETS09 is approximately one-third of that of the Caltech dataset, whereas the number of training samples of DroneFJU is approximately one-half of that of the Caltech dataset. This difference may affect the final network performance. Therefore, in this experiment, it was necessary to increase the number of training samples of the smaller datasets to the same or more than that of the Caltech dataset. We used two data augmentation methods to create two additional PETS09 training images and a DroneFJU training image. The first method flips the original image left and right, and the second method adds Gaussian noise to the original image. The Gaussian noise parameters had an average of 0 and a standard deviation of 0.01. Using these two data augmentation methods, we increased the number of training samples of PETS09 to 36,504, whereas that of DroneFJU increased to 35,632 via the addition of Gaussian noise. Finally, to balance the number of samples in each dataset, we used the number of training samples of the Caltech dataset as the standard and randomly extracted the same number of samples from the expanded dataset to obtain the final dataset.

Table 1. Number of dataset samples.

Dataset Name	Sample Count		
	Training	Validation	Testing
Caltech dataset	34,927	3346	33,043
PETS09	12,168	8813	14,474
DroneFJU	17,816	4286	6460

Eight types of multidomain datasets were used, and the abbreviations used for the new dataset combine the names of each. Therefore, for example, for Caltech + PETS09, the multidomain dataset is called cp, and for Caltech + PETS09 + DroneFJU, the dataset is called cpd. If the multidomain dataset used data augmentation to balance the number of samples, the symbol * was used next to the abbreviation. For example, Caltech + PETS09 with data augmentation is denoted as cp*, and Caltech + PETS09 with data augmentation + DroneFJU with data augmentation is denoted as cp*d*.

First, let us examine the evaluation results for different datasets when we use a single dataset, the so-called single-domain dataset, to train the model. We used the Caltech dataset, PETS09, and DroneFJU to train the MS-CNN model and to evaluate the three trained models, c, p, and d, on the test set of each dataset. The results are shown in Figure 13.

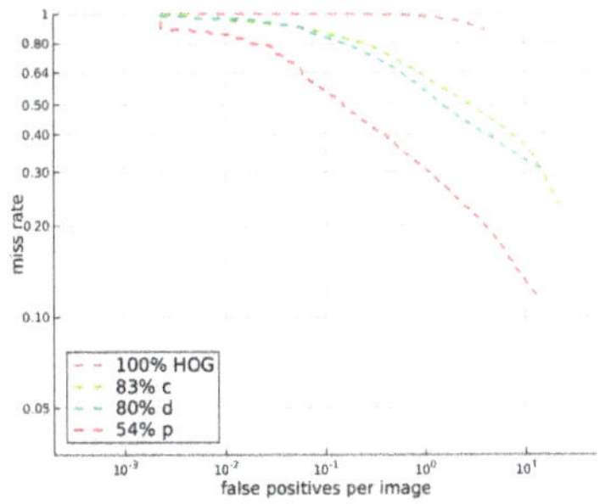

Caltech

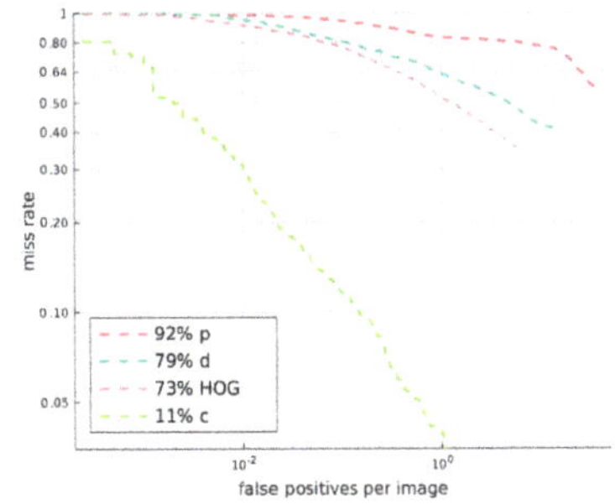

PETS09

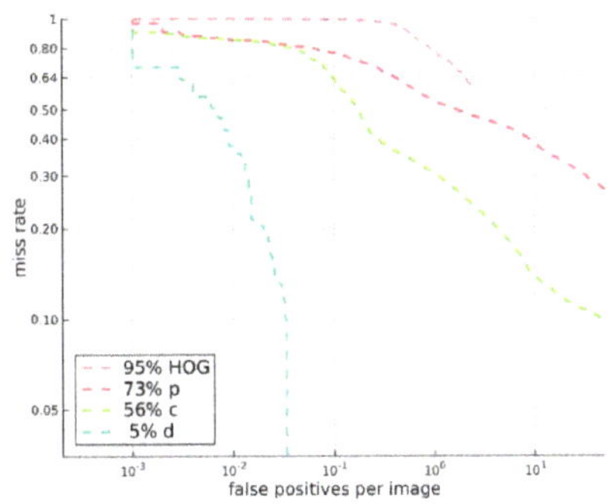

DroneFJU

Figure 13. Evaluation results of single-domain dataset model in different datasets.

Figure 13 shows that all the single-domain dataset models yield better evaluation results only for the dataset of the same domain and perform poorly on other domains. These results show that a model trained using a single dataset can be applied only to the same domain and cannot be generalized to other domains.

Table 2 lists the test results of the models trained using different multidomain datasets with appropriate settings of the test set of the Caltech dataset. We also include the evaluation results for the same setting when MS-CNN uses only the Caltech dataset as the baseline for comparison. The corresponding data are not shown for MDJL because they form a network designed for a multidomain dataset. Therefore, we did not consider the performance when using a single dataset for training. To perform a fair comparison, we selected only the multidomain dataset containing the Caltech dataset.

Table 2. Average miss rates of different multidomain dataset models for the Caltech test set.

Training Dataset	Average Miss Rate	
	MS-CNN	MDJL
cp*	0.12103	**0.114468**
cp*d*	0.115921	**0.113062**
cd*	0.11407	**0.113394**
cpd	0.113918	**0.106365**
cp	0.112784	**0.112716**
cd	0.112092	**0.108718**
c	0.110483	

The MS-CNN section is summarized in Table 2. When training was performed with the cp, cd, and cpd datasets without data augmentation, it was determined that simply adding a domain (such as cp and cd) to the dataset increased the average miss rate by approximately 0.2%, whereas adding cpd to the three domains increased it by approximately 0.5%. If we examine cp*, cd*, and cp*d* that use data augmentation, it is determined that the average miss rate further increased by approximately 1%. Thus, the overall average miss rate increased by 0.5%, with a standard deviation of 0.006. This indicates that the greater the number of domains involved in training the network with a multidomain dataset, the more significant the impact on the network performance. Moreover, using data augmentation to balance the number of samples increased the influence of the initially small dataset on the network training and also affected the accuracy of the network.

The experimental results of MDJL show that regardless of the number of domains contained in the multidomain dataset, the average miss rate increased by only approximately 0.2%. Furthermore, even if data augmentation was used, the average miss rate increased by only approximately 0.3%. Thus, the overall average increase in the average miss rate was 0.1%, and the standard deviation was 0.002. This indicates that MDJL is more stable when using multidomain dataset training and is not highly susceptible to the number of domains or the number of data. In addition, regardless of the multidomain dataset used for training, the average miss rate of MDJL was lower than that of MS-CNN. This implies that MDJL can improve the overall accuracy when training with a multidomain dataset.

In the following table, we compare the performance of the models trained using each multidomain dataset on the test set. Therefore, we indicate the dataset name used in the training after the method name to facilitate comparison. For example, the MDJL model trained with the cpd dataset is called MDJL-cpd.

We used the MS-CNN and MDJL models, which were trained on the eight multidomain datasets, to evaluate the test sets of the Caltech, PETS09, and DroneFJU datasets, and we recorded the average miss rates in Table 3. Two aspects were analyzed in this table. First, the performance of the MS-CNN and MDJL models that were trained using the same multidomain dataset for each evaluation dataset is identified by using a thick box. We investigated the network trained using the three datasets of cpd, cd, and pd, and it was determined that the average miss rate evaluated for PEST09 and DroneFJU using MDJL was higher than that of MS-CNN. None of these datasets used data augmentation to balance the sample size difference between the domains. Therefore, we speculate that even if our method can effectively divide the network into several small networks for fine-tuning on the basis of the response to each domain, as shown in Figure 14, shared and generalized neurons will still exist across the domains. These neurons are more susceptible to larger datasets (Caltech) during training and are dominated by these datasets, thus affecting the network performance on smaller datasets (PETS09 and DroneFJU). However, suppose we use the same domain combination of cp*d*, cd*, and p*d* for data augmentation. In this case, this problem is not encountered, which establishes that our assertion is reasonable. To optimize and balance the network's performance on each dataset, the number of samples of each domain in the multidomain dataset must also be balanced. Therefore, when discussing other phenomena, we will use the data augmentation model.

Second, the three blocks in Table 3 that are identified by using a gray color represent the network trained by each multidomain dataset. The performance on the test set of the dataset is interesting in each case. The results indicate that the model was trained using two datasets, cp* and cd*. It is foreseeable that the evaluation results of these models on the datasets, which have never been observed, are not satisfactory. However, the data show that MDJL-cp* exhibits an improvement of 2.7% in the average miss rate compared with MS-CNN-cp* and that MDJL-cd* is 1.7% better than MS-CNN-cd*. This improvement is not insignificant, so we believe that the model trained by MDJL has better generalizability.

Table 3. Average miss rate for each model for each dataset.

Model	Average Miss Rate		
	Caltech	PETS09	DroneFJU
MS-CNN-cpd	0.113918	**0.537132**	**0.080445**
MDJL-cpd	**0.106365**	0.546494	0.087827
MS-CNN-cp*d*	0.115921	0.498298	0.054163
MDJL-cp*d*	**0.113062**	**0.487953**	**0.051723**
MS-CNN-cp	0.112784	0.566055	0.578502
MDJL-cp	**0.112716**	**0.556025**	**0.516990**
MS-CNN-cp*	0.121030	0.523379	0.444892
MDJL-cp*	**0.114468**	**0.517952**	**0.417621**
MS-CNN-cd	0.112092	**0.764734**	**0.079890**
MDJL-cd	**0.108718**	0.794011	0.100103
MS-CNN-cd*	0.114070	0.799884	0.061482
MDJL-cd*	**0.113394**	**0.782861**	**0.055286**
MS-CNN-pd	**0.782319**	**0.517785**	**0.020215**
MDJL-pd	0.801515	0.538508	0.022154
MS-CNN-p*d*	**0.805207**	0.529371	0.024825
MDJL-p*d*	0.810766	**0.519591**	**0.018935**

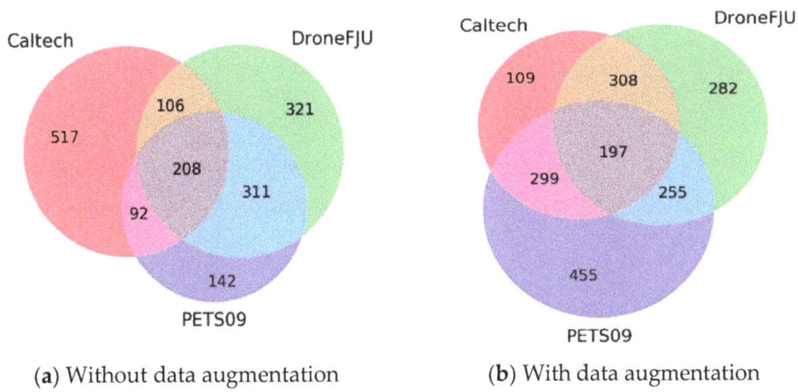

Figure 14. Degree of subnetwork overlaps of each domain.

However, if we examine the MS-CNN-p*d*and MDJL-p*d*models trained using the p*d* dataset, this phenomenon is not observed, and the evaluation performance of the two networks is similar. This result is not surprising given that the ability to generalize originates from similar situations that are learned or seen, from which knowledge is extended. The two datasets, PETS09 and DroneFJU, differ from the Caltech dataset used for the evaluation of the shooting angle and the shooting methods, which results in different image characteristics in the dataset. Therefore, neither MS-CNN nor MDJL can be used in PETS09, and DroneFJU learned how to correctly evaluate the Caltech dataset to obtain the presented results.

According to the results of this experiment, it is evident that the proposed method can effectively improve the impact of training networks using a multidomain dataset. Furthermore, using data augmentation to balance the number of samples in each domain in a multidomain dataset can increase the effectiveness of MDJL. Finally, MDJL appears to have better generalizability than MS-CNN does.

4.2. Analysis of Multidomain Joint Learning Method

In this experiment, the NIS calculated using MDJL is represented as a chart to analyze the relationship between the subnetworks after segmentation. First, we investigated whether there was a relationship between the response degree (NIS) of each neuron to different domains in the network. Three sets of comparisons were conducted, each of which selected two domains, A and B. We then sorted the NIS of domain A from small to large, arranged domain B in the same order, and represented the results as a line chart. As shown in Figure 15, domain A is the blue curve, and domain B is the red curve. The three results in the figure show that the NIS among the various domains does not exhibit a correlation, which means that the neurons in the network that strongly respond to each domain are not the same.

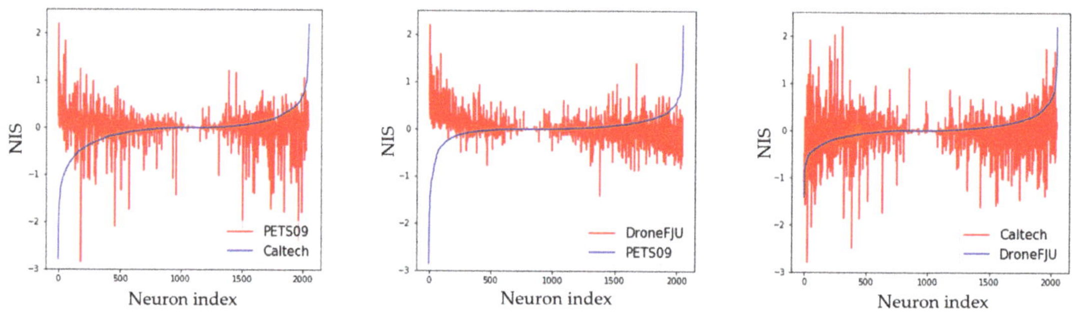

PETS09 vs. Caltech DroneFJU vs. PETS09 Caltech vs. DroneFJU

Figure 15. NIS correlation between different domains.

Next, we analyze the degree of overlap in the subnetworks divided according to the deterministic rule (NIS > 0), according to the NIS of each domain, and the results are represented as a Venn diagram. Figure 14 shows the distribution of neurons according to the models trained using the datasets with and without data augmentation.

First, in Figure 14a, a dataset without data augmentation is used to train the model, and the distribution of neurons is allocated to more or fewer neurons according to the training sample size of each domain.

Thus, a large domain (e.g., Caltech) is assigned to more neurons, and small domains (e.g., PETS09) are assigned to fewer neurons. This phenomenon is consistent with the observations summarized in Table 3. Therefore, large domains dominate the network. In Figure 14b, the neurons are distributed according to the complexity of the domain because of data augmentation. PETS09 was used because the pedestrians in the video were sometimes dense and sometimes loose. The shooting angle is a downward projection; it can easily cause pedestrian deformation, and the situation is complex. Therefore, when the network allocates neurons, it provides more neurons to the PETS09 domain. In contrast, although the scene at Caltech is relatively open because the camera's shooting angle is forward facing and the situation when pedestrians appear is relatively simple, the network allocates fewer neurons to this domain.

In addition, from Figure 15, it is evident that although there is no correlation between the NIS of each domain, there are still neurons in Figure 14 included in the multidomain subnets. We believe that these neurons are responsible for common information processing across domains. Moreover, the nonoverlapping parts are responsible for the unique information processing of the domain, thereby achieving the concept of divide and conquer. From the aforementioned two results, it is evident that MDJL can effectively divide the network into several subnetworks according to the degree of the neuron response to different domains. It includes dedicated neurons for each domain and neurons shared among the domains.

4.3. Domain-Oriented Fine-Tuning Experiments and Analysis

In this experiment, we used the previously trained model MDJL-cp*d* to perform further fine-tuning and obtain the model's evaluation results for each dataset, as summarized in Table 4. The gray background is the performance evaluated by the model trained using the single-domain dataset in the corresponding dataset. Therefore, for the evaluation, only Caltech assessed MS-CNN-c, PETS09 only assessed MS-CNN-p, and MS-CNN-d only used DroneFJU. We named this model MDJL-ft-cp*d*.

Table 4. Average miss rate evaluated in each dataset after fine-tuning.

Model	Average Miss Rate		
	Caltech	PETS09	DroneFJU
MS-CNN-c/-p/-d	0.110483	0.544825	0.053416
MDJL-cp*d*	0.113062	0.487953	0.051723
MDJL-ft-cp*d*	**0.107807**	**0.482616**	**0.046973**

According to Table 4, it is evident that the average miss rate of the model after fine-tuning for the evaluation using the Caltech dataset is 10.7807%, which is better than that of MS-CNN's 11.0483%. This is a slight improvement on PETS09 and DroneFJU. These results show that domain-oriented fine-tuning can effectively strengthen the domain-exclusive neurons and improve the network model. Therefore, we discuss in the next section how to adjust the parameters of domain-oriented fine-tuning.

There are three adjustable parameters for domain-oriented fine-tuning. The first is the T parameter in Equation (3), which we call the sigmoid scaler. The second is the R parameter in Equation (8), which we refer to as the reserve constant. The third is the number of iterations during fine-tuning training. We separately explore how the adjustment of these three parameters affects the performance of the network. The parameters that did not change in the following experiments were preset values: T = 1.0, R = 0, and iteration number = 20,000.

First, we discuss the T parameter in Equation (3), which is responsible primarily for adjusting the influence of the eNIS on the reserve rate of domain-exclusive neurons during fine-tuning. When $T \to 0$, stochastic DGD is converted to deterministic DGD because it can convert the result of Equation (3) to only 1 (eNIS $\geq$ 0) or 0 (eNIS < 0). In contrast, if $T \to \infty$, regardless of the value of eNIS, the result of Equation (3) is always 0.5, causing stochastic DGD to degenerate back to regular dropout. For convenience, we used the $1/T$ value as the adjustment target in the following experiments.

We adjusted $1/T$ from 0.1 to 1.0 and trained 10 models using 0.1 as the interval in the experiment. Each trained model was evaluated using the Caltech test set, PETS09, and the DroneFJU datasets, and the average of the results was calculated. The experimental results are shown in Figure 16a.

Figure 16a indicates that there is no consistent trend in the performance of different $1/T$ parameters for each dataset. However, if we observe the overall average performance in Figure 16a, we find that the relationship between the parameter T and the miss rate is slightly low, in the range of 0.2–0.5. Thus, the lowest point was at $1/T = 0.5$. According to Section 4.2, the maximum NIS value is approximately 2. Therefore, by applying Equation (3), the highest reserve rate of domain-exclusive neurons in fine-tuning is 73.11%.

The second parameter to be discussed is the R parameter in Equation (8), namely the reserve constant. This parameter is responsible for adjusting the reserve rate of neurons that do not belong to domain-exclusive neurons in domain-oriented fine-tuning and is a number less than 0. We adjusted the parameter R from 0 to -10, with -1 as the interval, and changed it 11 times. Each trained model was evaluated by using the Caltech test set, PETS09, and the DroneFJU datasets, and the average was calculated. The experimental results are shown in Figure 16b.

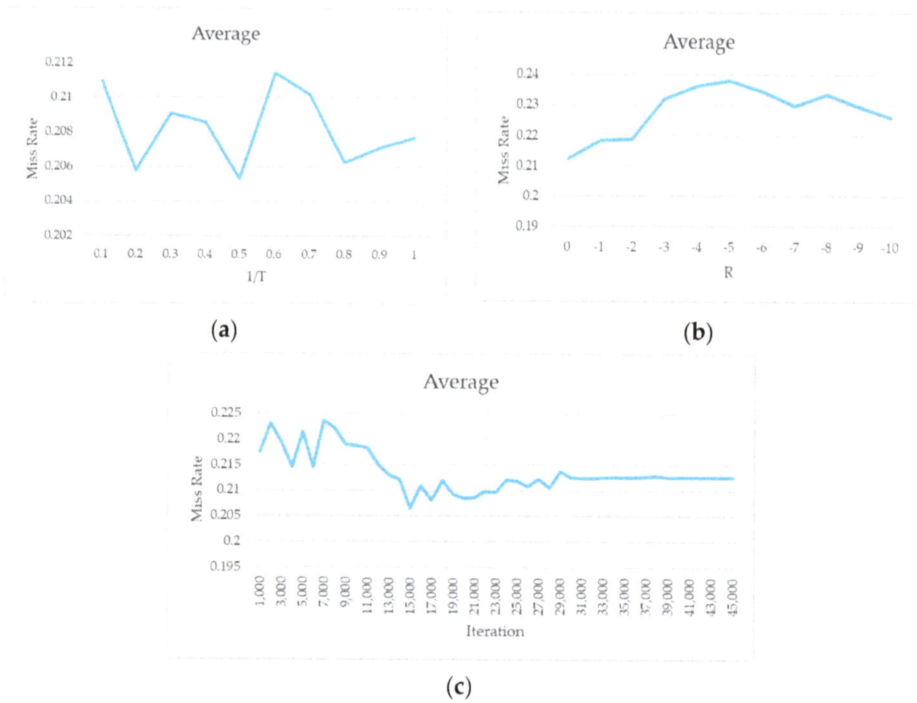

Figure 16. Miss rate curve obtained by (**a**) using different *1/T* fine-tuning parameters and (**b**) using different fine-tuning R parameters; (**c**) the number of iterations of fine-tuning.

In Figure 16b, we found that the average miss rate obtained using the evaluation had the lowest point: 0. When R = 0, the reserve rate of the non-domain-exclusive neurons was 50%. Thus, this value was too close to the reserve rate of the domain-exclusive neurons obtained in the T-parameter experiment (73.11%). It performed best on the PETS09 dataset and poorly on the other datasets. Therefore, it seems that R = −1 is the optimum value, with a reserve rate of 26.89%. This shows that the R parameters should not be too small and that non-domain-exclusive neurons should still have some fine-tuning space.

According to the aforementioned two experiments, the selection of the T and R parameters should be dominated by T and supplemented by R. The choice of T should refer to the maximum value of the NIS of each domain, such that the reserve rate is maintained at approximately 75%, which is $1/T = 0.5$ in this experiment. However, the choice of R should set the reserve rate of non-domain-exclusive neurons at approximately 25%, which is R = −1 in this experiment.

Finally, we discuss the iteration times of fine-tune training. Although fine-tuning has been widely used in many studies, there is no practical method for estimating the number of iterations that should be introduced in the fine-tuning stage. Therefore, in this experiment, we observed the performance of the model that was trained via domain-oriented fine-tuning for different iterations of each dataset on the basis of the investigation. In this experiment, 1000 iterations were used as units to increase the number of iterations to 45,000. A total of 45 models were trained. The base learning rate was 0.00002; for every 10,000 iterations, the learning rate was reduced by a factor of 0.1. Each trained model was evaluated using the Caltech test set, PETS09, and the DroneFJU datasets, and the average was calculated. The experimental results are shown in Figure 16c.

According to the experimental results shown in Figure 16c, we found that the miss rate of the model was minimal when the number of iterations was 15,000. If the training was continued, the miss rate started to increase. The miss rate remained constant after approximately 30,000 iterations. This phenomenon may be because of the reduction of the learning to 2×10^{-8}, which may be too low, rendering the network unable to effectively continue learning. The optimal fine-tuned iteration time was set as 15,000.

On the basis of the aforementioned three experiments, we selected the best parameters to train the new model and evaluated its performance on the test set for the three pedestrian detection datasets. The chosen parameters were $1/T = 0.5$ and $R = -2$, and the iteration number was 15,000. R was -2 because a value of 0.5 was chosen for $1/T$ to maintain a reserve rate of 26.89%; $R/T = -1$ must be satisfied. Using the above formula, we obtain a new $R = -2$. The model was named for the best fine-tuning parameter as MDJL-ft-bv-cp*d*, and the results are listed in Table 5. Thus, the evaluation results in the gray background are the same as those in Table 4. The evaluation results were obtained by training the model using the single-domain dataset in the corresponding dataset.

Table 5. Average miss rate measured in each dataset using the model with the best fine-tuning parameter.

Model	Average Miss Rate		
	Caltech	PETS09	DroneFJU
MS-CNN-c/-p/-d	0.110483	0.544825	0.053416
MDJL-cp*d*	0.113062	0.487953	0.051723
MDJL-ft-cp*d*	0.107807	0.482616	0.046973
MDJL-ft-bv-cp*d*	**0.103707**	**0.470637**	**0.045741**

Table 5 summarizes that the model trained with the best parameters improved by approximately 1% in each domain, which demonstrates the importance of choosing the correct parameters for fine-tuning.

4.4. Comparison with State-of-the-Art Methods

The experiment was conducted in two parts. The first part compares the performances of the models trained using single-domain and multidomain datasets to highlight the associated advantages. The second part compares the proposed method with several state-of-the-art methods.

Table 6 indicates that in the past, when training a machine-learning method or deep learning network, we often chose a training dataset that was commonly used in the research community, such as the Caltech or INRIA datasets, to train the model. Given that the video data contained in these datasets generally have the same characteristics, such as the shooting environment, shooting angle, and equipment used, this type of dataset is called a single-domain dataset.

Table 6. Average miss rate compared with that of the state-of-the-art methods.

Model	Classification	Training Dataset	Average Miss Rate
VJ [12]	AdaBoost	INRIA	0.947337
HOG [13]	Linear SVM	INRIA	0.733415
CCF + CF [19]	AdaBoost	Caltech	0.173247
Checkerboards+ [20]	AdaBoost	Caltech	0.170986
DeepParts [23]	DeepNet	Caltech	0.11889
MS-CNN-cp*d* [25]	DeepNet	cp*d*	0.115921
UDN+ [24]	DeepNet	Caltech	0.115217
MS-CNN [25]	DeepNet	Caltech	**0.110483**
MDJL-cp*d*	DeepNet	cp*d*	**0.113062**
MDJL-ft-cp*d*	DeepNet	cp*d*	**0.107807**
MDJL-ft-bv-cp*d*	DeepNet	cp*d*	**0.103707**

Figure 17 shows the results for the cross-domain evaluation between several pedestrian detection models that were trained using a single-domain dataset and pedestrian detection models trained using a multidomain dataset. The dotted lines in the figure represent the results for a part of the single-domain dataset. We used HOG trained using the INRIA dataset as the traditional baseline method. In addition, we used the Caltech dataset, PETS09, and the DroneFJU datasets trained using MS-CNN as the baseline of the deep learning network. The solid lines in the figure represent the results for a part of the multidomain dataset. The MS-CNN and MDJL models were used, which were trained on the cp*d* dataset. For the evaluation, we used the Caltech, PETS09, and DroneFJU test sets.

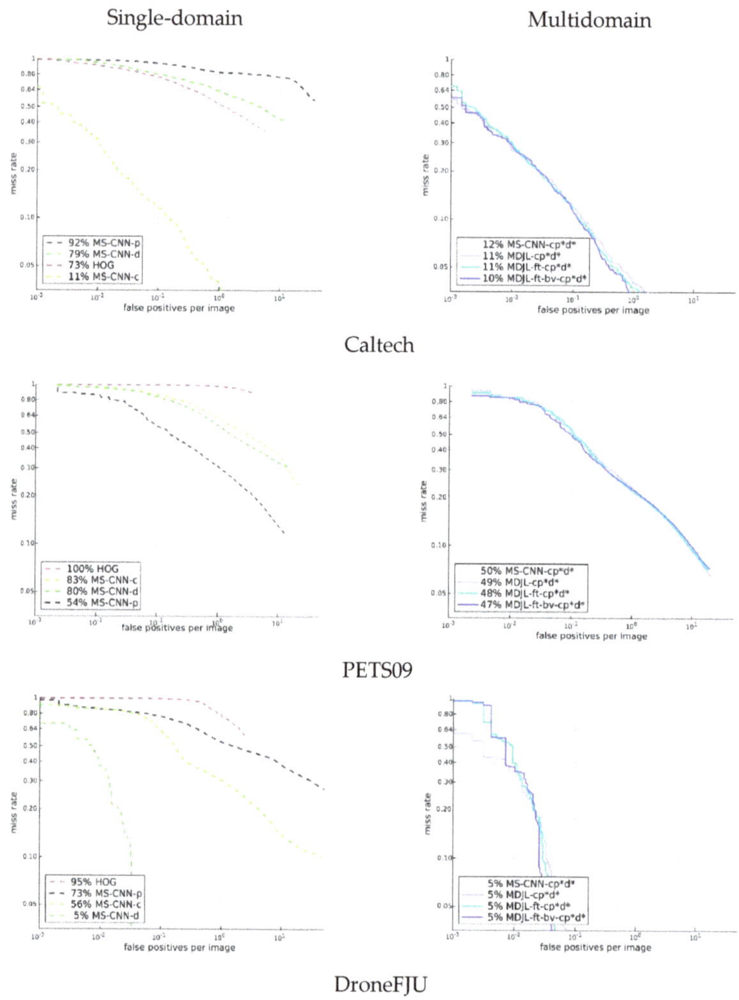

Figure 17. Single-domain dataset vs. multidomain dataset.

The results show that HOG was not observed in any dataset used in the evaluation in the training stage, so the effects on each dataset are not ideal. If we observe other models, we find that for the model trained using the single-domain dataset (dashed line), only the model trained with the training set of the same dataset performed better, and the rest were not ideal. For example, the average miss rate of MS-CNN for the Caltech dataset was 11%, but the average miss rates of MS-CNN-d and MS-CNN-p were as high as 79% and 92%, respectively. This result shows that it is difficult to achieve cross-domain generalization using single-domain dataset training. Therefore, we believe it is essential to use multidomain dataset training.

Figure 17 shows that the model (solid line) trained with the multidomain dataset has comparable performance to the model (dotted line) trained using the corresponding single-domain dataset, regardless of the dataset. For example, in the evaluation results for the DroneFJU, the average miss rates of the three models, MS-CNN-cp*d*, MDJL-cp*d*, and MS-CNN-d, were similar. If we examine the evaluation results of PETS09, the average miss rates of MS-CNN-cp*d* and MDJL-cp*d* using the multidomain dataset are 4% and 5% lower than that of MS-CNN-p, respectively. These results show not only that using multidomain dataset training allows the model to have cross-domain generalization capabilities but also that domains can occasionally have complementary effects.

Next, we compared the proposed method with current state-of-the-art approaches. Results were obtained for MS-CNN, MS-CNN-cp*d*, and MDJL-cp*d*. The data for the other algorithms were obtained by using the data on the Caltech pedestrian detection benchmark website. The classification methods used for each method and the dataset used for training are listed in Table 6.

Receiver-operating characteristic (ROC) analysis is a valuable tool for evaluating the performance of diagnostic tests and the accuracy of a statistical model's outcomes. We compared MDJL with state-of-the-art methods and drew the ROC curve shown in Figure 18. From Figure 18 and Table 6, it is evident that although MDJL used a multidomain dataset to train the network, it can still maintain the performance of the state-of-the-art pedestrian detection benchmark and can beat MS-CNN after fine-tuning.

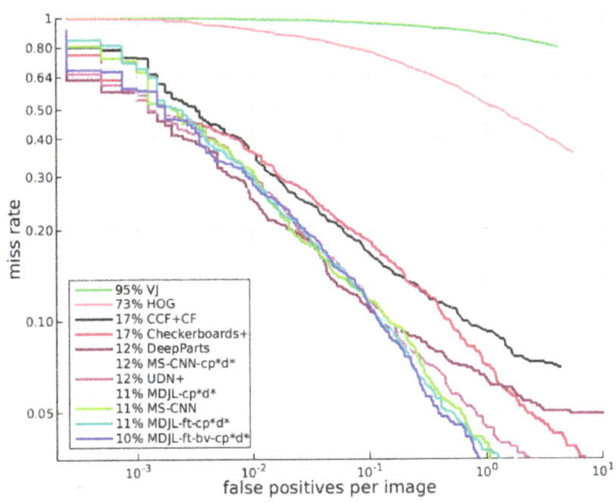

Figure 18. ROC line chart compared with state-of-the-art methods.

4.5. Pedestrian Tracking

In this experiment, we considered pedestrians as the primary tracking objects. In the experiment, we used the four videos from the MOT Challenge 2015 as the training and test data. The training part of the MDP tracker used S2L1 in PETS09 and TUD-Stadtmitte [38] (Table 7), whereas the experimental evaluation used S2L2 in PETS09 and TUD-Crossing as test videos.

Table 7. Experimental dataset.

	FPS	Resolution	Frame	Tracks
PETS09-S2L1	7	768 × 576	795	19
PETS09-S2L2	7	768 × 576	436	42
TUD-Crossing	25	640 × 480	201	13
TUD-Stadtmitte	25	640 × 480	179	10

In this experiment, two videos were selected in PETS09, namely S2L1 and S2L2, for training and testing, respectively. Both videos were acquired in an open outdoor environment, and the shooting angle was from the perspective of a fixed surveillance camera with a downward projection. In the PETS09 video set, several crowded groups of pedestrians walk freely, and there are numerous occlusions between the pedestrians, which leads to difficulty in detection or tracking. Figure 19a is a screenshot of a region of the pedestrian scene in the video set.

(**a**) (**b**)

Figure 19. (**a**) PETS09 dataset and (**b**) outdoor street scene.

The TUD-Stadtmitte and TUD-Crossing are outdoor street scenes, as shown in Figure 19b. The difference compared with PETS09 is that the shooting angle is fixed, and image acquisition is performed from a forward projection. Many pedestrians are observed walking along the street in the video. However, because of the forward projection, the occlusion between pedestrians is more significant compared with the surveillance view with a downward projection, which makes it more difficult to track the trajectory and associate the bounding box.

In the performance evaluation, multiple evaluation criteria [39] were used for the MOT benchmark [40], including multiobject tracking accuracy (MOTA), multiobject tracking precision (MOTP), ID switch, false positive (FP), false negative (FN), mostly tracked (MT), and mostly lost (ML).

FP denotes the total number of marking errors in the bounding box, and FN is the total number of unmarked ground truths. The ID switch is an important criterion for evaluating multiobject tracking, because the primary goal of this process is to obtain the object ID and monitor the trajectory. The definition of MOTA is given in Equation (11). MOTA is used primarily to evaluate the tracking accuracy of the object's trajectory. The referenced error-rate scores include the FN, FP, and ID switches, which are useful indicators. Compared with MOTA, MOTP does not consider the erroneous result of the ID switch but evaluates the tracker's accuracy in estimating the position and size of the object. The MOTP is calculated using Equation (12). First, the total error between the actual and estimated positions of all targets in the frames was counted and averaged. c_t is the number of objects in the ground truth of the t-th frame, and d_t^i is the overlapping rate of the bounding box of the i-th target in the t frames. Next, MT and ML were used to evaluate the integrity of the trajectory; however, the ID switch was not considered. If the trajectory time of an object was successfully tracked for more than 80%, then the trajectory was classified as MT; if it was only successfully tracked for less than 20%, then the trajectory was classified as ML.

$$OTA = 1 - \frac{\sum_{t=1}^{N_{frame}} (FN_t + FP_t + IDSW_t)}{\sum_{t=1}^{N_{frame}} GT_t} \tag{11}$$

$$MOTP = \frac{\sum_{i,t} d_t^i}{\sum_t c_t} \tag{12}$$

In this experiment, we determined the system's best parameters by observing the influence of each parameter on the final tracking result. We adjusted the parameters to filter the pedestrian detection results: confidence threshold, nonmaximum suppression

(NMS) threshold, and the bounding box threshold (parameter for pedestrian tracking). We used PEST09-S2L2 and TUD-Crossing as test videos and averaged the results.

The first parameter of pedestrian detection is the confidence threshold, which is used primarily to filter the bounding box obtained via pedestrian detection. This parameter affects the number of FP and FN. When the threshold is high, the detection results have a low FP, but it is relatively easy to obtain a higher FN. The effect of this threshold adjustment on the tracking results is shown in Figure 20a. The results show that this parameter has little impact on MOTP. However, when the threshold is too high or too small, MOTA and MT begin to decline, whereas ML increases. It is speculated that when FN is higher than a threshold value, the tracker is unable to associate with the target, and the trajectory is broken. Thus, the MT decreases, and the ML increases. The level of MOTA depends on the number of FN and FP. It was determined that the optimal parameter was 0.6, which resulted in the highest MOTA.

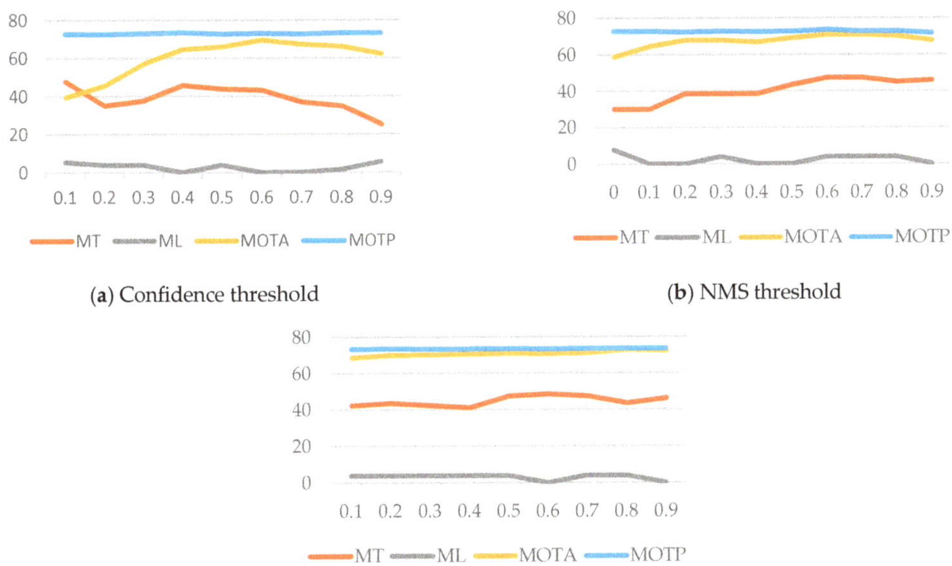

(**a**) Confidence threshold (**b**) NMS threshold

(**c**) Bounding box threshold

Figure 20. Influence of the adjustment of pedestrian detection parameters.

The second parameter for pedestrian detection was the NMS threshold. This parameter affects the accuracy and the quantity of the bounding box positions. The effect of this threshold adjustment on the tracking results is shown in Figure 20b. The results show that when the threshold is high, MOTP decreases slightly, and MT increases simultaneously. It is speculated that when the threshold is high, some bounding boxes have too many redundant detection results because of the inability to merge, which increases the number of FPs and decreases the value of MOTP. Conversely, when the threshold is lower, most bounding boxes obtain the best detection position. However, if multiple objects are too close, the bounding boxes of various objects may merge, thereby increasing the number of FN. Therefore, the MOTP will be higher, but the MT will be lower. Therefore, the optimum parameter was determined to be 0.6, which maintained the highest MOTA and MOTP.

Finally, we adjusted the pedestrian tracking parameter bounding box threshold. When the overlap rate between the pedestrian position predicted by the tracker and the bounding box was higher than this threshold, the algorithm associated the bounding box with the tracker and then continuously performed tracking. This parameter directly affects the tracking accuracy. The effect of this threshold adjustment on the tracking results is shown in Figure 20c. The results show that MT increases to a specific level when the threshold increases and then subsequently decreases. It is speculated that the tracker is more prone to associate errors when the threshold is low. Therefore, both MT and MOTA were relatively low. The tracker can associate the correct bounding box more accurately when the threshold increases. However, when the threshold is high, the tracker has a low error tolerance because of the threshold; therefore, both MT and MOTA decrease. Thus, the optimal parameter was chosen to be 0.8, which maintained the highest MOTA.

After adjusting the optimal parameters, the system was referred to as MDP_MDJL. Next, we compared our system with the current state-of-the-art methods in several experiments. The test videos used were PETS09-S2L2 and TUD-Crossing. Our method was compared with the six most advanced tracking techniques using the MOT benchmark, including two offline tracking methods, TSML_CDE [41] and DMT [42], and four online tracking methods, CDA_DDAL [43], AMIR [44], APRCNN [45], and MDP [33].

The results for PETS09-S2L2 are shown in Figure 21, which indicates that the proposed approach has the best MOTA and MOTP compared with the other methods. Compared with MDP, MOTA and MOTP increased by 18.8% and 5.3%, respectively. The result for the TUD-Crossing is shown in Figure 22; our approach has the best MOTA compared with the other methods. Compared with MDP, MOTA and MOTP increased by 9.7% and 4.7%, respectively.

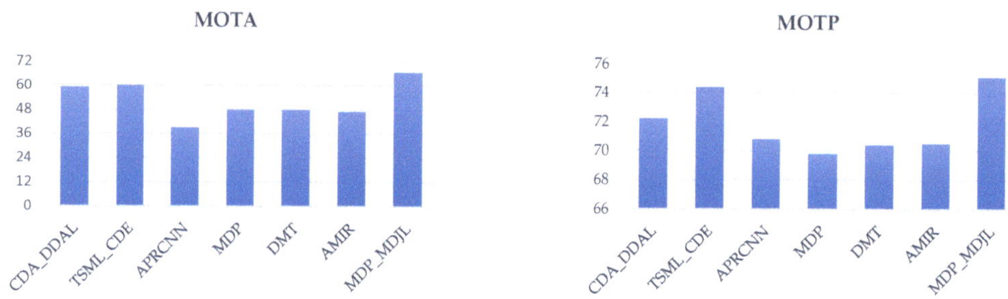

(**a**) MOTA (**b**) MOTP

Figure 21. PETS09-S2L2 analysis chart.

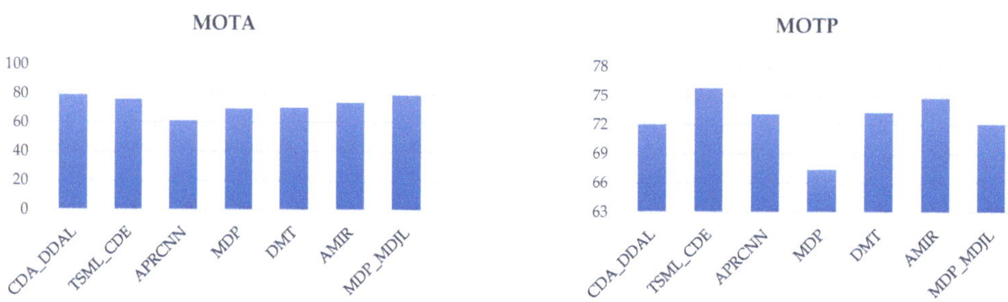

(**a**) MOTA (**b**) MOTP

Figure 22. TUD-Crossing analysis chart.

Overall, the proposed method performed best on the MOTA indicator in the two experimental video scenarios. This was primarily because of the high accuracy of MDJL in the case of multiple domains. In addition, optimizing the parameters allowed the detector to obtain the most accurate position, thus yielding the best performance.

5. Conclusions

This paper discussed the common problem of appropriate datasets often being unavailable in the implementation of machine learning to pedestrian detection and tracking. To address this issue, we proposed the MDJL method. This approach can effectively address the phenomenon in which the network performance in each domain is not ideal when multiple datasets are used for training. Moreover, it can improve the network's performance in similar domains. In addition, the MJDL model was more generalizable. When the number of samples in each dataset differed significantly, to prevent a large dataset from dominating the network and affecting the performance of the other domains, we used data augmentation to balance the difference in the number of samples between the domains. In addition, we also combined the MDJL and MDP trackers to create a multiobject tracking system that is suitable for flying drones. We used the characteristics of the MDJL model to enable the tracking system to effectively cope with a photography platform with a high degree of spatial freedom, such as an aerial camera. The experimental results also showed that MDJL can cope with different scenarios and significantly improve the systems' tracking performance.

When samples are expanded using data augmentation, it may be useful to consider referring to the characteristics of the original dataset; it may yield different training results. In addition, the drone's camera mobility and high shooting angle are not available in fixed-position surveillance and driving recorders. An inappropriate viewing angle leads to more misjudgments. Improvement in the pedestrian tracking recognition rate for horizontal viewing angles should be investigated in the future.

Author Contributions: Conceptualization, Y.-K.W. and J.G.; methodology, Y.-K.W. and J.G.; software, T.-M.P. and J.G.; validation, T.-M.P. and J.G.; formal analysis, Y.-K.W. and J.G.; data curation, J.G.; writing—original draft preparation, J.G.; writing—review and editing, T.-M.P. and Y.-K.W.; supervision, Y.-K.W. All authors have read and agreed to the published version of the manuscript.

Funding: This research received no external funding.

Conflicts of Interest: The authors declare no conflict of interest.

References

1. Boudjit, K.; Ramzan, N. Human detection based on deep learning YOLO-v2 for real-time UAV applications. *J. Exp. Theor. Artif. Intell.* **2022**, *34*, 527–544. [CrossRef]
2. Krishna, S.L.; Chaitanya, G.S.R.; Reddy, A.S.H.; Naidu, A.M.; Poorna, S.S.; Anuraj, K. Autonomous Human Detection System Mounted on a Drone. In Proceedings of the 2019 International Conference on Wireless Communications Signal Processing and Networking (WiSPNET), Chennai, India, 21–23 March 2019; pp. 335–338.
3. Hossain, S.; Lee, D.J. Deep learning-based real-time multiple-object detection and tracking from aerial imagery via a flying robot with GPU-based embedded devices. *Sensors* **2019**, *19*, 3371. [CrossRef] [PubMed]
4. Islam, M.M.; Newaz, A.A.; Gokaraju, B.; Karimoddini, A. Pedestrian detection for autonomous cars: Occlusion handling by classifying body parts. In Proceedings of the 2020 IEEE International Conference on Systems, Man, and Cybernetics (SMC), Toronto, ON, Canada, 11–14 October 2020; pp. 1433–1438.
5. Ahmed, S.; Huda, M.N.; Rajbhandari, S.; Saha, C.; Elshaw, M.; Kanarachos, S. Pedestrian and cyclist detection and intent estimation for autonomous vehicles: A survey. *Appl. Sci.* **2019**, *9*, 2335. [CrossRef]
6. Yan, Z.; Li, P.; Wang, B.; Ren, D.; Zuo, W. Towards Learning Multi-domain Crowd Counting. *IEEE Trans. Circuits Syst. Video Technol.* **2021**. [CrossRef]
7. Xiao, T.; Li, H.; Ouyang, W.; Wang, X. Learning Deep Feature Representations with Domain Guided Dropout for Person Re-identification. In Proceedings of the IEEE Conference on Computer Vision and Pattern Recognition, Las Vegas, NV, USA, 27–30 June 2016.
8. Priyanka, S.A.; Wang, Y.-K. Fully Symmetric Convolutional Network for Effective Image Denoising. *Appl. Sci.* **2019**, *9*, 778. [CrossRef]
9. Fan, C.T.; Wang, Y.-K.; Huang, C.R. Heterogeneous Information Fusion and Visualization for a Large-Scale Intelligent Video Surveillance System. *IEEE Trans. Syst. Man Cybern. Syst.* **2017**, *47*, 593–604. [CrossRef]
10. Pan, T.-M.; Fan, K.-C.; Wang, Y.-K. Object-based Approach for Adaptive Source Coding of Surveillance Video. *Appl. Sci.* **2019**, *9*, 2003. [CrossRef]
11. Wang, Y.-K.; Cheng, K.-Y. A Two-stage Bayesian Network Method for 3D Human Pose Estimation from Monocular Image Sequences. *EURASIP J. Adv. Signal Process.* **2010**, *2010*, 761460. [CrossRef]

12. Viola, P.; Jones, M. Robust Real-Time Face Detection. *Int. J. Comput. Vis.* **2004**, *57*, 137–154. [CrossRef]
13. Dalal, N.; Triggs, B. Histograms of Oriented Gradients for Human Detection. In Proceedings of the IEEE Computer Society Conference on Computer Vision and Pattern Recognition, San Diego, CA, USA, 20–25 June 2005.
14. Felzenszwalb, P.F.; Girshick, R.B.; McAllester, D.; Ramanan, D. Object Detection with Discriminatively Trained Part-Based Models. *IEEE Trans. Pattern Anal. Mach. Intell.* **2010**, *32*, 1627–1645. [CrossRef]
15. Cortes, C.; Vapnik, V. Support-vector networks. *Mach. Learn.* **1995**, *20*, 273–297. [CrossRef]
16. Benenson, R.; Mathias, M.; Timofte, R.; van Gool, L. Pedestrian detection at 100 frames per second. In Proceedings of the IEEE Conference on Computer Vision and Pattern Recognition, Providence, RI, USA, 16–21 June 2012.
17. Yang, B.; Yan, J.; Lei, Z.; Li, S.Z. Convolutional Channel Features. In Proceedings of the International Conference on Computer Vision, Santiago, Chile, 7–13 December 2015.
18. Zhang, S.; Benenson, R.; Schiele, B. Filtered channel features for pedestrian detection. In Proceedings of the IEEE Conference on Computer Vision and Pattern Recognition, Boston, MA, USA, 7–12 June 2015.
19. Chen, L.; Lin, S.; Lu, X.; Cao, D.; Wu, H.; Guo, C.; Liu, C.; Wang, F.Y. Deep neural network based vehicle and pedestrian detection for autonomous driving: A survey. *IEEE Trans. Intell. Transp. Syst.* **2021**, *22*, 3234–3246. [CrossRef]
20. LeCun, Y.; Bottou, L.; Bengio, Y.; Haffner, P. Gradient-based learning applied to document recognition. *Proc. IEEE* **1998**, *86*, 2278–2324. [CrossRef]
21. Sharma, V.; Mir, R.N. A comprehensive and systematic look up into deep learning based object detection techniques: A review. *Comput. Sci. Rev.* **2020**, *38*, 100301. [CrossRef]
22. Ren, S.; He, K.; Girshick, R.; Sun, J. Faster R-CNN: Towards Real-Time Object Detection with Region Proposal Networks. *IEEE Trans. Pattern Anal. Mach. Intell.* **2016**, *38*, C3. [CrossRef]
23. Tian, Y.; Luo, P.; Wang, X.; Tang, X. Deep Learning Strong Parts for Pedestrian Detection. In Proceedings of the International Conference on Computer Vision, Santiago, Chile, 7–13 December 2015.
24. Ouyang, W.; Zhou, H.; Li, H.; Li, Q.; Yan, J.; Wang, X. Jointly learning deep features, deformable parts, occlusion and classification for pedestrian detection. *IEEE Trans. Pattern Anal. Mach. Intell.* **2017**, *40*, 1874–1887. [CrossRef]
25. Cai, Z.; Fan, Q.; Feris, R.; Vasconcelos, N. A Unified Multi-scale Deep Convolutional Neural Network for Fast Object Detection. In Proceedings of the European Conference on Computer Vision, Amsterdam, The Netherlands, 11–14 October 2016.
26. Wang, D.; He, Y.; Liu, Y.; Li, D.; Wu, S.; Qin, Y.; Xu, Z. 3D object detection algorithm for panoramic images with multi-scale convolutional neural network. *IEEE Access* **2019**, *7*, 171461–171470. [CrossRef]
27. Srivastava, N.; Hinton, G.; Krizhevsky, A.; Sutskever, I.; Salakhutdinov, R. Dropout: A Simple Way to Prevent Neural Networks from Overfitting. *J. Mach. Learn. Res.* **2014**, *15*, 1929–1958.
28. Dehghan, A.; Assari, S.M.; Shah, M. GMMCP tracker: Globally optimal Generalized Maximum Multi Clique problem for multiple object tracking. In Proceedings of the IEEE Conference on Computer Vision and Pattern Recognition, Boston, MA, USA, 7–12 June 2015.
29. Gündüz, G.; Acarman, T. Efficient multi-object tracking by strong associations on temporal window. *IEEE Trans. Intell. Veh.* **2019**, *4*, 447–455. [CrossRef]
30. Leon, F.; Gavrilescu, M. A review of tracking and trajectory prediction methods for autonomous driving. *Mathematics* **2021**, *9*, 660. [CrossRef]
31. Yoon, K.; Kim, D.Y.; Yoon, Y.C.; Jeon, M. Data association for multi-object tracking via deep neural networks. *Sensors* **2019**, *19*, 559. [CrossRef] [PubMed]
32. Li, C.; Wang, Y.; Liu, X. A Multi-Pedestrian Tracking Algorithm for Dense Scenes Based on an Attention Mechanism and Dual Data Association. *Appl. Sci.* **2022**, *12*, 9597. [CrossRef]
33. Xiang, Y.; Alahi, A.; Savarese, S. Learning to Track: Online Multi-object Tracking by Decision Making. In Proceedings of the IEEE International Conference on Computer Vision, Santiago, Chile, 7–13 December 2015.
34. Venkateswara, H.; Chakraborty, S.; Panchanathan, S. Deep-Learning Systems for Domain Adaptation in Computer Vision. *IEEE Signal Process. Mag.* **2017**, *34*, 117–129. [CrossRef]
35. Ruder, S. An Overview of Multi-Task Learning in Deep Neural Networks. *arXiv* **2017**, arXiv:1706.05098.
36. Dollár, P.; Wojek, C.; Schiele, B.; Perona, P. Pedestrian Detection: An Evaluation of the State of the Art. *IEEE Trans. Pattern Anal. Mach. Intell.* **2012**, *34*, 743–761. [CrossRef] [PubMed]
37. Ferryman, J.; Shahrokni, A. PETS2009: Dataset and challenge. In Proceedings of the 2009 Twelfth IEEE International Workshop on Performance Evaluation of Tracking and Surveillance, Snowbird, UT, USA, 7–9 December 2009.
38. Mykhaylo, A.; Stefan, R.; Bernt, S. People-Tracking-by-Detection and People-Detection-by-Tracking. In Proceedings of the IEEE Conference on Computer Vision and Pattern Recognition, Anchorage, AK, USA, 23–28 June 2008.
39. Xu, Y.; Osep, A.; Ban, Y.; Horaud, R.; Leal-Taixé, L.; Alameda-Pineda, X. How to train your deep multi-object tracker. In Proceedings of the IEEE/CVF Conference on Computer Vision and Pattern Recognition, Seattle, WA, USA, 13–19 June 2020; pp. 6787–6796.
40. Leal-Taixé, L.; Milan, A.; Reid, I.; Roth, S.; Schindler, K. MOTChallenge 2015: Towards a Benchmark for Multi-Target Tracking. *arXiv* **2015**, arXiv:1504.01942.
41. Wang, B.; Wang, G.; Chan, K.L.; Wang, L. Tracklet Association by Online Target-Specific Metric Learning and Coherent Dynamics Estimation. *IEEE Trans. Pattern Anal. Mach. Intell.* **2017**, *39*, 589–602. [CrossRef]

42. Kim, H.-U.; Kim, C.-S. CDT: Cooperative Detection and Tracking for Tracing Multiple Objects in Video Sequences. In Proceedings of the European Conference on Computer Vision, Amsterdam, The Netherlands, 11–14 October 2016.
43. Bae, S.-H.; Yoon, K.-J. Confidence-Based Data Association and Discriminative Deep Appearance Learning for Robust Online Multi-Object Tracking. *IEEE Trans. Pattern Anal. Mach. Intell.* **2018**, *3*, 595–610. [CrossRef]
44. Sadeghian, A.; Alahi, A.; Savarese, S. Tracking the Untrackable: Learning To Track Multiple Cues with Long-Term Dependencies. In Proceedings of the IEEE International Conference on Computer Vision, Venice, Italy, 22–29 October 2017.
45. Chen, L.; Ai, H.; Shang, C.; Zhuang, Z.; Bai, B. Online Multi-Object Tracking with Convolutional Neural Networks. In Proceedings of the IEEE International Conference on Image Processing, Beijing, China, 17–20 September 2017.

drones

MDPI

Article

A Real-Time UAV Target Detection Algorithm Based on Edge Computing

Qianqing Cheng *, Hongjun Wang *, Bin Zhu, Yingchun Shi and Bo Xie

School of Electronic Countermeasures, National University of Defense Technology, Hefei 230037, China
* Correspondence: chengqianqing@nudt.edu.cn (Q.C.); wanghongjun17@nudt.edu.cn (H.W.); Tel.: +86-1552-803-7356 (Q.C.)

Abstract: Small UAV target detection plays an important role in maintaining the security of cities and citizens. UAV targets have the characteristics of low-altitude flights, slow speeds, and miniaturization. Taking these characteristics into account, we present a real-time UAV target detection algorithm called Fast-YOLOv4 based on edge computing. By adopting Fast-YOLOv4 in the edge computing platform NVIDIA Jetson Nano, intelligent analysis can be performed on the video to realize the fast detection of UAV targets. However, the current iteration of the edge-embedded detection algorithm has low accuracy and poor real-time performance. To solve these problems, this paper introduces the lightweight networks MobileNetV3, Multiscale-PANet, and soft-merge to improve YOLOv4, thus obtaining the Fast-YOLOv4 model. The backbone of the model uses depth-wise separable convolution and an inverse residual structure to simplify the network's structure and to improve its detection speed. The neck of the model adds a scale fusion branch to improve the feature extraction ability and strengthen small-scale target detection. Then, the predicted boxes filtering algorithm uses the soft-merge function to replace the traditionally used NMS (non-maximum suppression). Soft-merge can improve the model's detection accuracy by fusing the information of predicted boxes. Finally, the experimental results show that the mAP (mean average precision) and FPS (frames per second) of Fast-YOLOv4 reach 90.62% and 54 f/s, respectively, in the workstation. In the NVIDIA Jetson Nano platform, the FPS of Fast-YOLOv4 is 2.5 times that of YOLOv4. This improved model performance meets the requirements for real-time detection and thus has theoretical significance and application value.

Keywords: unmanned aerial vehicle (UAV); object detection; non-maximum suppression (NMS); weighted boxes fusion (WBF); lightweight network; multiscale

Citation: Cheng, Q.; Wang, H.; Zhu, B.; Shi, Y.; Xie, B. A Real-Time UAV Target Detection Algorithm Based on Edge Computing. *Drones* **2023**, *7*, 95. https://doi.org/10.3390/drones7020095

Academic Editor: Anastasios Dimou

Received: 17 December 2022
Revised: 15 January 2023
Accepted: 24 January 2023
Published: 30 January 2023

1. Introduction

In recent years, the rapid development of drone technology has resulted in increased convenience in people's daily lives. However, invasive and disorderly flying of UAVs are increasing, posing a serious threat to public safety [1]. UAV targets have the characteristics of low flying altitudes, slow speeds, and miniaturization [2]. This makes their detection by traditional radar and radio methods difficult and costly [3,4]. Detecting UAVs based on sound is greatly disturbed by noise, so their detection is not obvious [5]. In addition to the above methods, the deep learning method based on two-dimensional image data is gradually being applied to UAV detection and has achieved good results [6,7]. Therefore, it is promising to study a new object detection method based on UAV images to address the lack of suitable detection methods.

With the development of deep learning and updates to graphics computing devices, image-based target detection methods have become a hotspot in the field of target detection [8]. An increasing number of engineering practices are being applied to facial recognition, pedestrian target detection, and autonomous driving. The representative algorithms in these technologies include R-CNN and Faster R-CNN based on two-stage detection [9], as well as SSD (single-shot multi-box detector) and YOLO (you only look once) series algorithms based on one-stage detection [10]. The former methods are called two-stage algorithms because the target candidate regions are generated separately from the classification localization. The latter are called one-stage algorithms because the extraction of features and classification localization are carried out continuously. As the fourth version of the YOLO series of algorithms, YOLOv4 is the first to use CSPDarkNet-53 as the backbone network to extract image features [11], PANet (path aggregation network) as the feature fusion structure, and SPP (spatial pyramid pooling) to enhance feature extraction [12], which greatly improves the performance of the one-stage detection algorithm in model accuracy and reasoning speed.

At present, object detection applications are mostly divided into cloud computing and edge computing. Cloud computing is a mode of unifying the data collected on the edge side to the central processor for operation and then making decisions. This model has high latency, network instability, and low bandwidth problems, making it unsuitable for UAV detection tasks that require fast responses and have high error costs. On the contrary, edge computing can solve these problems well. Edge computing is a technology that provides cloud services and IT environment services for developers and service providers at the edge side of the network [13]. By integrating detection algorithms, edge computing platforms can partially replace the data processing functions of cloud devices and servers and directly store or detect the target data. This edge-side data processing method not only reduces the data transmission and communication time between devices but also saves on network bandwidth and energy consumption. Moreover, this method also frees target detection from its dependence on large servers and GPU devices and enables users to flexibly arrange edge computing devices to meet the needs of various detection tasks.

In target detection, accuracy and speed are the two most important metrics to measure a model's performance. Many scholars have also improved the detection algorithm based on these two points. In terms of accuracy, the proposed feature pyramid network and the application of various box-filtering algorithms enhance the detection capability of the model for multiscale targets. In terms of speed, various lightweight networks are proposed to greatly reduce the model parameters and accelerate the detection speed, so that the detection algorithm can be carried to the edge computing platform to complete the task. Aiming at the problems of slow detection speed and poor effect of multiscale target detection, we propose a UAV target detection algorithm Fast-YOLOv4 based on edge computing, which combines the lightweight network MobileNetV3, the improved Multiscale-PANet, and the soft-merge algorithm. To summarize, the innovations of the algorithm are as follows:

- Using the lightweight network MobileNetV3 to improve the backbone of YOLOv4 and introduce depth-wise separable convolution and inverse residual structures. These improvements can greatly reduce the number of model parameters and improve the detection speed.
- Adding the fourth feature fusion scale in the neck structure, PANet, of YOLOv4 and enhancing the flow superposition of high-dimensional image features and low-dimensional location features, improving the classification and localization accuracy of multiscale targets.
- Replacing NMS with the improved soft-merge algorithm. It is a predicted box-filtering algorithm and can fuse the predicted box information to obtain better-predicted boxes. The problems of missing detection and false detection are also reduced, enhancing the recognition effect.
- Combined with the edge-embedded platform NVIDIA Jetson Nano, the fast target detection algorithm is equipped to reduce the response time and energy consumption, and realize the real-time multi-scene accurate detection of UAV targets.

Through those improvements, Fast-YOLOv4 achieves high precision and fast detection of UAV targets on an edge computing platform, which provides a solution for research on UAV target detection.

2. Related Work

In this section, we mainly introduce work to four parts of our model: multiscale detection, box-filtering algorithms, lightweight networks, and edge computing about UAVs, as well as their related methods and applications. For a long time, multiscale detection has presented a major problem in target detection. From the initial image pyramid network and feature pyramid network method to the most commonly used Res2Net and PANet methods [12,14,15], multiscale detection is gradually maturing. For example, aiming at the significant differences in the scales of drone targets, Zeng et al. proposed a UAV detection network combining Res2net and a mixed feature pyramid structure, which achieved a more than 93% mAP in a self-built dataset [16]. However, the model is a little bloated and difficult to carry over to the edge platform. For mask target detection, Zhu et al. introduced a path aggregation network into the input feature layer of YOLOv4-Tiny, which improves the mAP of mask detection by 4.3% [17]. However, the method does not propose innovative improvements to the multiscale structure in that paper, which is slightly inadequate.

Another way to improve the model accuracy is to improve the box-filtering algorithm. Because the traditional NMS method is not effective in detecting occluded targets, many scholars have tried to study and improve it. To solve the problem of poor detection of occluded targets, Zhang et al. introduced position information into the NMS to adjust the final score of the box and proposed a distance-based intersection ratio loss function. The method was tested using classical datasets, and the accuracy was improved by 2% [18]. Ning et al. proposed an improved non-maximum suppression method called I-SSD (Inception-SSD) that can obtain weighted averages of the coordinates of the predicted boxes and improve the ability of the model's filtering results [19]. The proposed I-SSD algorithm achieves a 78.6% mAP on the Pascal VOC 2007 test. Roman et al. also proposed a weighted box fusion method by fusing predicted boxes of different detection models to obtain integration results. The method significantly improves the quality of the fused predicted rectangles for an ensemble and achieves the best results in dataset challenge competitions [20].

In addition to the detection accuracy of multiscale targets, the speed improvement in the detection model and the deployment of an edge platform are also current hot issues. To solve these problems, most scholars use lightweight networks to improve the target detection model. Zhou et al. proposed a lightweight YOLOv4 ship detection algorithm combined with MobileNetv2 to solve the problem that the large model could not run on the micro-platform and realized the high-precision and rapid detection of ship targets [21]. Liu et al. proposed a pruned-YOLOv4 model to overcome the problem of low detection speed and poor detection effect of small targets. The model achieves 90.5% mAP, and its processing speed is increased by 60.4%, which accomplishes the real-time detection of drones well [22]. However, the accuracy performance of the model is slightly inadequate. It can be seen from these papers that lightweight networks are very effective at improving target detection, which suggests ideas for their application to target detection in more ways.

Because of the broad application prospects of object detection combined with edge computing, many researchers try to carry the object detection algorithm onto the embedded platform to complete the detection task. Daniel et al. achieve an accurate detection of UAV targets by building YOLOv3 on the edge platform of NVIDIA Jetson TX2, with an average accuracy of 88.9% on the self-built dataset [23]. To solve the problem of dynamic obstacle avoidance in safe drone navigation, Adrian et al. propose a novel method of performing onboard drone detection and localization using depth maps. This method is integrated into a small quadrotor and reduces the maximum obstacle avoidance error distance to 10% [24]. In terms of expanding the extent of applicable mission scenarios of UAVs and coordinating the flight formations of fleets of UAVs, Roberto et al. proposed a YOLO object detection system integrated into an original processing architecture. This approach achieves a high level of accuracy and is robust against challenging conditions in terms of illumination, background, and target-range variability [25].

Combined with the ideas for improvement presented by related papers, the proposed Fast-YOLOv4 algorithm in this paper improves the detection accuracy in a multiscale structure and predicted box algorithm and improves the speed of the lightweight network. In the following experiments, the proposed algorithm shows higher accuracy and faster speed than other algorithms on the edge computing platform and realizes a highly accurate and fast detection of multiscale UAV targets.

3. System Model

This section mainly introduces the design idea of the system model. Figure 1 shows the system architecture of target detection using the edge computing platform and workstation. The target detection system consists of three parts: a high-speed intelligent camera, an edge computing platform NVIDIA Jetson Nano, and a workstation. Firstly, the system will set up a high-speed intelligent camera in a detection scene or on a mobile platform such as a robot. Then, the camera will monitor the surrounding environment and transmit the collected image data to the Jetson Nano. The Jetson Nano integrates the target detection algorithm Fast-YOLOv4, which can analyze the image data in real time. If there are UAV targets in the image, the algorithm will identify and locate the UAV targets. Finally, the Jetson Nano will send the abnormal results detected in real time to the workstation through a wireless connection. Administrators at the workstation can alert and respond to abnormal results. These are the process of real-time UAV target detection by the system.

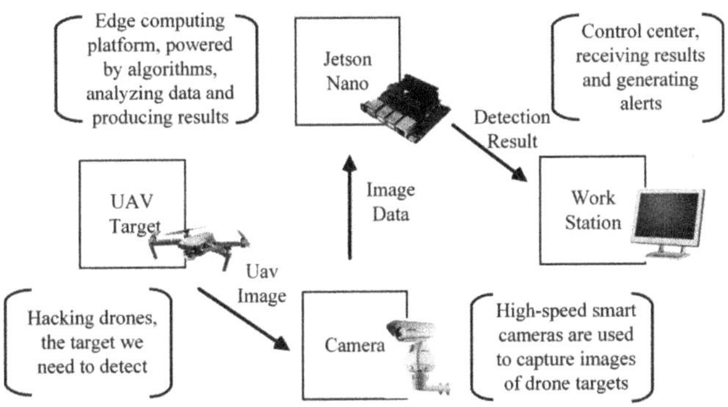

Figure 1. The object detection system model based on edge computing.

4. Algorithm Design

4.1. Fast-YOLOv4 Algorithm

This section mainly introduces the model architecture of Fast-YOLOv4, which is an improved algorithm based on YOLOv4. Figure 2 shows the architecture of Fast-YOLOv4 model and the structure of each module. The input image size of the model is $416 \times 416 \times 3$. The model architecture is mainly composed of MobileNet16, Multiscale-PANet, and YOLO-Head, including CBH module, MBN module, CBL module, and SPP module. CBH module is used as the initial layer of the backbone to process images, which is composed of a convolution block, normalization layer, and h-swish activation function. MBN module is a linear bottleneck inverted residual structure, which is mainly composed of a 1×1 convolution block, a 3×3 DW convolution block, and an h-swish activation function. When the input channel and output channel have the same number, the residual connection will be carried out to deepen the network. The CBL module is the basic module used for the detection and classification tasks, which is mainly composed of the convolution layer, normalization layer, and leaky ReLU activation function. The SPP module consists of the MaxPool layer and the full connection layer, which form a spatial pyramid pooling structure. This structure can obtain the same size feature vectors for any size of the input feature map, which can enhance the adaptability of the network to multiscale targets. MobileNet16 is stacked with CBH modules and MBN modules, with a total of 16 layers. Its function is to extract image features and output feature maps of different scales. Multiscale-PANet is mainly built by the CBL module and SPP module. Its function is to fuse and superposition the input feature map and enrich the feature map information. YOLO-Head is mainly built by a CBL module and a convolution layer. Its function is to detect the location and class of the target and input it into the soft-merge algorithm to filter out the final result.

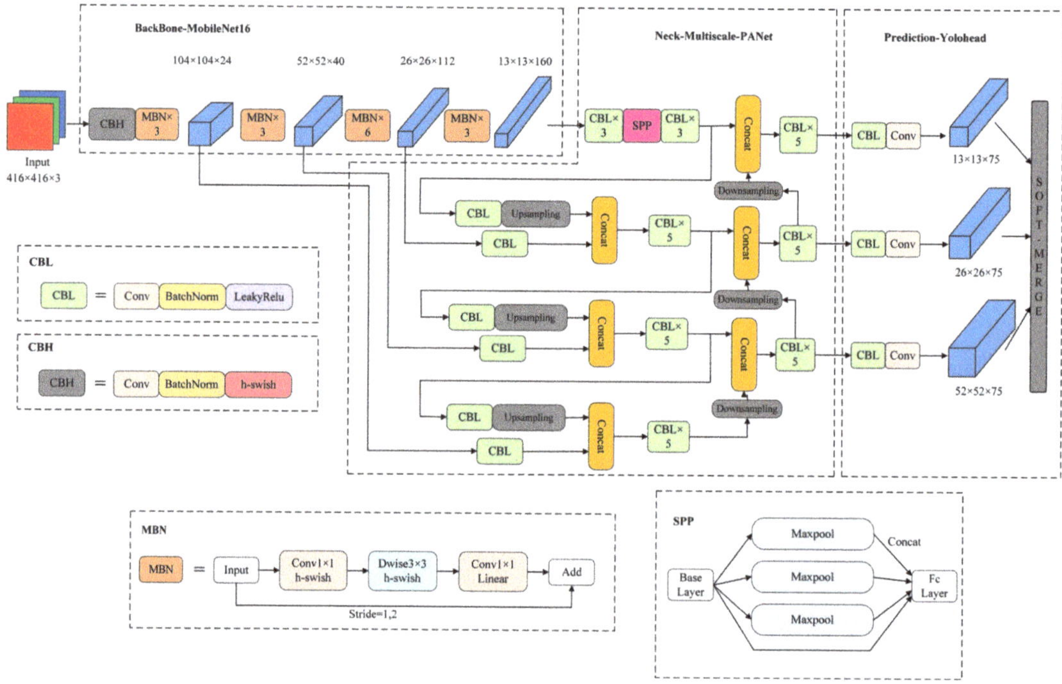

Figure 2. The architecture of the Fast-YOLOv4 model.

4.2. Lightweight Improvement of Backbone Network

The model inference speed determines the model detection speed. Therefore, reducing the number of model parameters and simplifying the network structure is the simplest way to speed up detection. Reducing the number of model parameters facilitates the integration of the algorithm into the edge computing platform and reduces energy consumption. Among the low-parameter models, the lightweight MobileNetV3 is a model with speed and accuracy advantages. MobileNetV3 is the third version of the lightweight network published by Google scientists such as Andrew Howard in 2019 [26]. It combines the deep separable convolution structure of V1 and the linear bottleneck inverse residual structure of V2 and introduces a new nonlinear activation function, h-swish, to save computational costs [27,28]. As a result, MobileNetV3 is a new lightweight network with higher performance.

Fast-YOLOv4 combines the lightweight structure of MobileNetV3 to transform the backbone network of YOLOv4. The backbone network uses the pruned MobileNetV3 as the starting part of the model, performs convolution operations on the input image, and extracts feature maps at four scales for feature fusion. The sizes of these feature maps are 104 × 104, 52 × 52, 26 × 26, and 13 × 13. These feature maps take into account the global and local features of various scale targets and provide sufficient information for subsequent classification and localization. The following Table 1 describes the backbone structure information of Fast-YOLOv4. In Table 1, the term "Operator" represents the module used at this layer and the size of the convolution kernel. The term "Exp size" represents the number of channels to be added to the MBN module. The term "Out size" represents the number of output channels. The term "SE" represents whether to add an SE (squeeze-and-excite) structure to the depth-wise separable convolution block. The term "s" stands for step size.

Table 1. Table of the Fast-YOLOv4 backbone.

Floor	Input	Operator	Exp Size	Out Size	SE	s
1	$416^2 \times 3$	CBH, 3×3	-	16	-	2
2	$208^2 \times 16$	MBN, 3×3	16	16	-	1
3	$208^2 \times 16$	MBN, 3×3	64	24	-	2
4	$104^2 \times 24$	MBN, 3×3	72	24	-	1
5	$104^2 \times 24$	MBN, 5×5	72	40	✓	2
6	$52^2 \times 40$	MBN, 5×5	120	40	✓	1
7	$52^2 \times 40$	MBN, 5×5	120	40	✓	1
8	$52^2 \times 40$	MBN, 3×3	240	80	-	2
9	$26^2 \times 80$	MBN, 3×3	200	80	-	1
10	$26^2 \times 80$	MBN, 3×3	184	80	-	1
11	$26^2 \times 80$	MBN, 3×3	184	80	-	1
12	$26^2 \times 80$	MBN, 3×3	480	112	✓	1
13	$26^2 \times 112$	MBN, 3×3	672	112	✓	1
14	$26^2 \times 112$	MBN, 5×5	672	160	✓	2
15	$13^2 \times 160$	MBN, 5×5	960	160	✓	1
16	$13^2 \times 160$	MBN, 5×5	960	160	✓	1

The convolution layer used in the YOLOv4 structure is the standard 3×3 convolution, which will increase the number of model parameters and the computation time in the operation process. Therefore, Fast-YOLOv4 uses depth-wise separable convolution instead of the standard convolution. Depth-wise convolution is a convolution block proposed in MobileNetV1 to replace traditional convolution [27]. Its structure consists of a light depth-wise layer for spatial filtering and a heavy 1×1 point-wise layer for feature generation. Separating spatial filtering and feature generation greatly reduces the number of convolution operations and saves on computational costs. Figure 3 shows the structure of the standard convolution and depth-wise separable convolution.

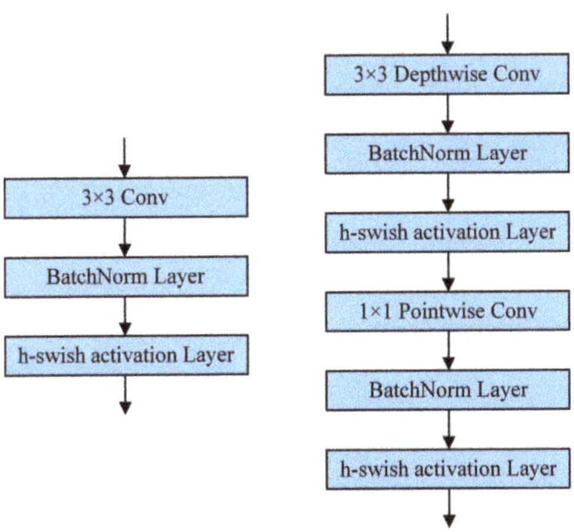

Figure 3. The structure of convolution.

The h-swish activation function is an approximate version of the swish activation function, which has a lower bound, no upper bound, and is not monotone, among other characteristics. The h-swish activation can make the neural network layer more expressive.

In order to adapt it to lightweight network use, the h-swish function is composed of common operators with small computational costs and is, thus, applicable to most frameworks. As a result, we use this function to improve the backbone network. The formula of the *h-swish* function can be written as:

$$h\text{-}swish[x] = x\frac{\mathrm{Re}LU6(x+3)}{6} \tag{1}$$

In Formula (1), x represents the value of the input activation function.

4.3. Multiscale Improvement of Neck Network

PANet is a path aggregation network applied to instance segmentation proposed by Shu [12]. Based on the top-down feature extraction of the FPN (feature pyramid network), PANet introduces a bottom-up path structure. Through feature fusion, the model can make full use of the local features of the shallow network and enrich the feature output of the deep network. An adaptive feature pooling layer is also added to the network to extract more features from the ROI (region of interest). Finally, the network uses the fully connected fusion layer to fuse the output of the convolution layer and the fully connected layer to obtain more accurate results.

The neck network of YOLOv4 adopts a PANet structure and extracts feature maps at three scales for fusion. This method enhances the network's ability of image feature extraction and integration but also results in losses of some features of small-scale targets. To strengthen the localization and feature extraction of small-scale targets, Fast-YOLOv4 improves the feature fusion structure of the neck network. Adding a feature output branch enriches the target feature extracted in the shallow network. By adding a feature output branch, the neck network can extract more target features from the shallow layer. Figure 4 shows the structure of the improved PANet.

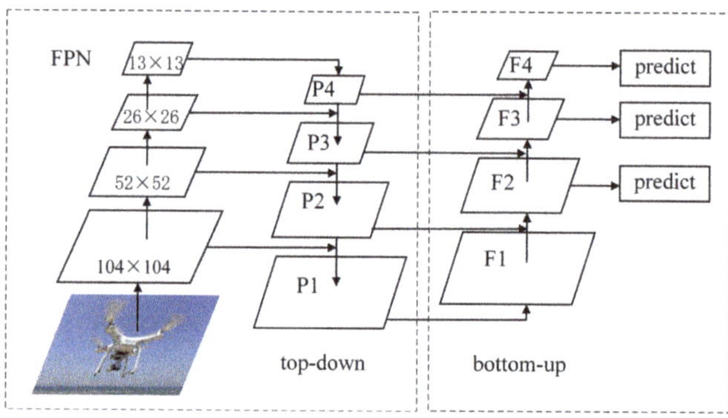

Figure 4. The diagram of the improved PANet structure.

4.4. Soft-Merge Algorithm

At present, NMS (non-maximum suppression) and Soft-NMS are commonly used in detection algorithms to filter predicted boxes [29]. These methods work well with a single model, but they can only filter relatively suitable predicted boxes and cannot effectively use the information of most predicted boxes. Different from NMS, Merge-NMS uses the confidence score and coordinate information of all predicted boxes to construct the fused predicted box, which is relatively closer to the real target boxes [21]. Figure 5 shows the difference in the results of Merge-NMS and NMS.

The operation of the traditional NMS is simple. In the filtering process, a predicted box larger than an IoU (intersection over union) threshold will be forcibly deleted, which may lead to missing detection or false detection in multi-UAV target scenarios. The IoU represents the area of the intersection ratio between two predicted bounding boxes, which can be used to measure the degree of overlap between bounding boxes. Our improved soft-merge algorithm combines the idea of Soft-NMS weight assignment, can weight and fuse all predicted boxes. This approach can make full use of the coordinate and confidence information of the predicted boxes and reduce problems such as missing detection and false detection. The algorithm process of soft-merge can be outlined as Algorithm 1.

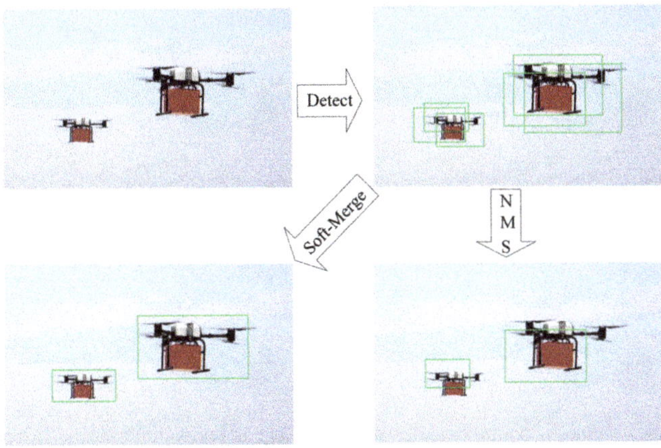

Figure 5. Comparison of Merge-NMS and NMS.

Algorithm 1: The algorithm process of soft-merge in Fast-YOLOv4

Input: Arrays $a_1, a_2, \ldots, a_m$, m represents the number of predicted boxes, a represents the information of predicted boxes, $a[confidence, X_1, X_2, Y_1, Y_2]$

Output: Array $F[f_1, f_2, \ldots, f_n]$, n represents the number of fusion boxes, and each f represents a fusion box result.

Assume that there are m predicted boxes in an image, and use soft-merge algorithm to obtain the fusion predicted box results. According to the confidence value of predicted box array a, all predicted boxes are sorted in descending order and added to array B to obtain $B[b_1, b_2, \ldots, b_m]$.;

for i = *1 to* m **do**
- **for** j = *1 to* n **do**
  - **if** $IoU(a_i, f_j) >= Thre$ **then**
    - Use the confidence update function for a_i and add it to the array L_j. L_j is the array corresponding to each fusion box in F, which is used to store the box to be fused;
  - **else**
    - $t \leftarrow t + 1$, t is used for counting, and the initial value is 0;
  - **end**
- **end**
- **if** $t = n$ **then**
  - $F[n+1] \leftarrow a_i$;
  - $n \leftarrow n + 1$;
- **end**
- **for** k = *1 to* n **do**
  - Use the weighted box fusion function to calculate the array L_k to obtain f_k;
  - $F[k] \leftarrow f_k$;
- **end**

end

The formula of the confidence-updating function is:

$$s_i = \begin{cases} s_i, IoU < Thre \\ s_i e^{-\frac{iou(M,b_i)^2}{\sigma}}, IoU > Thre \end{cases} \tag{2}$$

In Formula (2), s_i represents the score of the prediction box i, $Thre$ represents the threshold, M represents the selected predicted box, and b_i represents other predicted boxes.

The formula of weighted box fusion is:

$$X1,2 = \frac{\sum_{i=1}^{T} s_i \times X1,2_i}{\sum_{i=1}^{T} s_i}, \tag{3}$$

$$Y1,2 = \frac{\sum_{i=1}^{T} s_i \times Y1,2_i}{\sum_{i=1}^{T} s_i} \tag{4}$$

In Formulas (3) and (4), s_i represents the score of the prediction frame i, T represents the number of predicted boxes participating in the fusion, $X1,2$ and $Y1,2$ represent the coordinates of the fused box, and $X1,2_i$ and $Y1,2_i$ represent the coordinates of the predicted box i. During the soft-merge calculation, the confidence scores of other boxes will be suppressed according to the IoU of the selected box. Boxes with high confidence are given a higher weight in the fusion formula, and the influence of the low-confidence boxes will be reduced in the final result.

5. Dataset and Experimental Preparation

5.1. Dataset Information

The datasets used in the experiment include the PASCAL VOC 07+12 dataset and UAV dataset. Firstly, the algorithm is trained on the VOC 07+12 dataset, and then the algorithm is trained and tested on the UAV dataset via transfer learning. The PASCAL VOC 07+12 dataset is a public competition dataset used by the PASCAL VOC Competition in 2007 and 2012. The dataset contains 20 classes of targets with a total of 21,504 images, of which 16,551 are training set images and 4952 are test set images for target detection. UAV dataset is constructed by combining the public dataset [30] and autonomously taking images of UAV targets. It includes three scales of large, medium, and small UAV targets, with 3698 images and 3719 targets. The division of the experimental dataset first randomly selects 20% of the images as the test set and then selects 90% of the remaining 80% of images as the training set and the rest as the validation set. Table 2 shows detailed information on the UAV dataset.

Table 2. The table of UAV dataset information.

Class	Number	Large	Medium	Small
train	2662	1621	547	509
validation	296	176	60	63
test	740	434	171	138
total	3698	2231	778	710

5.2. Experimental Environment and Hyperparameter

The experimental environment includes the workstation and the edge computing platform NVIDIA Jetson Nano. The workstation is used to train and validate the algorithm, and Jetson Nano is used to test the final performance of the algorithm. The configuration of the workstation is Intel (R) Core(TM) I9-10900F CPU @ 2.80 GHz 2.81 GHz 64 G memory based on Windows 10 operating system. The GPU is an NVIDIA Quadro P4000 with 8 G of video memory. The configuration of NVIDIA Jetson Nano is a 128-core CUDA Maxwell GPU, a quad-core ARM A57 1.43 GHz CPU with 4 GB LPDDR4 memory and 472 GFLOPS of processing power. The deep learning framework used in the experiment is PyTorch 1.8.1, and the GPU acceleration library is CUDA10.2.

Hyperparameters are a set of parameters related to model training. They are also related to the length of training time and training effect. Table 3 shows the hyperparameter information of model training.

Table 3. The table of model training hyperparameter setting.

Type	Parameter	Note
Image size	416 × 416	Image input size
Epoch	200	Total training times
Batch size	16	Number of images per iteration
Learning rate1	0.01	Initial learning rate
Learning rate2	0.0001	Minimum learning rate
Momentum	0.937	Momentum of optimizer
Weight decay	0.0005	Decay of weights

5.3. Evaluation Metrics

The evaluation metrics of the experiment include object detection accuracy AP (average precision), mAP (mean average precision), IoU, detection speed FPS (frames per second), model parameters, and model volume size.

The mAP can comprehensively evaluate the localization and classification effect of the model for multi-class and multi-target tasks. Calculating the mAP requires calculating the AP for each class in the recognition task and then taking its average. The formula is as follows:

$$mAP = \frac{\sum_{i=1}^{C} AP_i}{C}, \tag{5}$$

In Formula (5), C represents the number of total classes, and AP_i represents the AP value of class i.

Calculating AP requires knowing the values of P (precision) and R (recall). The formulas for these three metrics are as follows:

$$P = \frac{TP}{TP + FP}, \tag{6}$$

$$R = \frac{TP}{TP + FN}, \tag{7}$$

$$AP = \int_0^1 P(R)dR, \tag{8}$$

In Formulas (6)–(8), TP (true positive) means that the input is a positive sample and the predicted result is also a positive sample; FP (false positive) means that the input is a negative sample and the predicted result is a positive sample; FN (false negative) means that the input is a positive sample and the prediction result is a negative sample; and TN (true negative) means that the input is a negative sample and the prediction result is a negative sample.

The IoU metric is used to calculate the ratio of the intersection and union of two bounding boxes. In essence, it converts the accuracy of object detection division into a comparison of the area between detection results and the true values. Under different IoU conditions, we can calculate different detection accuracies to comprehensively measure the accuracy of the model. Assuming that the areas of the two bounding boxes are A and B, the formula of the IoU metric can be written as:

$$IoU = \frac{\text{Area of Overlap}}{\text{Area of Union}} = \frac{A \cap B}{A \cup B} \tag{9}$$

The FPS metric is the time that a model takes to detect a picture or the number of pictures detected in one second. The larger the FPS, the faster the model is detecting, which can be used to measure the detection speed of the model. The model parameters and model volume size are both metrics of model complexity. They all represent the size of the model, which can directly reflect the model size.

6. Experimental Results and Analysis

6.1. Performance Experiment on PASCAL VOC 07+12

To investigate the improvements of Fast-YOLOv4, we performed a comparative experiment on the PASCAL VOC 07+12 dataset. This dataset contains 20 categories of large-, medium-, and small-scale targets, which can comprehensively reflect the detection performance of the algorithm. The comparative algorithms used in the experiment include Faster R-CNN, YOLOv4, YOLOv3, SSD, YOLOv4-MobileNetV3, and Fast-YOLOv4. All algorithms will be trained and validated in the workstation. In addition, the metrics used in this experiment are mAP (IoU = 0.5), FPS, parameter, and volume. Table 4 shows the detection results of each algorithm on PASCAL VOC 07+12.

Table 4. The performance comparison of algorithms on PASCAL VOC 07+12 (in the workstation).

Model	Backbone	Input Size	mAP (%)	FPS (f/s)	Parameter ($\times 10^6$)	Volume (MB)
Faster R-CNN	ResNet50	416 × 416	77.02	12	137.10	522.99
YOLOv4	CSPDarkNet53	416 × 416	84.29	29	63.94	243.90
YOLOv3	DarkNet53	416 × 416	80.24	35	61.63	235.08
SSD	VggNet16	300 × 300	78.27	43	26.29	100.27
YOLOv4-MobileNetV3	MobileNetV3	416 × 416	79.75	60	11.30	43.51
Fast-YOLOv4	MobileNetV3	416 × 416	81.45	54	13.47	51.39

As can be seen from Table 4, in terms of detection accuracy, our approach has good performance compared with the other algorithms. Its mAP reaches 81.45%, second only to YOLOv4. Compared with YOLOv4-MobileNetV3 and YOLOv3, Fast-YOLOv4's accuracy increased by 1.7% and 1.21%, respectively. In terms of the detection speed, FPS, and model parameters, the performance of Fast-YOLOv4 is slightly worse than YOLOv4-MobileNetV3, but better than the other algorithms. The FPS metric reaches 55 f/s, and the model parameters and volume are only 13.47 M and 51.39 MB, which are better than the other models. In addition, compared with the low detection speed and the large number of parameters in YOLOv4 and Faster R-CNN, the proposed Fast-YOLOv4 has greater real-time detection performance and more lightweight parameters.

To illustrate the detection performance of the proposed algorithm on different categories of targets, we select Fast-YOLOv4 and YOLOv4 to compare their AP scores in six categories of targets in the PASCAL VOC 07+12 dataset. Figure 6 shows the comparison results.

It can be seen from Figure 6 that compared with YOLOv4, Fast-YOLOv4 improved the detection accuracy in most categories, especially for small-scale targets, such as potted plants, boats, and birds. The mAP has improved in these categories by 20%, 15%, and 7%, respectively. There are also some improvements in categories of medium-scale targets, such as dog, sheep, and cow, with increased mAPs of 8%, 5%, and 4%, respectively. This indicates that the proposed algorithm has enhanced the detection ability of small targets after adding scale fusion branches from shallow feature maps.

To sum up, Fast-YOLOv4 combines MobileNetV3's lightweight network, the improved Multiscale-PANet structure, and the soft-merge box fused algorithm. This combination greatly reduces the number of parameters in the model, speeds up the detection, and obtains good detection results on the PASCAL VOC 07+12 dataset, especially for small-scale targets. Compared with common one-stage and two-stage detection algorithms, such as YOLOv3, YOLOv4, SSD, and Faster R-CNN, Fast-YOLOv4 achieves better performance, reflecting the feasibility of the algorithm and its effectiveness in small-scale target detection.

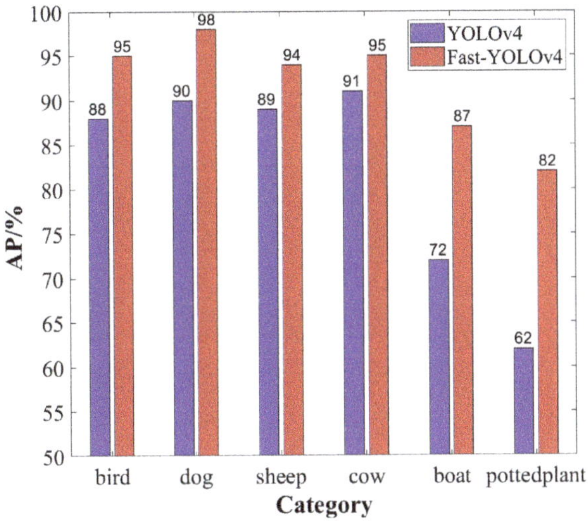

Figure 6. The comparison of AP of different categories.

6.2. Performance Experiment on the UAV Dataset

To demonstrate the effectiveness of Fast-YOLOv4 in UAV target detection, we conduct an experiment on the UAV dataset using Faster R-CNN, SSD, YOLOv3, YOLOv4, YOLOv4-MobileNetV3, and Fast-YOLOv4. All algorithms are trained and validated in the workstation. The metrics used in this experiment are mAP (IoU = 0.5), AP_S, AP_M, and AP_L (IoU = 0.5:0.95), where AP_S, AP_M, and AP_L are the average accuracies of the three scales of the targets. Table 5 shows the detection results of each algorithm on the UAV dataset. Figure 7 shows the comparison of the partial detection results of YOLOv4 and Fast-YOLOv4.

Table 5. The performance comparison of algorithms on the UAV dataset (in the workstation).

Model	mAP	AP_S	AP_M	AP_L
Faster R-CNN	88.12	12.89	38.06	61.68
YOLOv4	90.18	20.70	39.03	57.28
YOLOv3	88.92	17.93	33.18	53.41
SSD	86.88	8.36	38.82	65.50
YOLOv4-MobileNetV3	88.81	16.73	34.92	55.33
Fast-YOLOv4	90.62	23.88	37.53	59.71

From the experimental results in Table 5, the mAP (IOU = 0.5) of the proposed approach reaches 90.62%, which is superior to the other algorithms. Meanwhile, Fast-YOLOv4 performs well on the AP accuracy metrics of three scales, with the accuracy of small-scale targets to large-scale targets being 23.88%, 37.53%, and 59.71%, respectively. In particular, compared with YOLOv4 and YOLOv4-MobileNetV3, the APs of small-scale targets have increased by 3.18% and 7.15%, respectively, reflecting improvements in the multiscale structure and soft-merge brought to small target detection. Furthermore, our approach performs fairly well in medium-scale target detection. Compared with YOLOv3 and YOLOv4-MobileNetV3, Fast-YOLOv4 has a certain improvement. However, due to model size limitations, the AP is slightly inferior to YOLOv4, SSD, and Faster R-CNN. In large-scale target detection, our approach performs slightly worse than SSD and Faster R-CNN. The main reason is that Fast-YOLOv4 has a big gap in model volume compared with the other algorithms, and the large-scale model has more advantages in feature extraction and feature fusion, so it performs better in large-scale target detection.

Moreover, it can be seen from Figure 7 that Fast-YOLOv4 detected more small-scale UAV targets than the other models. It also reduces the number of missing detection and false detection, which shows that our approach has better detection performance than the original YOLOv4. In view of all performance metrics, the proposed Fast-YOLOv4 model combines the advantages of high accuracy, fast speed, and fewer model parameters, making it very suitable for integration with edge computing platforms for improving UAV target detection.

Figure 7. The partial detection results comparison of YOLOv4 and Fast-YOLOv4.

6.3. Ablation Experiment

To directly observe the improvement of Fast-YOLOv4, we conduct an ablation experiment in the workstation platform. The experimental method will train and validate each improved part of Fast-YOLOv4 on the UAV dataset and record its mAP (IoU = 0.5), APS, APM, APL (IoU = 0.5:0.95), FPS, and model parameters to check the improvement effect. Table 6 shows the performance results of the original YOLOv4 and Fast-YOLOv4 improvements.

From the experimental results in Table 6, improvement can be seen in each step of Fast-YOLOv4. Among all improvements, the lightweight network based on the backbone has significantly improved the performance of the model. Compared with YOLOv4, the YOLOv4-MobileNetV3 model improves the FPS to 60 f/s and reduces the number of parameters to 11.30M at the cost of only a 1.37% reduction in mAP. Multiscale-PANet improves the mAP and APS of the model by 1.37% and 4.05%, respectively, but the FPS decreases by 3 f/s, and the number of parameters increases by 3.64M. The soft-merge method improves the mAP of the YOLOv4 model by 0.86%, and other AP metrics are improved to varying degrees; however, the FPS decreases by 2 f/s. Finally, the proposed approach, Fast-YOLOv4, which combines these three methods, achieves an mAP of 90.62%, an APS of 23.88%, an FPS of 54 f/s, and 11.54M parameters. Compared with YOLOv4, this is an overall improvement and reflects the success of the improved method in UAV target detection tasks.

Table 6. The performance comparison of YOLOv4 and Fast-YOLOv4's improvements based on the UAV dataset (in the workstation).

Model	mAP	AP_{Small}	AP_{Medium}	AP_{Large}	FPS	Parameter
YOLOv4	90.18	20.70	39.03	57.28	29	63.94
YOLOv4-MobileNetV3	88.81	16.73	34.92	55.33	60	11.30
YOLOv4-Multiscale	91.55	24.75	39.40	56.19	26	64.89
YOLOv4-Soft-Merge	90.73	20.94	39.66	59.76	27	63.94
Fast-YOLOv4	90.62	23.88	37.53	59.71	54	11.54

6.4. Edge Computing Application

To investigate the real-time detection performance of Fast-YOLOv4 in the edge computing platform, all detection models are trained in the Quadro P4000 GPU and integrated into the Jetson Nano edge platform for target detection of UAVs in a video. The experiment tests the real-time detection effect of the model by comparing the speed and confidence of detection of each model in the edge platform. Meanwhile, we monitor the CPU temperature and memory footprint of the embedded platform to observe their changes when running different algorithms. Figure 8 shows the running results of different algorithms in the Jetson Nano. Figure 9 shows the comparison of Fast-YOLOv4's and YOLOv4's performances in detecting UAV targets in the videos.

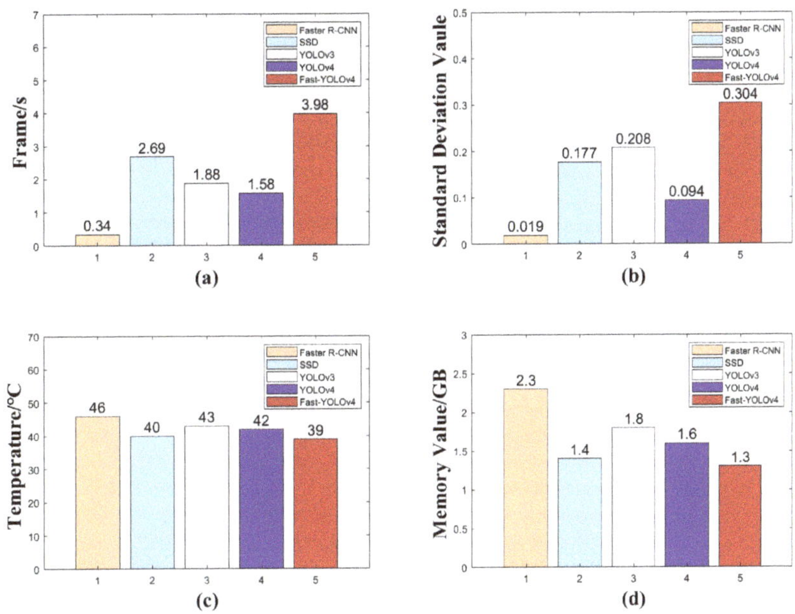

Figure 8. The comparison of running results of different algorithms (in the Jetson Nano). (**a**) The first panel is the average FPS. (**b**) The second panel is the standard deviation of FPS. (**c**) The third panel is CPU temperature. (**d**) The fourth panel is memory footprint.

In Figure 8, we select the average FPS, the standard deviation of FPS, CPU temperature, and memory footprint metrics to compare the algorithms. It can be seen that the average FPS of Fast-YOLOv4 is 3.98 f/s, which is the best result in all algorithms. In CPU temperature and memory footprint metrics, Fast-YOLOv4 shows lower heat dissipation requirements and memory usage. Although Fast-YOLOv4's FPS stability is slightly inadequate, its performance in Jetson nano is better than other algorithms on the whole. In addition, Figure 9 also shows the detection effect of YOLOv4 and Fast-YOLOv4 on UAV videos. In

Figure 9, the FPS of the improved model is 2.5 times higher than that of YOLOv4, and the confidence is comparable. Compared with other algorithms, our approach greatly improves the detection speed and accuracy. In terms of the results, the application results show that Fast-YOLOv4 has good performance in the edge computing platform and can achieve high-precision detection in real time.

YOLOv4

Fast-YOLOv4

Figure 9. The comparison of YOLOv4 and Fast-YOLOv4 detection results (in the Jetson Nano).

7. Conclusions

Aiming to address the problems in the detection of the multiscale UAV target, we present a novel real-time target detection algorithm called Fast-YOLOv4 based on edge computing. The proposed algorithm is based on YOLOv4 and uses the lightweight network MobileNetV3 to improve the backbone, which can reduce structural complexity and model parameters to speed up detection. Then, our approach combines the Multiscale-PANet to enhance the use and extraction of shallow features, strengthen the detection of small-scale targets, and improve the model's accuracy. In view of missing detection and false detection of multiscale targets, the proposed approach adopts an improved weighted boxes fusion algorithm, called soft-merge, to perform the weighted fusion of the predicted boxes instead of the traditional NMS. Finally, Fast-YOLOv4 achieves the best results in the fast detection of UAV targets in the edge computing platform Jetson Nano. To summarize, Fast-YOLOv4 provides a practical and feasible research method for real-time UAV target detection based on edge computing, which is worthy of further research.

Author Contributions: Conceptualization, Q.C., H.W. and B.Z.; methodology, Q.C.; software, Q.C.; validation, Q.C.; formal analysis, Q.C. and H.W.; investigation, Q.C.; resources, H.W. and B.X.; data curation, Q.C.; writing—original draft preparation, Q.C.; writing—review and editing, Q.C., H.W. and Y.S.; visualization, Q.C.; supervision, H.W.; project administration, H.W., B.Z. and Y.S.; funding acquisition, H.W., B.Z. and Y.S. All authors have read and agreed to the published version of the manuscript.

Funding: This research was funded by the Theoretical Research Project (KY20S011), and the Fusion Special Project of Anhui Province (KY21C008), China.

Data Availability Statement: The data used to support the findings of this study are available from the corresponding author upon request.

Acknowledgments: The author would like to thank all contributors to this study.

Conflicts of Interest: The authors declare no conflict of interest.

References

1. Yuan, C.X.; Li, G.; Su, D.C. Development status of anti-low-slow and small UAV systems. *Electron. World* **2021**, *23*, 138–140. (In Chinese) [CrossRef]
2. Jiang, R.Q.; Bai, R.K.; Peng, Y.P. Review of low and slow small UAV target detection technology. *Maneuverable Missile* **2020**, *49*, 100–105. (In Chinese) [CrossRef]
3. Mendis, G.J.; Randeny, T.; Wei, J.; Madanayake, A. Deep learning based doppler radar for micro UAS detection and classification. In Proceedings of the Military Communications Conference, Baltimore, MD, USA, 1–3 November 2016; pp. 924–929. MILCOM.2016.7795448. [CrossRef]
4. Bisio, I.; Garibotto, C.; Lavagetto, F.; Sciarrone, A.; Zappatore, S. Unauthorized Amateur UAV Detection Based on WiFi Statistical Fingerprint Analysis. *IEEE Commun. Mag.* **2017**, *56*, 106–111. [CrossRef]
5. Bougaiov, N.; Danik, Y. Hough Transform for UAV's Acoustic Signals Detection. *Adv. Sci. J.* **2015**, *6*, 65–68. [CrossRef]
6. Briese, C.; Guenther, L. Deep learning with semi-synthetic training images for detection of non-cooperative UAVs. In Proceedings of the 2019 International Conference on Unmanned Aircraft Systems (ICUAS), Atlanta, GA, USA, 11–14 June 2019; IEEE: Piscataway, NJ, USA, 2019; pp. 981–988. [CrossRef]
7. Mejias, L.; McFadyen, A.; Ford, J J. Sense and avoid technology developments at Queensland University of Technology. *IEEE Aerosp. Electron. Syst. Mag.* **2016**, *31*, 28–37. [CrossRef]
8. Xie, F.; Zhu, D J. Survey on Deep Learning Object Detection. *Comput. Syst. Appl.* **2022**, *2*, 1–12. . (In Chinese) [CrossRef]
9. Ren, S.; He, K.; Girshick, R.; Sun, J. Faster R-CNN: Towards Real-Time Object Detection with Region Proposal Networks. *IEEE Trans. Pattern Anal. Mach. Intell.* **2017**, *39*, 1137–1149. [CrossRef] [PubMed]
10. Liu, W.; Anguelov, D.; Erhan, D.; Szegedy, C.; Reed, S.; Fu, C.Y.; Berg, A.C. SSD: Single Shot MultiBox Detector. In Proceedings of the European Conference on Computer Vision, Amsterdam, The Netherlands, 11–14 October 2016; pp. 21–37. [CrossRef]
11. Bochkovskiy, A.; Wang, C.Y.; Liao, H.Y.M. YOLOv4: Optimal Speed and Accuracy of Object Detection. In Proceedings of the IEEE Conference on Computer Vision and Pattern Recognition, Seattle, WA, USA, 14–19 June 2020.
12. Liu, S.; Qi, L.; Qin, H.; Shi, J.;Jia, J. Path aggregation network for instance segmentation. In Proceedings of the IEEE Conference on Computer Vision and Pattern Recognition, Salt Lake City, UT, USA, 18–23 June 2018; pp. 8759–8768. [CrossRef]
13. Ai, Y.; Peng, M.; Zhang, K. Edge computing technologies for internet of things: A primer. *Digit. Commun. Netw.* **2018**, *2*, 77–86. [CrossRef]
14. Lin, T.; Dollár, P.;Girshick, R.; He, K.; Hariharan, B.; Belongie, S. Feature pyramid networks for object detection. In Proceedings of the IEEE Conference on Computer Vision and Pattern Recognition, Honolulu, HI, USA, 21–26 July 2017; pp. 2117–2125. [CrossRef]
15. Gao, S.; Cheng, M.M.; Zhao, K.; Zhang, X.Y.; Yang, M.H.; Torr, P. Res2net: A new multiscale backbone architecture. *IEEE Trans. Pattern Anal. Mach. Intell.* **2019**, *43*, 652–662. [CrossRef] [PubMed]
16. Zeng, Z.; Wang, Z.; Qin, L.; Li, H. Drone Detection Based on Multi-scale Feature Fusion. In Proceedings of the 2021 International Conference on UK-China Emerging Technologies (UCET), Chengdu, China, 4–6 November 2021; pp. 194–198. [CrossRef]
17. Zhu, J.; Wang, J.L.; Wang B. Lightweight mask detection algorithm based on improved YOLOv4-tiny. *Chin. J. Liq. Cryst. Disp.* **2021**, *36* 1525–1534. (In Chinese) [CrossRef]
18. Zhang, C.; Zhang, C.; Wang, H.; He, Q.; Liu, Y. Research on non-maximum suppression based on attention mechanism in object detection. *Electron. Meas. Technol.* **2021**, *44*, 82–88. (In Chinese) [CrossRef]
19. Ning, C.; Zhou, H.; Song, Y.; Tang, J. Inception single shot multibox detector for object detection. In Proceedings of the 2017 IEEE International Conference on Multimedia and Expo Workshops (ICMEW), Hong Kong, China, 10–14 July 2017; IEEE: Piscataway, NJ, USA, 2017; pp. 549–554. [CrossRef]
20. Solovyev, R.; Wang, W.; Gabruseva, T. Weighted boxes fusion: Ensembling boxes from different object detection models. *Image Vis. Comput.* **2021**, *107*, 104117. [CrossRef]
21. Zhou, L. Piao, J. A Lightweight YOLOv4 Based SAR Image Ship Detection. In Proceedings of the 2021 IEEE 4th International Conference on Computer and Communication Engineering Technology (CCET), Beijing, China, 13–15 August 2021; pp. 28–31. [CrossRef]
22. Liu, H.; Fan, K.; Ouyang, Q.; Li, N. Real-Time Small Drones Detection Based on Pruned YOLOv4. *Sensors* **2021**, *21*, 3374. [CrossRef] [PubMed]
23. Xun, D.T.W.; Lim, Y.L.; Srigrarom, S. Drone detection using YOLOv3 with transfer learning on NVIDIA Jetson TX2. In Proceedings of the 2021 Second International Symposium on Instrumentation, Control, Artificial Intelligence, and Robotics (ICA-SYMP), Bangkok, Thailand, 20–22 January 2021; IEEE: Piscataway, NJ, USA, 2021; pp. 1–6. [CrossRef]
24. Carrio, A.; Tordesillas, J.; Vemprala, S.; Saripalli, S.; Campoy, P.; How, J.P. Onboard detection and localization of drones using depth maps. *IEEE Access* **2020**, *8*, 30480–30490. [CrossRef]
25. Opromolla, R.; Inchingolo, G.; Fasano, G. Airborne visual detection and tracking of cooperative UAVs exploiting deep learning. *Sensors* **2019**, *19*, 4332. [CrossRef] [PubMed]

26. Howard, A.; Sandler, M.; Chu, G.; Chen, L.C.; Chen, B.; Tan, M.; Wang, W.; Zhu, Y.; Pang, R.; Vasudevan, V.; et al. Searching for MobileNetV3. In Proceedings of the IEEE Conference on Computer Vision and Pattern Recognition (CVPR), Long Beach, CA, USA, 15–20 June 2019; pp. 1314–1324. [CrossRef]
27. Howard, A.G.; Zhu, M.; Chen, B.; Kalenichenko, D.; Wang, W.; Wey, T.; Andreetto, M.; Adam, H. Mobilenets: Efficient convolutional neural networks for mobile vision applications. *arXiv* **2017**, arXiv:1704.04861.
28. Sandler, M.; Howard, A.; Zhu, M.; Zhmoginov, A.; Chen, L.C. Mobilenetv2: Inverted residuals and linear bottlenecks. In Proceedings of the IEEE Conference on Computer Vision and Pattern Recognition, Salt Lake City, UT, USA, 18–22 June 2018; pp. 4510—4520. [CrossRef]
29. Bodla, N.; Singh, B.; Chellappa, R.; Davis, L.S. Improving object detection with one line of code. *arXiv* **2017**, arXiv:1704.04503.
30. Hao, Y.J.; Teck, L.K.; Xiang, C.Y.; Jeevanraj, E.; Srigrarom, S. Fast Drone Detection using SSD and YoloV3. In Proceedings of the 2021 21st International Conference on Control, Automation and Systems (ICCAS), Jeju, Repulic of Korea, 12–15 October 2021; pp. 1172–1179. [CrossRef]

drones

MDPI

Article

Improved Radar Detection of Small Drones Using Doppler Signal-to-Clutter Ratio (DSCR) Detector

Jiangkun Gong [1], Jun Yan [1,*], Huiping Hu [2], Deyong Kong [3] and Deren Li [1]

[1] State Key Laboratory of Information Engineering in Surveying, Mapping and Remote Sensing, Wuhan University, Wuhan 430072, China; gjk@whu.edu.cn (J.G.); drli@whu.edu.cn (D.L.)
[2] Wuhan Geomatics Institute, Wuhan 430022, China; huhuiping@whu.edu.cn
[3] School of Information Engineering, Hubei University of Economics, Wuhan 430205, China; kdykong@hbue.edu.cn
* Correspondence: yanjun_pla@whu.edu.cn; Tel.: +86-027-68778527

Abstract: The detection of drones using radar presents challenges due to their small radar cross-section (RCS) values, slow velocities, and low altitudes. Traditional signal-to-noise ratio (SNR) detectors often fail to detect weak radar signals from small drones, resulting in high "Missed Target" rates due to the dependence of SNR values on RCS and detection range. To overcome this issue, we propose the use of a Doppler signal-to-clutter ratio (DSCR) detector that can extract both amplitude and Doppler information from drone signals. Theoretical calculations suggest that the DSCR of a target is less dependent on the detection range than the SNR. Experimental results using a Ku-band pulsed-Doppler surface surveillance radar and an X-band marine surveillance radar demonstrate that the DSCR detector can effectively extract radar signals from small drones, even when the signals are similar to clutter levels. Compared to the SNR detector, the DSCR detector reduces missed target rates by utilizing a lower detection threshold. Our tests include quad-rotor, fixed-wing, and hybrid vertical take-off and landing (VTOL) drones, with mean SNR values comparable to the surrounding clutter but with DSCR values above 10 dB, significantly higher than the clutter. The simplicity and low radar requirements of the DSCR detector make it a promising solution for drone detection in radar engineering applications.

Keywords: drone detection; Doppler signal-to-clutter ratio (DSCR); missed target; signal-to- noise ratio (SNR)

Citation: Gong, J.; Yan, J.; Hu, H.; Kong, D.; Li, D. Improved Radar Detection of Small Drones Using Doppler Signal-to-Clutter Ratio (DSCR) Detector. *Drones* **2023**, *7*, 316. https://doi.org/10.3390/drones7050316

Academic Editors: Bo Li, Chunwei Tian, Daqing Chen and Ming Yan

Received: 17 April 2023
Revised: 4 May 2023
Accepted: 8 May 2023
Published: 10 May 2023

1. Introduction

Detecting drones (or unmanned aerial vehicles, UAVs) using radar poses a significant challenge due to their slow velocities and small radar cross-section (RCS) values [1–6], which can result in a high number of missed targets. Common drone types include multirotor drones, helicopters, and unmanned fixed-wing aircraft [7–10]. Furthermore, the issue of drone swarms is expected to pose a significant challenge for the counter-drone industry in the future [11–14]. This vulnerability can lead to drones being used as ghost spies, silently invading borders and threatening coastal border defenders, highlighting the importance of developing counter-drone technologies.

Small drones are characterized by their small RCS, slow speeds, and low altitudes. These drones typically fall into Group 1 (Table 1) as reported by the U.S. Department of Defense, which requires rotating blades for aerial flight [10]. Group 1 drones generally have an RCS of 0.01–0.1 m^2, which is approximately 1/10,000 to 1/1000 the size of a typical airplane. Current radar detectors use the amplitude detection, Doppler detection, and other improved detectors, but traditional signal-to-noise ratio (SNR) detectors often result in a missed target due to weak radar signals from the small RCS values of drones [15]. Researchers have developed different methods to suppress background clutter and improve the SNR of targets, such as a constant false alarm rate (CFAR) and modified CFAR detectors [16–19]. Additionally, Doppler detectors, including moving target detection (MTD) [20] and track-before-detection (TBD) technology, have been developed for detecting drones [21–24], although the latter requires a long tracking period and may struggle with sudden drone hovering.

Table 1. Drone classification according to the U.S. Department of Defense (DoD) [1,2].

Category	Size	Maximum Gross Takeoff Weight (Pounds)	Normal Operating Altitude (ft)	Airspeed (Knots)
Group 1	Small	<20	<1200 AGL [3]	<100
Group 2	Medium	21–55	<3500 AGL	<250
Group 3	Large	<1320	<18,000 MSL [4]	<250
Group 4	Larger	>1320	<18,000 MSL	Any airspeed
Group 5	Largest	>1320	>18,000 MSL	Any airspeed

[1] Source: "Eyes of the Army", U.S. Army Roadmap for UAS 2010–2035 [10]. https://home.army.mil/rucker/index.php, accessed on 4 May 2023; [2] If the drone has even one characteristic of the next level, it is classified in that level. [3] AGL = Above Ground Level. [4] MSL = Mean Sea Level.

Due to the limitations of current detectors that rely on SNR values, which are affected by the small RCS values of drones, there is a need to develop a detector that is independent of RCS values. To address this issue, we propose a new detector that extracts the Doppler signal-to-clutter ratio (DSCR) of targets. The DSCR detector is capable of detecting radar signals in almost real time and does not require tracking. In this study, we present the algorithm for the DSCR detector in Section 2 and compare its performance with actual radar data in Section 3. We consider different detection ranges, background clutter, and drone types. The experimental results demonstrate that the DSCR detector outperforms the traditional SNR detector. In Section 4, we analyze the difference between missed targets and false alarms. Finally, we conclude our findings in Section 5.

2. Materials and Methods

2.1. Theoretical Model

The classical method of radar detection involves detecting signals in noise along the range dimension and using the SNR to determine the ratio between the target signal power and the mean noise level. Figure 1 shows three possible outcomes: "False Alarm", "Detected Target", and "Missed Target". A "False Alarm" occurs when the detected signals are not from the target, while a "Missed Target" represents a situation where the signal power is below the detection threshold of the SNR detector. Although radar detection does not explicitly involve an identification process, it still incorporates it implicitly because it is necessary to distinguish between "False Alarm" and "Missed Target" (or" Missed Alarm"). In the case of small-slow targets, "Missed Alarm" is a common problem due to their low signal power, which can be similar to the mean noise level. Therefore, reducing "Missed Target" should be a priority when detecting small-slow targets to increase detection probability. Once the "Missed Target" are minimized, automatic target recognition (ATR) technology can be employed to classify between a "False Alarm" and a "Detected Target" to lower false alarms rate and maintain high detection probability.

The traditional radar detector assumes that a target is a point object with a mean RCS. This detector calculates the SNR, which is a measure of a radar's ability to detect a specific target at a given range, by comparing the prioritized scattering power of a target over the background. The SNR is an essential parameter used to maintain the detection probability, which is calculated by [17]:

$$SNR\,(\mathrm{dB}) = 10log10\left(\frac{E_r}{E_n}\right) = 10log10\left(\frac{P_t G_r G_t \lambda^2 \sigma}{(4\pi)^3 R^4 K T_s B_n L}\right) \quad (1)$$

where E_r = the power of radar echoes from a target and E_n = the power of noise. T_s = the system noise temperature, B_n = the noise bandwidth of the receiver, L = the total system losses, K = the Boltzmann's constant. P_t = the transmitted power, G_r = the received gain, G_t = the transmitted gain, R = the measured range, σ = the radar cross section (RCS) of the target, λ = the radar wavelength. For a certain radar, its transmitted parameters are given, and then Equation (1) can be revised into

$$SNR \propto \frac{\sigma}{R^4} \tag{2}$$

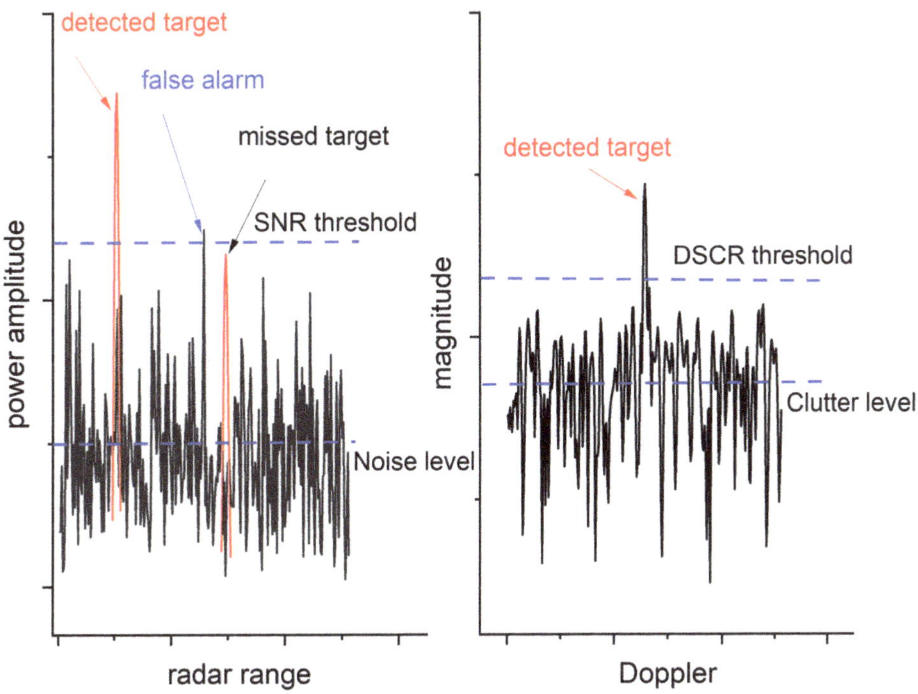

Figure 1. The detection of signals in noise and clutter (**Left**: SNR detector, **Right**: DSCR detector).

Equation (2) demonstrates that the SNR of a target is a function of its RCS and the detection range. The basic radar detector is the threshold detection, which can be represented as

$$P_{SNR} \geq SNR_{thr} \tag{3}$$

where SNR_{thr} = the threshold of the SNR detector, and P_{SNR} = the measured SNR value of the target. For single pulse detection, if the detection probability exceeds 50%, the target's SNR should be at least 13.1 dB, and to achieve a 95% probability, the SNR should be 16.8 dB [17]. In simpler terms, a smaller target with a lower RCS will have a lower SNR, resulting in a shorter detection range. Thus, radar systems often encounter the problem of "Missed Target" when detecting drones and other objects with small RCS values.

It is widely acknowledged that targets are not merely point objects, but rather distributed ones. This notion is supported by the observation of micro-Doppler phenomena, as described by micro-Doppler theory [25–27]. For instance, when it comes to a drone, its spectrum contains body Doppler, micro-Doppler, and clutter Doppler components. The body Doppler in the spectrum is induced by the Doppler effect, which is expressed as follows:

$$\overline{f_{bd}} = -\frac{2V_b}{\lambda} \tag{4}$$

where V_b = the flying speed of the target, λ = the radar wavelength. Secondly, micro-motion can be characterized as a simple periodic motion superimposed on the main movement of the target, and the micro-Doppler shifts in the spectrum is given by

$$\overline{f_{md}(t)} = \frac{L}{\lambda}\omega cos\beta\, cos(\omega t) + \overline{f_{bd}} \tag{5}$$

where L = the length of the micro-structure, β = the elevation angle, and ω = the rotating speed. Thus, the Doppler spectra of targets always contain generalized Doppler peaks, which include both micro-Doppler, body Doppler shifts and clutter shifts, as given by

$$\overline{f_d(t)} = \overline{f_{bd}} + \overline{f_{md}(t)} + \overline{f_{cd}(t)} \tag{6}$$

where $\overline{f_{bd}}$ = the body Doppler vector, $\overline{f_{md}(t)}$ = the body micro-Doppler vector, and $\overline{f_{cd}(t)}$ = the clutter Doppler vector, which are related to the clutter. Here, we use the term "generalized Doppler" to describe Doppler, including body Doppler and micro-Doppler.

We introduce a novel parameter to quantify the strength of the generalized Doppler signal in a radar spectrum. Specifically, given radar data $X(n)$ within a single radar resolution cell, and the corresponding Doppler spectrum $F(k)$, we define the Doppler signal-to-clutter ratio (DSCR) of a Doppler frequency D within this cell as follows:

$$DSCR(D)(dB) = 10log10[\frac{F(D)}{\frac{\sum_1^N F(K)}{N}}] \tag{7}$$

where $F(K)$ = the amplitude of frequencies in the spectrum of the current radar bin; K = the Doppler frequency in the spectrum; and N = the length of the spectrum. D could be either the body Doppler or micro-Doppler, or even any Doppler in the spectrum. Radar echoes in one radar bin include the scattering power of the target and the background clutter, and the Doppler effect can extract the scattering part of the target (i.e., $F(D)$) from the background (i.e., $F(K)$). It should be noted that while Equation (4) only considers the body Doppler shift components in the spectrum, the DSCR in Equation (7) takes into account both the magnitude of the Doppler component and the Doppler speed. Compared to Equations (1) and (2), because the DSCR uses no value related to the RCS value of the target, it is independent of the detection range.

2.2. Algorithm Description

Here, we propose a concise detector using the above DSCR value for detecting radar signals of drones, which combines amplitude detection and Doppler detection, as shown in Algorithm 1. The steps for the DSCR detector are as follows:

(1) Obtain the radar echoes raw data, $X(n)$, in one radar resolution cell, and calculate its Doppler spectrum, $F(K)$.
(2) Search the spectrum, $F(K)$, to locate the strongest Doppler shift of D starting from the beginning point of M. Generally, we assume that the Doppler points below M belong to the background clutter. Calculate the maximum DSCR of the strongest Doppler shift of D.
(3) Compare the value of the maximum DSCR with the detection threshold T. If the DSCR is above the threshold, then it is a target. Otherwise, it is not.

We believe that this method is effective for detecting drones and can provide near-real-time results without the need for a tracking period.

Before using the DSCR detector, two parameters are required: the spreading width of the background clutter (i.e., the beginning point of M) and the detection threshold (i.e., the detection threshold of T). These parameters are determined based on statistical data from the specific scenario. The spreading width of the background clutter (or the beginning point) depends on the type of clutter. For a ground clutter, a fixed length of M can be used, but for a sea clutter, an adaptive method to adjust the width of M is preferable due to its time–space variance. The detection threshold of the DSCR detector can be either hard-style or soft-style. The hard threshold is based on the bottom noise level of the radar system and is typically set at twice the bottom noise level. For example, if the radar system has a bottom noise level of 5 dB, the DSCR detector threshold can be set at 8 dB. The soft threshold requires mean statistical data from the scenario and allows the DSCR detector to automatically adjust the threshold to adapt to new scenarios.

Algorithm 1: Calculating the maximum DSCR value.

```
Function begins:
Compute the Fourier transform of X(n) using fast Fourier transform (FFT), store it in F(k).
Initialize a variable Ma to zero.
while each index i from M to N-M, do
    if F(i) > Ma, then
        update Ma to F(i) and D to i.
    end if
end while
Compute the mean value of F(k) and store it in a variable Me.
Compute the spectral contrast ratio (DSCR), Ds, as the ratio of Ma to Me.
Translate the Ds into a dB value using 10log10(Ds).
Return the values of Ds and D.
Function end.
```

2.3. Experimental Tests

In addition to the theoretical calculations, we use real radar data to verify the comparison between the SNR and DSCR values of drones.

(1) Ku-band radar test

We first conducted experiments to evaluate the performance of the DSCR detector using data collected by a Ku-band pulsed-Doppler phased array radar (surface surveillance radar) with a coherent pulse integration (CPI) of approximately 30 ms and a pulse repetition frequency (PRF) of 6 kHz. The radar has a range resolution of 15 m. The data were collected at the Yellow River wetland bird protection area in China, which includes various environments such as roads, rivers, farmlands, and woods with moving targets such as vehicles, people, flocks of sheep, and birds. We used a DJI Phantom 3 drone, manufactured by DJI Inc. (Shenzhen, China) in our experiment, which is a quadcopter with propellers in a tractor configuration. The drone's body and propellers are primarily composed of plastic, while the individual rotors are made of carbon fiber and plastic. The drone weighs 1.280 kg, has a maximum horizontal flight speed of 57.6 km/h, and a maximum rotor blade rotary speed of 150°/s. It can fly up to a maximum height of 6000 m and has a flight time of 23 min. It should be noted that the drone used in our experiment is not the same as the one shown in Figure 2.

Figure 2. The test environment using the Ku-band radar.

(2) X-band radar test

We also utilized a pulse-Doppler phased array radar (maritime surveillance radar) operating in the X-band, in addition to the Ku-band radar, to validate the DSCR detector. The radar had a PRF of approximately 5 kHz, a CPI of about 20 ms and a range resolution of 12 m. It had an active electronically scanned phased array antenna, and the raw data collected were float complex data. To evaluate the DSCR detector, we used three drones as cooperative targets, and their specifications are provided in Table 2, with their images displayed in Figure 3. We collected data from various coastal areas, including the cities of Qidong and Rizhao in China. Figure 3 indicates the relative positions of the drones with respect to the radar. During the tests, the Qidong sea had a lower sea scale than Class 3, while the Rizhao sea had a sea scale of approximately Class 5. To distinguish between the two sea scales, we referred to the former as the "Qidong sea" and the latter as the "Rizhao sea" in this study. However, these names are not indicated on any official maps. Additionally, we collected radar data in the ground clutter background in Nanjing city.

Table 2. Parameters of the drones.

Drone Type	Hybrid VTOL Fixed-Wing	Multirotor	Fixed-Wing
Model	TX25A	Phantom 4	Albatross 1
Manufacturer	Harryskydream Inc.	DJI Inc.	Homemade
Flight weight	26 kg	1.38 kg	0.3 kg
Wingspan	360 cm	40 cm	108 cm
Body size	197 cm	40 cm	80 cm
Blade length	30 cm	20 cm	10 cm
Rotor number	5	4	2
Cruise speed	25 m/s	15 m/s	10 m/s
Aero-frame materials	FRP (Fiber reinforced plastic)	PC (Polycarbonate)	EPP (Expanded polypropylene)

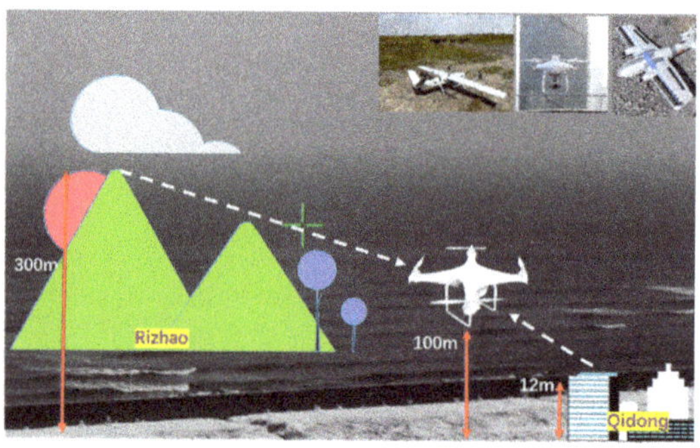

Figure 3. The relative locations between the drones and the radar in the tests using the X-band radar.

3. Results

3.1. Simulated Result

We conducted simulations to demonstrate the DSCR of radar signals from a drone. Figure 4a shows the simulated raw data of radar signals from a drone where $x(n)$ is the time series and $f(k)$ is the spectrum. $x(n)$ is given by:

$$x(n) = 8e^{2\pi\times0.5n\times j} + 4e^{2\pi\times20n\times j} + 2e^{2\pi\times40n\times j} + 3\times randn(size(n)) \tag{8}$$

where $n =$ the sample data from 1 to 256; $size(n)$ = the MATLAB function calculating the size of a variable. $randn(n)$ = the MATLAB function providing the normally distributed pseudorandom numbers. The sampling frequency is 256. Figure 4a shows three digital frequencies of 0.5, 20, and 40, with magnitudes of 5.06, 3.84, and 1.94, respectively. We assume the clutter Doppler to be 0.5 + *randn* and the body Doppler of the drone to be 20, while the micro-Doppler of the drone produced by the blade is 40. Many sources contribute to the clutter, such as environmental and biological clutter (e.g., birds) when the radar detects drones. Thereby, clutters could have low Dopplers (e.g., environmental clutter) around 0 and high Dopplers (e.g., biological clutter) such as 78 in Figure 4b.

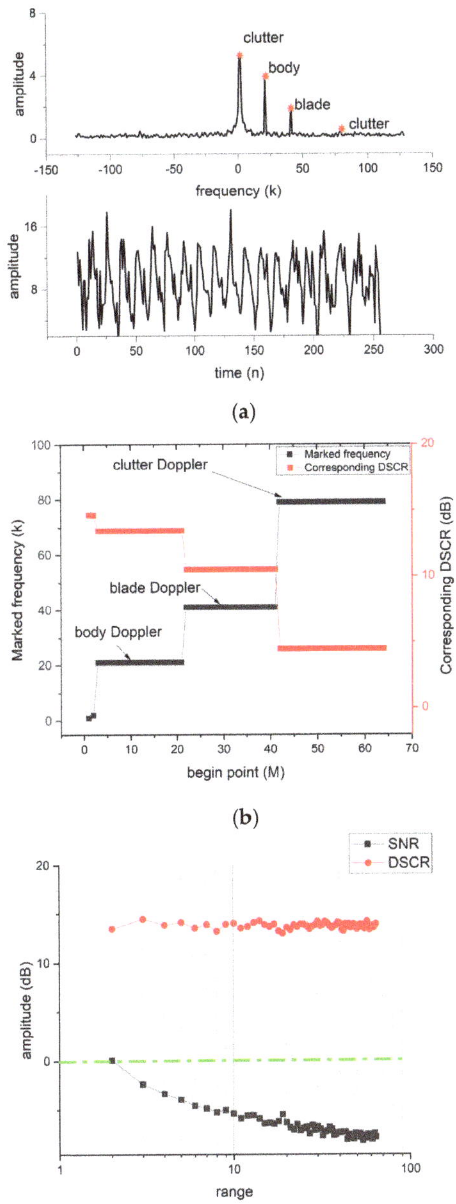

Figure 4. Simulated radar data to calculate the DSCR of signals: (**a**) raw signals and their spectrum, (**b**) the relationship between the beginning point from 1 to 64 and the DSCR value, as well as the marked digital frequency, (**c**) comparison of SNR and DSCR values given different range factors.

The DSCR of a Doppler is closely related to the choice of the beginning point of M in Step (2). Figure 4b illustrates the impact of varying the beginning point of M on the calculated DSCR values. As M is changed from 1 to 64, the marked Doppler frequency D and the corresponding DSCR value also change. In most cases, the body Doppler of 20 and the blade Doppler of 40 are correctly identified as D, and their corresponding DSCR values exhibit only slight fluctuations. Therefore, it is important to select M carefully to accurately identify the body Doppler of 20. If M is either too small or too large, the marked body Doppler frequency will be incorrect.

The impact of the detection range on the DSCR of radar signals is minimal compared to that of the SNR. Figure 4c compares the SNR and DSCR of radar signals under different detection ranges. The radar signals affected by the detection range are represented as follows:

$$y(n) = \mathrm{A} \times x(n) \tag{9}$$

where A = the transmission loss of radar signals related to the range [28]. With the noise threshold level set as 64, M as 8, and D as 20, Figure 4c illustrates the decay in amplitude of $y(n)$ with an increasing detection range. The values are dimensionless. The green dotted line denotes the detection threshold of 0 dB, and as the detection range factor increases, the SNR values of the signals also decrease from 0.08 dB to −8.09 dB. In contrast, the DSCR values remain relatively stable, fluctuating between 14.46 dB and 12.97 dB. The standard deviation of DSCR is only 0.28, which accounts for just 19% of the 1.46 SNR value. Equation (7) shows that the denominator of the DSCR is the mean value of the entire spectrum; therefore, the DSCR is largely unaffected by the detection range.

3.2. Real Ku-Band Radar Data

The DSCR detector effectively separates the radar signals of the drone from the background clutter in the frequency spectrum. Figure 5 displays the normalized data and Doppler spectra of both the drone and the clutter. MATLAB's signal processing tools, specifically the FFT function, were utilized to obtain the spectra from the raw time data. The drone's spectrum reveals a dominant bulk Doppler, corresponding to the drone's body Doppler, with a maximum DSCR value of 17.24 dB and a speed of −2.7 m/s indicating that the drone was flying away from the radar. Conversely, there is no bulk Doppler evident in the clutter spectrum, with the Doppler magnitudes being uniform, suggesting the absence of any objects in the clutter radar data.

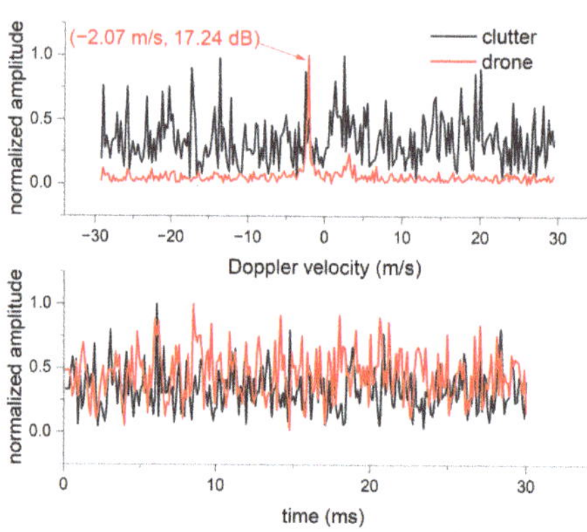

Figure 5. Ku-band raw radar data and Doppler spectra of a quad-rotor drone and clutter in one radar cell.

The results of the tracking demonstrate the ability of the DSCR detector to detect weak radar signals of the drone in a clutter. The tracking results of the SNR, DSCR, and Doppler speeds of the drone are presented in Figure 6. In most cases, the SNR of the drone is similar to that of the clutter within the same range. The tracking range is approximately 1.3 km from the radar location, with a mean SNR of the drone of about 0.93 dB, similar to the clutter's 0.003 dB. The dynamic range of the drone's clutter is 3.45 dB, larger than the clutter's 2.41 dB, indicating that the RCS of the drone changes with radar observation angles and the drone's radar signals are weak, similar to the clutter level. Therefore, traditional SNR detectors may have difficulty in detecting radar echoes from the drone.

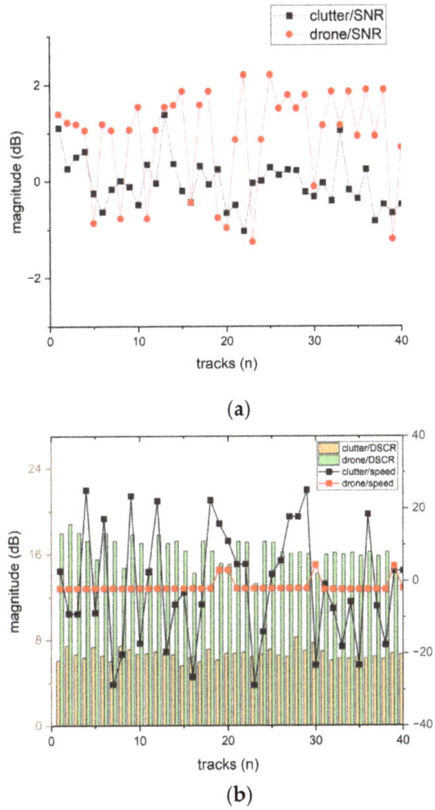

Figure 6. Ku-band tracking SNR and DSCR values of a quad-rotor drone, (**a**) SNR comparison of the clutter and the drone, (**b**) DSCR comparison of the clutter and the drone.

However, the DSCR detector can detect such weak radar signals from the drone. In most cases, the DSCR of the drone is much larger than that of the clutter in Figure 6b. The mean DSCR of the drone is approximately 16.28 dB, about 10 dB larger than the clutter's 6.67 dB, and approximately 13 dB larger than the SNR of the drone. This suggests that the detection threshold can be lowered by at least 10 dB. According to the radar Equation (1), a 10 dB improvement means that the radar distance of the same target will be enhanced by approximately 77% when given the false alarm probability and detection probability. Alternatively, a different target with a much smaller RCS value of 90% can be detected. Therefore, the DSCR detector is superior to the SNR detector in detecting radar signals from the drone because it has a lower detection threshold. Furthermore, Figure 6b shows that the DSCR detector also extracts the body Doppler speeds of the drone. The dynamic range of the drone's Doppler is only 2.3 m/s, much smaller than the clutter's 27.83 m/s, with a standard deviation of 0.50 m/s, much smaller than the clutter's 8.57 m/s. The body Dopplers of the drone have a pattern, whereas the Dopplers of the clutter appear random.

3.3. Real X-Band Radar Data

The DSCR detector analyzes the spectrum of radar signals over time to detect targets. Figure 7 shows the raw X-band data and Doppler spectra of drones. The raw data's amplitude is normalized and dimensionless. We processed the raw time data using signal processing tools, including the FFT function in MATLAB, to obtain the spectrum. We collected radar data from hybrid vertical take-off and landing (VTOL) and quad-rotor drones in a sea clutter background and from a fixed-wing drone in a ground clutter background. The quad-rotor drone, hybrid VTOL drone, and fixed-wing drone have max DSCR values of 9.41 dB, 9.91 dB, and 13.18 dB, respectively, with Doppler velocities of −6.9 m/s, −15.6 m/s, and 8.7 m/s. A negative Doppler indicates flying away from the radar. Sea clutter or ground clutter can cause a spreading Doppler, which may suppress a drone's body Doppler if it is flying at a low speed.

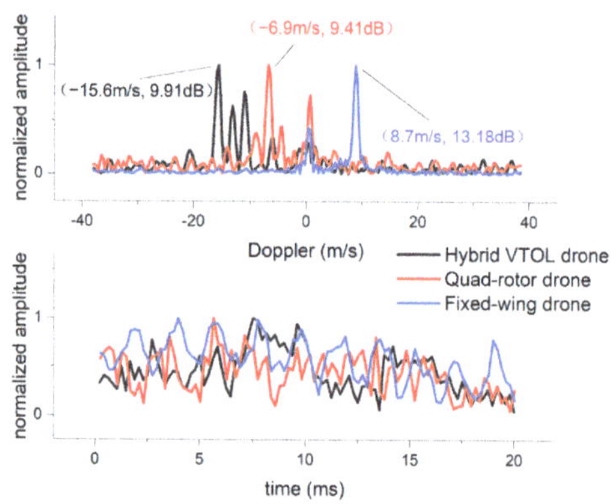

Figure 7. X−band raw radar data and Doppler spectra of drones in one radar cell.

The DSCR detector outperforms the SNR detector in detecting targets. In Figure 8, we present the detection results of a quad-rotor drone in a range window comprising 20 radar bins. This area has a length of 240 m since the range resolution is 12 m. The drone was located in radar bin #12. The DSCR values denote the maximum DSCR in each radar cell, whereas the SNR values are true numbers. In the Qidong case, the quad-rotor drone was about 10 km away from the radar, and the range was roughly 5.7 km in the Rizhao case. The SNR of the drone was almost zero in the Qidong data. However, the SNR value was higher (approximately 6 dB) when the detection range was shorter. Additionally, the elevation angle beams and different sea classes are also considered, as shown in Figure 3.

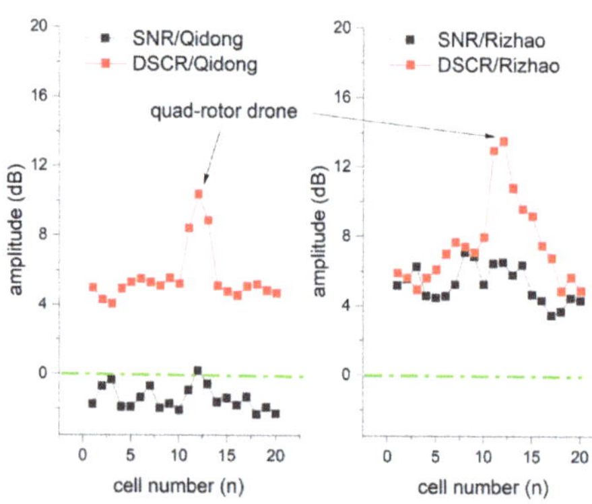

Figure 8. The SNR and DSCR values of the 20 radar bins detected at ranges of 10 km in the Qidong case and 5 km in the Rizhao case, where the drone is in radar cell #12.

The clutter in the Rizhao case has higher SNR values than that in the Qidong case. Three reasons may explain this difference. Firstly, the SNR values represent the received power when using the same radar system, and they are not the SNR values when using a CFAR detector. Secondly, the data in the Rizhao case were detected at approximately half of the detection range of that in the Qidong case, and the SNR of targets and clutter could be larger in the Qidong case based on the radar equation. Thirdly, the data in the Rizhao case were detected with a look-down angle, and the clutter mainly came from the sea clutter. On the other hand, the data in the Qidong case were detected with a look-up angle, and the clutter data were generated by the air. Therefore, the SNR values in the Rizhao case were much higher than those in the Qidong case.

However, the SNR values of the drone are comparable to the surrounding clutter SNR values in each case. The SNR values of both the air clutter and the drone in the Qidong case were around 0 dB, while those of both the sea clutter and the drone were approximately 6 dB in the Rizhao case. Sometimes, the SNR of the drone was smaller than that of the surrounding clutter. For instance, the SNR in radar bin #6 in the Rizhao case was higher than that of the drone in radar bin #12. Therefore, the SNR detector failed to detect the drone in both cases. In contrast, the DSCR values of the drone (i.e., 10.3 dB and 13.5 dB) were more stable in the two cases, and the DSCR detector successfully detected the drone in both cases.

4. Discussion

A "Missed Target" occurs when the signal power of the target is lower than the SNR detector's detection threshold, while a "False Alarm" indicates that the detected signals are not from the target. Although the identification process is not explicitly stated in radar detection, it is necessary to classify a missed target and false alarm. "Missed Targets" are the most common situations in detecting drones because they have small RCS values resulting in low signal power, similar to the RMS noise level. Therefore, reducing missed targets to zero should be a priority in drone detection.

The DSCR detector has the advantage of having a lower missed target rate but the disadvantage of a high false alarm rate, without the ability to differentiate between the clutter and the target. To reduce the false alarm rate, automatic target recognition (ATR) technology can be used after detection by the DSCR detector. The Swerling-I model suggests that a target's SNR should be at least 13.1 dB to achieve a detection probability of over 50% and 16.8 dB for a probability of 95%. However, in most cases of detecting radar signals from drones, the SNR values of drones are significantly lower than the required values. Table 3 shows the statistical results of radar detection, where "MIN" and "MAX" represent the minimum and maximum values, respectively. "MEAN" is the mean value, while "STDEV" represents the standard deviation. Additionally, "STDEV/MEAN" is the ratio between the standard deviation and the mean value. The mean SNR values of drones are lower than the corresponding DSCR values, with a difference of 8.55, 10.05, and 11.43 for the VTOL drone, quad-rotor drone, and fixed-wing drone, respectively. This suggests that the DSCR detector can detect signals with a lower threshold than the SNR detector, or reduce the missed target rate compared to the SNR detector when using the same threshold. Furthermore, the "STDEV/MEAN" of the three drones' SNR is almost 50%, whereas that of the DSCR is only 10%, as shown in Figure 9. It is important to note that these numbers were measured at a fixed range, and if different ranges were considered, the "STDEV/MEAN" values of the SNR could be worse as the SNR is dependent on the detection range. Overall, in addition to the DSCR values of a drone being higher than the SNR, the DSCR is also much more stable than the SNR.

The DSCR detector offers two distinct advantages over the SNR. Firstly, it extracts relative values that demonstrate relativity both in space and time. By utilizing radar data within one range unit across different distance bins, the DSCR detector effectively limits the scattering energy of the background clutter while preserving the scattering power of the target, resulting in an increased detection probability. Additionally, the short radar dwell time of one range bin during one CPI significantly reduces interference from the background clutter and removes much of the time-varying background clutter. This relativity results in DSCR values that are almost independent of the detection range, and the detector combines the advantages of amplitude detection and Doppler detection. By transforming radar signals from the time domain into the frequency domain using Fourier transformation, the Doppler frequency contains both the amplitude and velocity of the target, providing a powerful tool for detection.

Table 3. Performance of different detection methods.

Drone Types		Hybrid VTOL Drone	Quad-Rotor Drone	Fixed-Wing Drone
Detection Background		Sea	Sea	Ground
Detection Ranges		8~14 km	10 km	5 km
SNR	MIN (dB)	2.58	0.20	0.70
	MAX (dB)	8.07	4.28	5.13
	MEAN (dB)	4.83	1.76	2.59
	STDEV (dB)	1.86	1.18	1.19
	STDEV/MEAN	0.38	0.67	0.46
DSCR	MIN (dB)	11.24	10.06	11.09
	MAX (dB)	17.08	14.13	16.83
	MEAN (dB)	13.38	11.81	14.02
	STDEV (dB)	1.38	1.15	1.52
	STDEV/MEAN	0.10	0.09	0.11

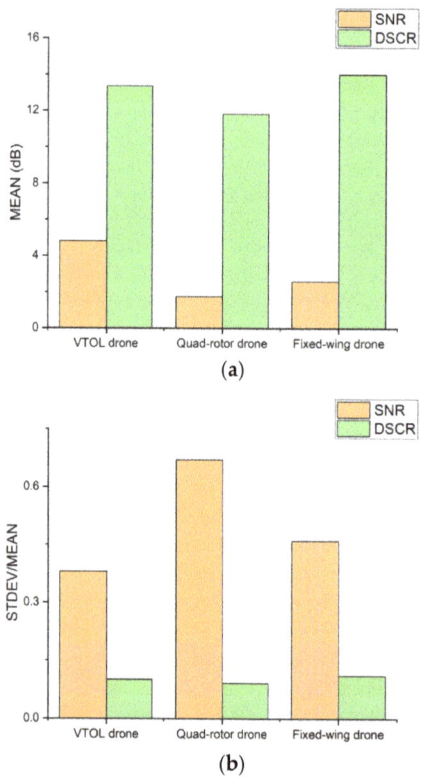

Figure 9. The comparion of SNR and DSCR values of the three drones: (**a**) the mean values, (**b**) the ratio between the standard deviation and the mean value.

Secondly, Fourier transformation enables the classification of different scattering sources into distinct spectral parts. While the clutter tends to concentrate in the low-frequency part, targets fall into other parts of the spectrum. Even the body Doppler of drones due to their slow speed can fall into the clutter part, but the micro-Doppler due to the rotating blades can jump out of the clutter. The DSCR detector can effectively extract radar signals from targets such as drones, making it a useful and versatile tool in engineering applications.

In addition to the basic DSCR detector described in Section 2, the DSCR value can be utilized in other improved solutions for radar detection and classification. Table 4 provides a comparison of different solutions for detecting drones, highlighting the importance of designing effective detection and recognition algorithms to minimize both missed targets and false alarms. A combined approach using the DSCR detector and automatic target recognition (ATR) technology can effectively address the radar challenge of countering drones by reducing missed targets and false alarms. Therefore, reducing missed targets in drone detection is crucial, and the DSCR detector, together with ATR technology, can offer an effective solution.

Table 4. Different solutions for detecting drones.

Solutions	Detection	Recognition
Solution 1	SNR detector	Kinetic features (e.g., trace)
Solution 2	SNR detector	Signals signatures (e.g., micro-Doppler)
Solution 3	DSCR detector	Kinetic features (e.g., trace)
Solution 4	DSCR detector	Signals signatures (e.g., micro-Doppler)

5. Conclusions

The primary challenge in radar surveillance of drones is the high rate of "Missed Targets" due to their small RCS values, which limit the effectiveness of traditional SNR detectors. In this study, we propose a comparative method based on the DSCR value, which combines the advantages of amplitude and Doppler detection. Our results with Ku-band and X-band radars indicate that the SNR detector is ineffective in detecting drones due to cases of almost 0 dB-level SNR values. Although the DSCR detector may increase the number of false alarms, it can effectively reduce missed targets. However, an effective ATR solution is required to reduce false alarms. The DSCR detector is a detection tool, and effective ATR technology can recognize the objects detected by the DSCR detector and reduce false alarms. In our future work, we plan to investigate Solutions 3 and 4 in Table 4. Additionally, we will research adaptive methods using the DSCR detector.

Author Contributions: Conceptualization, J.Y.; methodology, J.G.; software, J.G.; validation, J.Y.; formal analysis, J.G.; investigation, J.Y.; resources, D.L.; data curation, J.Y.; writing—original draft preparation, J.G.; writing—review and editing, H.H.; visualization, H.H.; supervision, J.Y.; project administration, D.L.; funding acquisition, D.K. All authors have read and agreed to the published version of the manuscript.

Funding: This research received some support from the Natural Science Foundation of Hubei Providence (General Program: 2021CFB309).

Institutional Review Board Statement: Not applicable.

Informed Consent Statement: Not applicable.

Data Availability Statement: Some of the data presented in this study may be available on request from the corresponding author. The data are not publicly available due to the internal restriction of the research group.

Acknowledgments: We appreciate both the testers during the collection of the data, and we also want to thank the authors whose photographs are reproduced in this study.

Conflicts of Interest: The authors declare that they have no conflict of interest.

References

1. Čisar, P.; Pinter, R.; Čisar, S.M.; Gligorijević, M. Principles of Anti-Drone Defense. In Proceedings of the 2020 11th IEEE International Conference on Cognitive Infocommunications (CogInfoCom), Online on MaxWhere 3D Web, 23–25 September 2020; pp. 19–26.
2. Shi, X.; Yang, C.; Xie, W.; Liang, C.; Shi, Z.; Chen, J. Anti-Drone System with Multiple Surveillance Technologies: Architecture, Implementation, and Challenges. *IEEE Commun. Mag.* **2018**, *56*, 68–74. [CrossRef]
3. Gong, J.; Yan, J.; Li, D.; Kong, D.; Hu, H. Interference of Radar Detection of Drones By Birds. *Prog. Electromagn. Res. M* **2019**, *81*, 1–11. [CrossRef]
4. Markow, J.; Balleri, A.; Catherall, A. Statistical analysis of in-flight drone signatures. *IET Radar Sonar Navig.* **2022**, *16*, 1737–1751. [CrossRef]
5. Svanström, F.; Alonso-Fernandez, F.; Englund, C. Drone Detection and Tracking in Real-Time by Fusion of Different Sensing Modalities. *Drones* **2022**, *6*, 317. [CrossRef]
6. Ezuma, M.; Anjinappa, C.K.; Funderburk, M.; Guvenc, I. Radar Cross Section Based Statistical Recognition of UAVs at Microwave Frequencies. *IEEE Trans. Aerosp. Electron. Syst.* **2022**, *58*, 27–46. [CrossRef]
7. HECHT, E. Drones in the Nagorno-Karabakh War: Analyzing the Data. Available online: https://www.militarystrategymagazine.com (accessed on 3 May 2023).
8. Park, S.; Kim, H.T.; Lee, S.; Joo, H.; Kim, H. Survey on Anti-Drone Systems: Components, Designs, and Challenges. *IEEE Access* **2021**, *9*, 42635–42659. [CrossRef]
9. Wellig, P.; Speirs, P.; Schuepbach, C.; Oechslin, R.; Renker, M.; Boeniger, U.; Pratisto, H. Radar systems and challenges for C-UAV. In Proceedings of the International Radar Symposium, Bonn, Germany, 20–22 June 2018; pp. 1–8.
10. U.S. Army UAS Center of exCellenCe. *"Eyes of the Army" U.S. Army Roadmap for Unmanned Aircraft Systems 2010–2035*; U.S. Army: Arlington County, VA, USA, 2010.
11. Zheng, J.; Chen, R.; Yang, T.; Liu, X.; Liu, H.; Su, T.; Wan, L. An Efficient Strategy for Accurate Detection and Localization of UAV Swarms. *IEEE Internet Things J.* **2021**, *8*, 15372–15381. [CrossRef]
12. Zheng, J.; Yang, T.; Liu, H.; Su, T.; Wan, L. Accurate Detection and Localization of Unmanned Aerial Vehicle Swarms-Enabled Mobile Edge Computing System. *IEEE Trans. Ind. Inform.* **2021**, *17*, 5059–5067. [CrossRef]
13. Qi, D.; Zhang, J.; Liang, X.; Li, Z.; Zuo, J.; Lei, P. Autonomous Reconnaissance and Attack Test of UAV Swarm Based on Mosaic Warfare Thought. In Proceedings of the 2021 6th International Conference on Robotics and Automation Engineering (ICRAE), Guangzhou, China, 19–22 November 2021; pp. 79–83.
14. Zhou, Y.; Rao, B.; Wang, W. UAV Swarm Intelligence: Recent Advances and Future Trends. *IEEE Access* **2020**, *8*, 183856–183878. [CrossRef]
15. Dale, H.; Baker, C.; Antoniou, M.; Jahangir, M.; Atkinson, G.; Harman, S. SNR-dependent drone classification using convolutional neural networks. *IET Radar Sonar Navig.* **2022**, *16*, 22–33. [CrossRef]
16. VizcarroCarretero, M.; Harmanny, R.I.A.; Trommel, R.P. Smart-CFAR, a machine learning approach to floating level detection in radar. In Proceedings of the EuRAD 2019—2019 16th European Radar Conference, Paris, France, 2–4 October 2019; pp. 161–164.
17. Skolnik, M.I. *Introduction to Radar Systems*; McGraw Hill Book Co.: New York, NY, USA, 1999; Volume 19.
18. Humphreys, E.; Antoniou, M.; Baker, C.; Stafford, W. Cognitive Approaches to Detection of Small Targets. In Proceedings of the EuRAD 2020—2020 17th European Radar Conference, Utrecht, The Netherlands, 10–15 January 2021; pp. 366–369.
19. Roldan, I.; del-Blanco, C.R.; Duque de Quevedo, Á.; Ibañez Urzaiz, F.; Gismero Menoyo, J.; Asensio López, A.; Berjón, D.; Jaureguizar, F.; García, N. DopplerNet: A convolutional neural network for recognising targets in real scenarios using a persistent range–Doppler radar. *IET Radar Sonar Navig.* **2020**, *14*, 593–600. [CrossRef]
20. Tait, P. *Introduction to Radar Target Recognition*; Institution of Electrical Engineers: London, UK, 2006; ISBN 9781849190831.
21. Amrouche, N.; Khenchaf, A.; Berkani, D. Multiple target tracking using track before detect algorithm. In Proceedings of the 2017 19th International Conference on Electromagnetics in Advanced Applications, ICEAA 2017, Verona, Italy, 11–15 September 2017; pp. 692–695.
22. Buzzi, S.; Lops, M.; Venturino, L. Track-before-detect procedures for early detection of moving target from airborne radars. *IEEE Trans. Aerosp. Electron. Syst.* **2005**, *41*, 937–954. [CrossRef]
23. Kim, K.; Uney, M.; Mulgrew, B. Estimation of Drone Micro-Doppler Signatures via Track-Before-Detect in Array Radars. In Proceedings of the 2019 International Radar Conference (RADAR), Toulon, France, 23–27 September 2019. [CrossRef]
24. Guerraou, Z.; Khenchaf, A.; Comblet, F.; Leouffre, M.; Lacrouts, O. Particle Filter Track-Before-Detect for Target Detection and Tracking from Marine Radar Data. In Proceedings of the 2019 IEEE Conference on Antenna Measurements & Applications (CAMA), Kuta, Indonesia, 23–25 October 2019; pp. 1–4.
25. Chen, V.C. *The Micro-Doppler Effect in Radar*; Artech House: Norwood, MA, USA, 2011; ISBN 9781608070572.
26. Palamà, R.; Fioranelli, F.; Ritchie, M.; Inggs, M.; Lewis, S.; Griffiths, H. Measurements and discrimination of drones and birds with a multi-frequency multistatic radar system. *IET Radar Sonar Navig.* **2021**, *15*, 841–852. [CrossRef]

27. Hanif, A.; Muaz, M.; Hasan, A.; Adeel, M. Micro-Doppler Based Target Recognition With Radars: A Review. *IEEE Sens. J.* **2022**, *22*, 2948–2961. [CrossRef]
28. Skolnik, M. *Radar Handbook*, 3rd ed.; McGraw-Hill Education: New York, NY, USA, 2008.

drones

MDPI

Article

Research on the Intelligent Construction of UAV Knowledge Graph Based on Attentive Semantic Representation

Yi Fan [1], Baigang Mi [1,*], Yu Sun [2] and Li Yin [1]

[1] School of Aeronautics, Northwestern Polytechnical University, Xi'an 710072, China; fan_yionline@163.com (Y.F.); 13880626200@163.com (L.Y.)
[2] School of Foreign Studies, Northwestern Polytechnical University, Xi'an 710072, China; sunyu@nwpu.edu.cn
* Correspondence: mibaigang@163.com

Abstract: Accurate target recognition of unmanned aerial vehicles (UAVs) in the intelligent warfare mode relies on a highly standardized UAV knowledge base, and thus it is crucial to construct a knowledge graph suitable for UAV multi-source information fusion. However, due to the lack of domain knowledge and the cumbersome and inefficient construction techniques, the intelligent construction approaches of knowledge graphs for UAVs are relatively backward. To this end, this paper proposes a framework for the construction and application of a standardized knowledge graph from large-scale UAV unstructured data. First, UAV concept classes and relations are defined to form specialized ontology, and UAV knowledge extraction triples are labeled. Then, a two-stage knowledge extraction model based on relational attention-based contextual semantic representation (UASR) is designed based on the characteristics of the UAV knowledge extraction corpus. The contextual semantic representation is then applied to the downstream task as a key feature through the Multilayer Perceptron (MLP) attention method, while the relation attention mechanism-based approach is used to calculate the relational-aware contextual representation in the subject–object entity extraction stage. Extensive experiments were carried out on the final annotated dataset, and the model F1 score reached 70.23%. Based on this, visual presentation is achieved based on the UAV knowledge graph, which lays the foundation for the back-end application of the UAV knowledge graph intelligent construction technology.

Keywords: UAV; knowledge graph; knowledge extraction; attention mechanism; semantic representation

Citation: Fan, Y.; Mi, B.; Sun, Y.; Yin, L. Research on the Intelligent Construction of UAV Knowledge Graph Based on Attentive Semantic Representation. *Drones* **2023**, *7*, 360. https://doi.org/10.3390/drones7060360

Academic Editor: Diego González-Aguilera

Received: 29 April 2023
Revised: 22 May 2023
Accepted: 27 May 2023
Published: 30 May 2023

1. Introduction

With the penetration of artificial intelligence technology into the total factor of warfare, intelligent warfare is gradually becoming the mainstream combat style [1]. UAVs play the role of both a material foundation and a main support of combat capability in the context of intelligent warfare, which can realize more accurate perception, swifter decision-making, and more efficient action on the battlefield. In the rapidly changing intelligent warfare environment, the generation of large-scale UAV knowledge data leads to changes in knowledge processing, storage, query, and integration technologies, bringing new challenges and opportunities for the application of UAVs in practical mission scenarios [2].

As a new paradigm of knowledge processing, the development of knowledge graphs brings new possibilities for UAV knowledge management [3]. Objectively, knowledge graphs have more powerful data synthesis governance capabilities, and on a large scale, multi-source and different forms of UAV knowledge data can be deeply mined and represent the semantic relation and knowledge systems the graph contains. Subjectively, the knowledge graph is part of the development process of expert systems. However, the knowledge graph construction research on specific domains mainly focuses on medical [4], financial [5], and energy [6] fields. Compared with other domains, there are two main challenges for UAV knowledge graph construction:

- At the level of UAV knowledge and data, several challenges need to be addressed. Firstly, the authority of UAV data is often lacking. These data primarily originate from encyclopedias and news pages, where a direct binding connection between data producers and data credibility is absent. Secondly, the accuracy of UAV knowledge suffers from deficiencies, leading to conflicts and noticeable errors among different UAV data sources. Thirdly, there is a scarcity of systematic data within the UAV domain, and publicly available UAV data are in unstructured text format. Extracting fine-grained knowledge directly from such data becomes challenging, thus hindering comprehensive system research in this field.
- At the level of UAV knowledge graph construction and application, knowledge extraction is the core task in the construction process. The classical knowledge extraction methods can be commendable in the general domain. However, in the face of UAV domain data, it is necessary to develop an algorithmic extraction model adapted to the domain properties, to improve the efficiency and accuracy of extraction, and thus to enhance the generalization ability of the traditional methods. In the face of downstream UAV task applications, there is no direct reference case for the development of domain-oriented system engineering for user requirements and visual interaction. The overall architecture of the system needs to be explored and developed under the guidance of the requirements, in combination with application models from other domains.

Based on the above problems, this paper proposes a knowledge graph construction and application framework for the UAV domain. Firstly, a fine-grained ontology knowledge system model of UAV classes and their relations is defined, and the UAV knowledge extraction dataset is labeled according to the ontology. Then, a two-step UASR knowledge extraction model is designed according to the characteristics of the UAV triple corpus, divided into two subtasks of relation prediction and subject–object entity identification. Then, we design a two-step UASR knowledge extraction model based on the characteristics of the UAV triple corpus. The main contributions of this paper are as follows:

- A fine-grained knowledge ontology is formed by defining the concept and relation attributes of UAVs based on a collection of unstructured UAV data of a significant scale. From this ontology, a UAV knowledge extraction dataset is created by selecting high-quality texts that align with predefined UAV ontology entities and relation annotations.
- A UASR knowledge extraction model is proposed, taking into account the characteristics of UAV knowledge extraction data. The BERT pre-trained language model is utilized to generate character feature encoding. In the decoder stage, the model incorporates the MLP attention mechanism to enhance the representation of relation types in the text for relation prediction. Additionally, a relationship-aware attention approach is employed to assign higher weights to tokens closely associated with relation classification and entity recognition tasks, thus enhancing the contextual semantic representation of subject–object entities.
- The UASR knowledge extraction model undergoes extensive comparison and ablation experiments using a self-built dataset. The experimental results demonstrate the model's effectiveness in solving knowledge extraction challenges within the UAV corpus. Furthermore, the knowledge graph generated by the UASR model's extraction enables visual storage applications. These quantitative and qualitative experiments substantiate the efficacy and validity of the UASR framework.

2. Related Works

In this section, we mainly introduce two types of related work: the results of macro-construction of UAV knowledge graphs and micro-knowledge extraction methods.

2.1. Construction of UAV Knowledge Graph

The knowledge graph is a network composed of entities and the relations between them. These entities include entity types, entity type properties, relation types, and relation type properties, all of which require a unified semantic specification, known as the knowledge graph schema [7]. In the context of the Semantic Web, the knowledge graph schema is often referred to as an ontology, primarily aimed at abstracting, semantifying, and conceptualizing the content of knowledge graphs. However, there is no mature ontology in the UAV domain, while more mature ontologies have been constructed in the aviation domain on top of UAVs. For aircraft maintenance faults, Wang et al. [8] proposed an ontology for aircraft faults in order to solve the problem that multi-source aviation maintenance data lead to syntactic and semantic conflicts, thus limiting the integration and sharing of aircraft fault information, which provides a unified and specific description of multi-source data and eliminates the semantic heterogeneity. For aviation intelligence and operational safety, Mi et al. [9] used natural language processing techniques and clustering algorithms to intelligently extract ontologies for navigational announcement information.

Based on the top-level ontology model, the knowledge graph can be formed by combining the underlying data instances. At present, the research on the construction and application of knowledge graphs in the UAV domain is basically in the initial stage, and there are limitations in the specialty and the scale is far from the practical application requirements. Qiu Ling et al. [10] constructed a knowledge graph with more than 900 entities and 1800 relations for UAV faults, but its ontology construction is too simple and lacks certain authority. Nie Tongpan et al. [11] constructed a knowledge graph containing 74 entities and 98 relations for UAV power system fault diagnosis, and although the original fault manual documents are available on the data to ensure professionalism, the graph storage capacity is too small.

Although the above two typical UAV knowledge graphs basically present the form of the knowledge graph in terms of the results, they essentially directly replicate the general domain methods in terms of the methods and cannot meet the needs of the domain for knowledge graphs, while in other specific domains, knowledge graph construction techniques are relatively mature. Li et al. [12] applied the BERT-BiLSTM-CRF model for information extraction in the military domain to achieve extraction of multi-source military intelligence information. In the geographic domain, Molina-Villegas et al. [13] selected the Mexican geological exploration news text as the research object and applied the general word-embedding approach to accomplish information extraction and disambiguation at the same time. In the energy domain, Wang et al. [14] proposed a recognition method applicable to electric power text, which improves the performance of the BiLSTM-CRF decoding recognition model by fusing character-level pre-training models, left-neighbor entropy, and lexical feature encoding. In the aviation domain, Bao et al. [15] used the traditional BiLSTM-CRF framework and incorporated an attention mechanism to extract named entities related to aviation design. Wang Hong et al. [16] used a self-attentive mechanism and BiLSTM to extract the triples of the accident occurrence process and used scenario reproduction to analyze the causality of aviation safety accidents.

2.2. UAV Knowledge Extraction Approach

The UAV knowledge extraction task is essentially an automated process of extracting triples from large-scale unstructured texts. Currently, it is primarily tackled using deep learning methods, which involve training deep neural networks to learn relevant features for knowledge classification. This process can be divided into three main components: an embedding layer, a network layer, and classification.

Since sentences are not directly input into the neural network for computation, textual word element-embedding representations are required, and common approaches include the static word representations Word2vec [17,18] and GloVe [19], and the dynamic pre-training models ELMo [20], BERT [21], and GPT [22] for the input-embedding layers. The main approaches of the network layer include the convolutional neural network (CNN) [23], firstly used for text classification and then further extended to knowledge extraction, mainly implemented using multiple convolutional kernels and multiple windows. Based on this, Zeng et al. [24] proposed a maximum convergent segmental convolution based on location information, which can capture more structured information of entity pairs. Similarly, Recurrent Neural Network (RNN) can also extract relations, such as the SDP-LSTM framework [25], which applies a multi-channel LSTM to the shortest sequence path of entity pairs. In independent sentence representation, graph convolution networks (GCN) [26] can be utilized to encode sentences with relational knowledge, and the multi-headed attention mechanism selects weights corresponding to different relational edges, further proposing a knowledge-adaptive coarse- and fine-grained attention mechanism combined with an information-filtering relation extraction model. The output layer of the model is processed with SoftMax or sigmoid for single and multiclassification cases, respectively.

According to the model structure, there are pipeline and joint extraction models [27]. The biggest drawback of the pipeline model is that it cannot solve the relation overlap problem, which is manifested in the case of one entity corresponding to multiple relations, the case of two relations between the principal and object entities, and the case of multiple nesting within the principal and object entities. These problems are commonly found in UAV knowledge extraction tasks, such as the PURE model proposed by Chen et al. [28]. However, Yu et al. [29] first proposed the ETL model for the overlapping problem of relations, where the main entity is extracted first as a priori information, and then the object entities and relations are classified by sequence annotation. However, this method cannot solve the situation that there are two relations between the main and object entities. The CasRel model, proposed by Wei et al. [30] for this purpose, first identifies all possible principal entities in the sentence as a priori information and extracts both object entities and relations using sequence annotation. However, the method requires judging a large number of redundant relations and can only handle one principal entity at a time, which is less efficient in engineering practice. Wang et al. [31] proposed the TPLinker model to construct the global information matrix, a matrix used for extracting all the principal and object entities, and additionally, for each relation, to construct two matrices for the beginning and end positions of the principal and object entities. Finally, the alignment of entities and relations is achieved by using the corresponding annotation method to decode the triples. The relation extraction model based on potential relations and global correspondence (PRGC) proposed by Zheng et al. [32] can not only solve the overlap phenomenon in the corpus, but can also significantly reduce the number of relation judgments due to the presence of a priori potential relation judgments and significantly improve the training and computational efficiency.

3. Construction of UAV Knowledge Graph Based on UASR

3.1. UAV Ontology Definition

Currently, ontology construction usually adopts a seven-step approach [33], in which the core aspects include listing important terms in the domain, defining classes and their hierarchical relation, defining class attributes and their relation, etc. As for UAV ontology construction, it is more oriented to unstructured text data, and there is currently no ontology that can be directly reused to learn from. Therefore, combined with the specific UAV knowledge graph construction and application scenario requirements, the UAV ontology definition is simplified into the following two steps:

1. Analyze the core knowledge concepts of the UAV domain and sort out the knowledge system of the UAV system. This paper is oriented to UAV ontology construction and was completed under the guidance of domain expert knowledge, which defines UAV-related concepts, attributes, and relations, and prepares "raw materials" for the subsequent steps. This process does not require a completely clear and conflict-free classification of the above elements, but only a list of as many desired elements as possible, for example, for the UAV system components enumerated in the flight control, weaponry, and landing systems, etc. Although there is no UAV ontology that can be directly reused, it can be borrowed from the ontology model of the military domain [34], which has a strong relevance to the UAV studied in this paper and can serve a complementary role in the process of the knowledge system formation. This contains the main basic information of the aircraft and

knowledge of derived systems, such as warfare, equipment systems, design manufacturers and facilities, and equipment, which are basically the same concepts as the UAV.

2. Define the UAV classes and their hierarchical and relational attributes. After determining the UAV-related elements, the next step is to filter out the representative elements and further constitute the framework of the UAV classification system with its hierarchical relation, in which the concepts represented by the upper-level categories must fully encompass the concepts represented by the lower-level categories. Relations generally correspond to the categories that need to be interacted with, and attributes refer to the inherent qualities of the categories themselves. However, the UAV relation and attribute constraints can be transformed into each other to some extent. For example, the UAV production manufacturer can be defined as an attribute inherent to the UAV in the description metrics, but often the manufacturer latently has a strong interaction with categories such as the location country. Therefore, this paper provides a unified definition of relation and attribute constraints. According to the above ontology construction method, the first level of the UAV classification system includes two subsystems, UAV equipment and UAV events. Three third-level subsystems, of the composition system, description attributes, and technical indicators, are included under UAV equipment, while UAV events include eleven third-level subsystems, such as design, manufacturing, and operation. The fourth-level subsystems include 70 subsystems, such as appearance, the control system, and duties. The layers of the UAV composition system are shown in Figure 1.

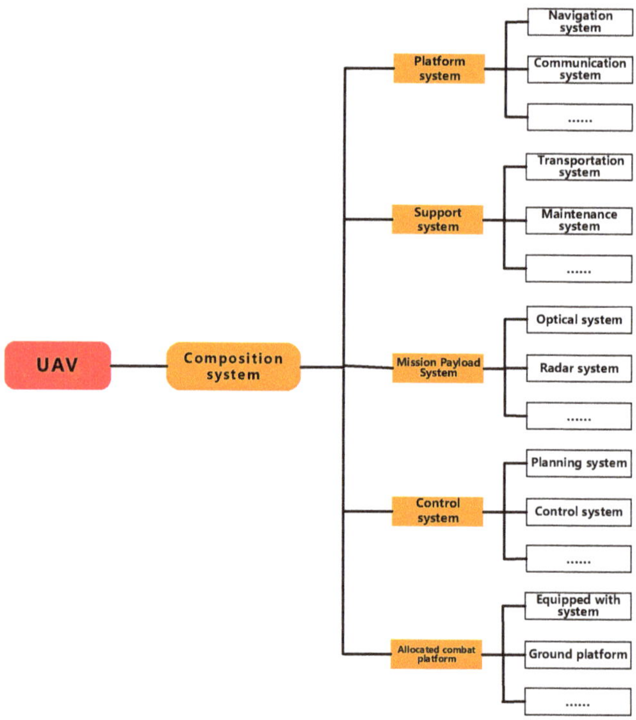

Figure 1. UAV ontology of the composition system.

3.2. UASR Knowledge Extraction Model

3.2.1. Problem Formulation

The task of UAV knowledge extraction model is to get the UAV knowledge triples $T(s) = \{(s, r, o) | s, o \in E, r \in R\}$ from the given sentences with n tokens $S = \{x_1, x_2, \ldots, x_n\}$. Among them, E and R denote the set of entities and the set of relations, respectively. In the encoding part, in order to obtain better results for token and relation embedding, the BERT [35] large-scale language pre-training model is used to encode the input sentences and obtain the contextual representation of each word, and the result is represented as: $Y_{enc}(S) = \left\{h_1, h_2, \ldots, h_n \middle| h_i \in \mathbb{R}^{d\times 1}\right\}$. In the decoding part, it can be split into two stages: relation prediction and subject–object entity identification. For a given sentence S, the goal of the relation extraction subtask is to predict the potential relations contained in the sentence, and the output is:$Y_r(S) = \{r_1, r_2, \ldots, r_m \in R\}$. Among them, m denotes the number of relations in the subset of potential relations. The entity recognition subtask annotates each token corresponding to a given sentence S and the potential relation r_i, predicted by the entity recognition task according to the predefined scheme and BIO rules, and the output is: $Y_e(S, r_i | r_i \in R = \{t_1, t_2, \ldots, t_n\})$. Among them, t_i denotes each label of the annotation. A given sentence S produces a corresponding score for each subject–object entity start token, which corresponds to a higher score for the start tokens and lower scores for the other tokens in a triadic pair. The output results are: $Y_s(S) = M \in \mathbb{R}^{n\times n}$. Among them, M represents the global correspondence matrix. Combined with the specific UAV knowledge extraction context, Figure 2 shows that considering the input text: "RQ-4 is deployed in the Western Pacific", the final output triple is: RQ-4, Position (Deployment), Western Pacific. The relation judgment subtask determines the potential relation from the five predefined relations, such as "Position (Deployment)". The entity recognition subtask predicts "RQ-4" and "Western Pacific" with the labels "B-SUB", "I-SUB", "B-OBJ", and "I-OBJ". The global correspondence matrix for the subject–object alignment subtask predicts higher correspondence scores for the tokens starting with "Seahawk" and ending with "West Pacific", and lower correspondence scores for other tokens.

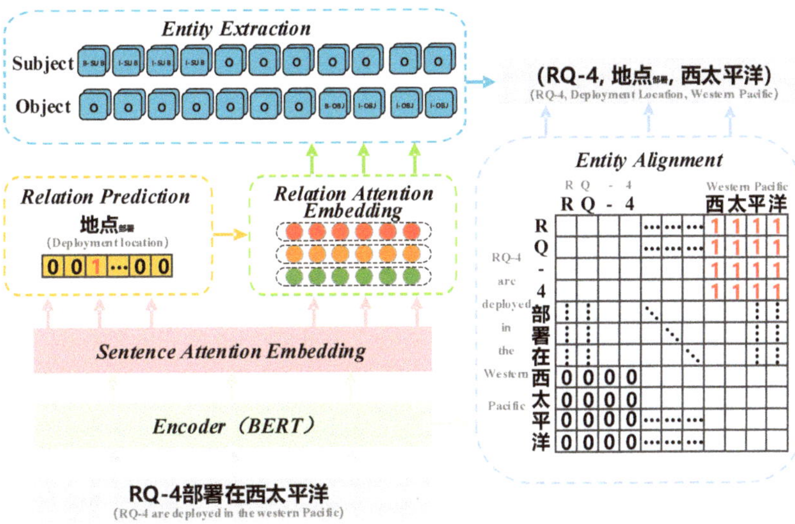

Figure 2. The overall structure of our proposed model.

3.2.2. Relation Prediction

For the relation prediction stage, MLP attention [36] is used to generate contextual semantic representations based on tokens and relation representations to further extract key information, and the extraction method is shown in Equation (1), where the weight of each token in the sentence is generated by multi-layer perceptron attention. If token x_k has a tight semantic association in the subject–object entity extraction or relation classification tasks, x_k receives a higher weight.

$$\begin{aligned} \mathcal{V}_k &= \mathrm{MLP}_k(x_k) \mathrm{s.t.} k \in [1, n] \\ \alpha_{\mathbf{k}} &= \frac{\exp(\mathcal{V}_k)}{\sum\limits_{m=1}^{n} \exp(\mathcal{V}_m)} \\ \mathcal{F}_{\mathrm{s}} &= \sum_{m=i}^{n} \alpha_m x_m \end{aligned} \tag{1}$$

From the MLP attention method, the sentence semantic representation obtained in the relation prediction stage is $\mathcal{F}_{\mathrm{s}}$, while the relation classification is simplified into a multi-label binary classification task, σ denotes the *sigmoid* function, $\mathrm{W}_r \in \mathbb{R}^{d\times 1}$ denotes the training parameters, and P_{rel} denotes the probability of the sentence relation classification, as the corresponding relation of the binary classification task will be labeled as 1 (when the relation probability is exceeded); otherwise, it will be 0. Therefore, the subsequent tasks only need to label the sequences with potential relation sequences, and not all relational sequences need to be labeled.

$$\mathrm{P}_{rel} = \sigma(\mathrm{w}_r \mathcal{F}_{\mathbf{s}} + \mathrm{b}_r) \tag{2}$$

3.2.3. Subject and Object Extraction

Different from the conventional knowledge extraction model, this paper further incorporates the attention mechanism [37] to obtain the contextual semantic representation. For the general entity–relation extraction method, the subject and object entities are first extracted, and then the relation vectors are spliced and fused with the subject and object entity vectors. Thus, the extraction of entities and relations is inseparable from the semantics of the context, where the relation embedding is represented by a learnable network, and the specific relation attention mechanism is formulated as follows:

$$\begin{aligned} a_{ij} &= \alpha^T\left[W_1 x_i + b_1; W_2 r_j + b_2\right], \\ \lambda_{ij} &= softmax(a_{ij}), \\ h_s &= r_j + \sum^{n} \lambda_{ij}(W_3 x_j + b_3), \end{aligned} \tag{3}$$

To overcome the overlap phenomenon in the UAV knowledge extraction corpus, this paper utilizes the relation-embedding approach to label object entities. Based on the semantic association of entities and relations, the main entity and relational embeddings are input to a deep neural network model to predict the object entities. In order to enhance the effect of relational embedding in entity classification, the weight of each token in the sentence is calculated based on the relational attention approach under each specific class of relations. In the subject–object entity recognition task, different relational representations for each sentence are generated through relation-aware contextual semantics with the following prediction formula for each token, specifically:

$$\begin{aligned} \mathrm{P}_{i,j}^{sub} &= Softmax(\mathrm{w}_{sub}(\mathrm{h}_i + \mathrm{u}_j) + \mathrm{b}_{sub}) \\ \mathrm{P}_{i,j}^{obj} &= softmax(\mathrm{w}_{obj}(\mathrm{h}_i + \mathrm{u}_j) + \mathrm{b}_{obj}) \end{aligned} \tag{4}$$

Among them, $\mathrm{u}_j \in \mathbb{R}^{d\times 1}$ denotes the j-th relation representation in the learning-embedding matrix, n_r denotes the total number of relations in the set of relations, $\mathrm{h}_i \in \mathbb{R}^{d\times 1}$ denotes the encoding of the i-th token, $\mathrm{W}_{sub}, \mathrm{W}_{obj} \in \mathbb{R}^{d\times 3}$ denote the learnable weights, and the three dimensions correspond to the annotation set $\{B, I, O\}$.

In the process of subject–object entity alignment, the correct subject–object entity pairs are defined using the global correspondence matrix. The prediction of a potential relation can also be simultaneously performed during the learning process of the global correspondence matrix. The process is as follows: first, number the possible subject–object entity pairs, then check the corresponding score of each subject–object entity pair in the global correspondence matrix, and keep the score if it exceeds the threshold λ_2; otherwise, filter and delete it.

For a given global correspondence matrix $M \in \mathbb{R}^{n\times n}$, n denotes the number of tokens in the sentence, each element of the matrix corresponds to the starting position of the subject–object entity, and the correspondence score is the confidence of matching the subject and object entities. The higher the score, the higher the confidence of matching the two entities. The corresponding score of each element in the matrix is calculated as follows:

$$\mathrm{P}_{i_{sub},j_{obj}} = \sigma(\mathrm{w}_g[\mathrm{h}_i^{sub};\mathrm{h}_j^{obj}] + \mathrm{b}_g) \tag{5}$$

Among them, $\mathrm{h}_i^{sub}, \mathrm{h}_j^{obj} \in \mathbb{R}^{d\times 1}$ denote the input encoding of the i-th and j-th tokens formed by the subject–object entity, $\mathrm{W}_g \in \mathbb{R}^{2d\times 1}$ denotes the learnable weights, and σ denotes the *sigmoid* function.

3.2.4. Training and Inference

The cross-entropy loss function is used in the training inference phase to calculate the final loss to optimize the training model, while sharing the parameters encoded by the BERT model. The loss function can be decomposed into three components:

$$\begin{aligned} \mathcal{L}_{rel} &= -\frac{1}{n_r}\sum_{i=1}^{n_r}(y_i \log \mathrm{P}_{rel} + (1-y_i)\log(1-\mathrm{P}_{rel})) \\ \mathcal{L}_{seg} &= -\frac{1}{2\times n\times n_r^{pot}}\sum_{t\in\{sub,obj\}}\sum_{j=1}^{n_r^{pot}}\sum_{i=1}^{n}\mathrm{y}_{i,j}^t \log \mathrm{P}_{i,j}^t \\ \mathcal{L}_{global} &= -\frac{1}{n^2}\sum_{i=1}^{n}\sum_{j=1}^{n}(y_{i,j}\log \mathrm{P}_{i_{sub},j_{obj}}) + (1-y_{i,j}\log(1-\mathrm{P}_{i_{sub},j_{obj}})) \end{aligned} \tag{6}$$

Among them, n_r denotes the number of all relations and n_r^{pot} denotes the number of predicted potential relations.

The weighted sum of the three-part loss function is:

$$\mathcal{L}_{total} = \alpha\mathcal{L}_{rel} + \beta\mathcal{L}_{seg} + \gamma\mathcal{L}_{global} \tag{7}$$

Among them, α, β, γ denote the adjustable hyperparameters, which can be set to 1 to simplify the model calculation.

4. UAV Knowledge Extraction Experiments

4.1. Experimental Setup

4.1.1. Dataset

The experimental dataset was generated based on the UAV ontology defined in Section 3.1 using the manual annotation method, with a total number of 377,670 characters and 8063 sentences. The core elements of relation labeling still revolve around the UAV entity, corresponding to the relation between UAV and UAV, including homologation, modifications before and after production design; operation between the UAV and the organization; UAV and its related locations such as the production, origin and user country etc.; the direct correspondence between the UAV and the engine; the relations between the UAV and event participation, exhibition, and occurrence of accidents, etc. The final annotated corpus entity data and representatives are shown in Table 1.

Table 1. Statistics and examples of each type of relation.

Relation	Examples	Number
UAV and Engine	RQ-4 Global Hawk—Rolls-Royce AE 3007	88
UAV and UAV	RQ-4 Global Hawk—ACTD Prototype	974
UAV and Events	RQ-4 Global Hawk—Afghanistan War	135
UAV and Country	RQ-4 Global Hawk—United States	363
UAV and Organization	RQ-4 Global Hawk—Northrop Grumman	702

4.1.2. Model Setting

The above constructed UAV knowledge extraction dataset was divided into train, validation, and test dataset data models in the ratio of 8:1:1 in the experiment. The accuracy, precision (P), recall (R), and the F1 score were selected as the evaluation index of knowledge extraction task:

$$\begin{aligned} P &= T_P/(T_P + F_P) \\ R &= T_P/(T_P + F_n) \\ F1 &= 2 * P * R/(P + R) \end{aligned} \tag{8}$$

where T_P denotes the number of entities whose true class is positive and that were predicted to be positive as well, F_P denotes the number of entities whose true class is negative but were predicted to be positive, and F_n denotes the number of entities whose true class is positive but that were predicted to be negative.

The RoBERTa-wwm-ext language pre-training model proposed by Cui et al. [38] was chosen as the basis for character-level vector coding, and the experimental base hyperparameters were set as shown in Table 2. The deep learning framework was Pytorch V1.6.0, the basic running environment was Python V3.7.0, and the hardware configuration was 32G RAM, NVIDIA RTX A4500 (20G).

Table 2. Relation extraction model hyperparameter setting.

Hyperparameter	Value
Batch size	8
Sequence length	256
λ_1	0.5
λ_2	0.1
Warmup proportion	0.05
Decay rate	0.5
Learning rate	1×10^{-3}
Embedding learning rate	1×10^{-4}
Dropout	0.3
Loss	CE
Optimizer algorithm	Adam

4.2. Experimental Results

4.2.1. Pre-Experiment Result

Before applying the UASR model on the in-house UAV knowledge extraction dataset, a pre-experiment on the public dataset NYT was conducted to justify the feasibility. As depicted in Table 3, the UASR model demonstrated superior performance compared to other models on the NYT dataset. Our model outperformed the baselines in terms of precision, recall, and F1 score, with a notable increase of 0.34% in the F1 score over PRGC. While the performance on the public dataset was relatively satisfactory, it should be noted that there is still room for improvement. Nevertheless, this pre-experiment serves as a strong indication that the UASR model is viable for knowledge extraction tasks, providing a solid foundation for subsequent experiments on our in-house UAV knowledge extraction dataset.

Table 3. Comparison of the results of different models for the pre-experiment.

Model	P (%)	R (%)	F1 (%)
CasRel	87.71	90.53	89.10
TPLinker	89.35	90.67	90.01
PGCN	93.54	91.62	92.57
UASR (Our Model)	93.58	92.24	92.91

4.2.2. Overall Comparison

The UAV knowledge extraction results are presented in Table 4. Our proposed UASR model demonstrated the best performance compared to other models, achieving an impressive F1 score of 70.23%. In comparison to two other entity–relation joint extraction models, namely CasRel and TPLinker, the CasRel model exhibited exposure bias and error propagation issues. It overlooked the interaction between entities and relations, resulting in a decrease of 6.37% in the F1 score compared to our model. On the other hand, the TPLinker model performed relatively better, with a decrease of 5.49% in the F1 score. This improvement is primarily attributed to the adoption of a handshake labeling strategy for end-to-end sequence labeling. This strategy connects entity heads with entity tails, head entity heads with tail entity heads, and head entity tails with tail entity tails, using a matrix-based approach for labeling. By training the model to predict the matrix under each relation, the TPLinker model achieved simultaneous entity–relation learning, avoiding the inconsistency between training and prediction orders. Consequently, it demonstrated superior performance on the UAV dataset. Additionally, the PGCN model, which shares closer similarities with our approach, attained an F1 score of 66.59%, showing a decrease of 3.64%. This outcome further emphasizes the advantages of our attention-based relation-aware method in acquiring contextual semantic information.

Table 4. Comparison of the results of different models for UAV knowledge extraction.

Model	P (%)	R (%)	F1 (%)
CasRel	67.12	60.16	63.86
TPLinker	62.13	67.58	64.74
PGCN	64.93	68.35	66.59
UASR (Our Model)	67.79	72.86	70.23

4.2.3. Ablation Experiments

In order to further demonstrate the effectiveness of the improvement strategies of each part of the UASR model, further ablation experiments were conducted in this paper. The experiments removed the MLP attention module in the relation prediction phase and the relation attention module in the subject–object entity identification phase, respectively. The two main improvement modules of the model shown in Table 5 had significant effects on the overall performance. When the MLP attention module was removed, the F1 score of the model decreased by 3.22%, presumably because the relation prediction stage enhanced the model's perception of relation classification to a certain extent for the relation-embedding representation. When the relation attention module was removed, the F1 score of the model decreased by 2.68%, presumably because the main object entity recognition stage above the semantic relationship-aware representation further enhanced the information fusion of task associations.

Table 5. Comparison of the results of UAV knowledge extraction ablation experiments.

Model	P (%)	R (%)	F1 (%)
UASR (Our Model)	67.79	72.86	70.23
MLP attention	66.21	67.83	67.01
Relation attention	62.51	73.49	67.55

The impact of different components within the loss functions on the performance of our proposed UASR model was analyzed through a comprehensive experiment. The results, presented in Table 6, provide valuable insights into the significance of three components. Upon removing the relation loss function, we observed a decrease in the performance of the UASR model by 5.67%. This validates the crucial role of the relation prediction component, which aims to accurately predict potential relation subsets. Moreover, Table 6 reveals that the removal of the sequence loss function led to the inability of the UASR model to correctly extract UAV knowledge. This can be attributed to the inclination of the model to solely memorize the entity positions rather than comprehending the underlying semantics. Additionally, the limited size of our in-house dataset may have contributed to this limitation. Furthermore, the absence of the global correspondence loss function significantly impacted the performance of the UASR model, resulting in an increased number of predicted triples with numerous mismatched pairs. This emphasizes the importance of the constraint imposed by this component.

Table 6. Comparison of the results of UASR loss function ablation experiments.

Model	P (%)	R (%)	F1 (%)
UASR (Our Model)	67.79	72.86	70.23
Relation loss	57.32	73.90	64.56
Sequence loss	Null	Null	Null
Global loss	39.26	28.70	33.16

To assess the impact of different BERT models on our proposed model, we conducted a comparative analysis of three prominent BERT models: BERT-base, BERT-wwm, and RoBERTa-wwm. These models have demonstrated state-of-the-art performance in related knowledge extraction tasks. The experimental results, presented in Table 7, indicate that RoBERTa-wwm outperformed the other models, exhibiting F1 score improvements of 2.10% and 1.00%, respectively. However, as the focus of this paper is not specifically on pre-training language models, the influence of these models, as indicated by the results of our ablation experiments, can be considered negligible.

Table 7. Comparison of the results of UASR BERT ablation experiments.

Model	P (%)	R (%)	F1 (%)
RoBERTa-wwm (Our Model)	67.79	72.86	70.23
BERT-base	66.71	69.62	68.13
BERT-wwm	65.22	73.77	69.23

4.3. UAV Knowledge Graph Visualization

UAV knowledge extraction was implemented through the UASR knowledge extraction model, and the extracted relation needed to be stored and visualized. In this paper, Neo4j was chosen as the relation storage tool in view of the data structure of UAV and the convenience of interactive visualization. In this paper, the UAV knowledge graphs are represented as a triad of <entity, relation, entity> and <entity, attribute, attribute value>, such as <RQ-4, deployment location, Western Pacific> and <MQ-9, endurance, 42 h>. In this paper, we adopted the LOAD CSV method in Cypher query language to import the above knowledge triples and visualize them using Cypher statements. Figure 3a shows the UAV knowledge graph of 4000 entities and more than 10,000 relations at a macro-level. Figure 3b shows the UAV knowledge graph of a single entity "RQ-4" and its relation attributes at a micro-level.

(**a**) Macro-Display (**b**) Micro-display

Figure 3. UAVKG visualization results.

5. Conclusions

As a basis for intelligent target identification and detection, UAV knowledge graphs can provide fine-grained structured knowledge for downstream situational awareness tasks. In this paper, a knowledge graph construction and application framework for the UAV domain was proposed for the problems at the data, system, technology, and application levels in the process of UAV knowledge graph construction. The UAV ontology was formed by combining the characteristics of the UAV knowledge structure system, and a certain scale of the UAV knowledge extraction triple dataset was labeled according to the ontology model, which solved the difficulties at the data and knowledge levels. Based on the UAV data and schema layers, a two-step UASR knowledge extraction model was proposed, and the contextual semantic representation of the two stages of relation prediction and subject–object entity recognition was optimized using attention representation and a relation-aware attention mechanism, respectively, based on the word meta-embedding of the large-scale language pre-training model, with full consideration of the UAV corpus characteristics. Finally, extensive experiments were conducted on the labeled dataset to demonstrate the effectiveness of the UASR model, and the extracted UAV knowledge was implemented for visual storage and intelligent question-and-answer (Q&A) applications.

However, the proposed framework for UAV knowledge graph construction and application still has some limitations:

(1) Due to the limited UAV data, there is still room for improvement in the knowledge extraction accuracy of the UASR model. In the future, under the condition that the dataset is large enough, we could divide the sub-datasets according to the categories of overlapping relations and the length of entity texts in the relations. We could then conduct experiments on the sub-datasets, analyze them from multiple perspectives based on the experimental results, and finally, conclude the reasons for the poor performance of the model on the UAV dataset and the directions for improvement.

(2) Although the application of the UAV knowledge graph was simply realized in this paper, there is still a gap in the application technology between intelligent Q&A and UAV target recognition and detection applications. In the future, the method of knowledge graph embedding can be used to input UAV knowledge in the form of pre-trained templates into the downstream situational game perception task to realize the development of UAV knowledge graphs for knowledge understanding and reasoning applications in actual aerospace situation.

Author Contributions: Conceptualization, B.M.; methodology, Y.F.; writing—review and editing, Y.S.; writing—original draft, L.Y. All authors have read and agreed to the published version of the manuscript.

Funding: This research was funded by the National Natural Science Foundation of China, grant number 12202363, and Fundamental Research Projects in Characteristic Disciplines, grant number G2023WD0124.

Data Availability Statement: The data are unavailable due to privacy reasons.

Acknowledgments: The authors would like to acknowledge the support of the National Natural Science Foundation of China (Grant No. 12202363) and the support of Fundamental Research Projects in Characteristic Disciplines (Grant No. G2023WD0124).

Conflicts of Interest: The authors declare no potential conflict of interest with respect to the research, authorship, and/or publication of this article.

References

1. Mohsan, S.A.H.; Khan, M.A.; Alsharif, M.H.; Uthansakul, P.; Solyman, A.A.A. Intelligent Reflecting Surfaces Assisted UAV Communications for Massive Networks: Current Trends, Challenges, and Research Directions. *Sensors* **2022**, *22*, 5278. [CrossRef] [PubMed]
2. Baigang, M.; Yi, F. A review: Development of named entity recognition (NER) technology for aeronautical information intelligence. *Artif. Intell. Rev.* **2022**, *56*, 1515–1542. [CrossRef]
3. Huo, C.; Ma, S.; Liu, X. Hotness prediction of scientific topics based on a bibliographic knowledge graph. *Inf. Process. Manag.* **2022**, *59*, 102980. [CrossRef]
4. Wu, X.; Duan, J.; Pan, Y.; Li, M. Medical Knowledge Graph: Data Sources, Construction, Reasoning, and Applications. *Big Data Min. Anal.* **2023**, *6*, 201–217. [CrossRef]
5. Cheng, D.; Yang, F.; Wang, X.; Zhang, Y.; Zhang, L. Knowledge Graph-based Event Embedding Framework for Financial Quantitative Investments. In Proceedings of the 43rd International ACM SIGIR Conference on Research and Development in Information Retrieval (SIGIR), Electr Network, online, 25–30 July 2020; pp. 2221–2230.
6. Chen, Z.; Deng, Q.; Ren, H.; Zhao, Z.; Peng, T.; Yang, C.; Gui, W. A new energy consumption prediction method for chillers based on GraphSAGE by combining empirical knowledge and operating data. *Appl. Energy* **2022**, *310*, 118410. [CrossRef]
7. Shao, B.; Li, X.; Bian, G. A survey of research hotspots and frontier trends of recommendation systems from the perspective of knowledge graph. *Expert Syst. Appl.* **2021**, *165*, 113764. [CrossRef]
8. Wang, Y.; Li, Q.; Sun, Y.; Chen, J. Aviation equipment fault information fusion based on ontology. In Proceedings of the 2014 International Conference on Computer, Communications and Information Technology (CCIT 2014), Beijing, China, 16–17 January 2014; pp. 8–10.
9. Mi, B.; Fan, Y.; Sun, Y. Ontology Intelligent Construction Technology for NOTAM. In Proceedings of the 2021 11th International Conference on Intelligent Control and Information Processing (ICICIP), Dali, China, 3–7 December 2021; pp. 137–142.
10. Qiu, L.; Zhang, A.; Zhang, Y.; Li, S.; Li, C.; Yang, L. An application method of knowledge graph construction for UAV fault diagnosis. *Comput. Eng. Appl.* **2023**, *59*, 280–288.
11. Nie, T.; Zeng, j.; Cheng, Y.; Ma, L. Knowledge graph construction technology and its application aircraft power sys tem fault diagnosis. *Acta Aeronaut. Et Astronaut. Sin.* **2022**, *43*, 46–62.
12. Li, Z.; Dai, Y.; Li, X. Construction of sentimental knowledge graph of Chinese government policy comments. *Knowl. Manag. Res. Pract.* **2022**, *20*, 73–90. [CrossRef]
13. Molina-Villegas, A.; Muñiz-Sanchez, V.; Arreola-Trapala, J.; Alcántara, F. Geographic named entity recognition and disambiguation in Mexican news using word embeddings. *Expert Syst. Appl.* **2021**, *176*, 114855. [CrossRef]
14. Chen, Y.; Wang, J.; Zhu, S.; Gu, Y.; Dai, H.; Xu, J.; Zhu, Y.; Wu, T. Knowledge Graph Construction for Foreign Military Unmanned Systems. In Proceedings of the CCKS 2022-Evaluation Track: 7th China Conference on Knowledge Graph and Semantic Computing Evaluations, CCKS 2022, Qinhuangdao, China, 24–27 August 2022; Revised Selected Papers. pp. 127–137.
15. Bao, Y.; An, Y.; Cheng, Z.; Jiao, R.; Zhu, C.; Leng, F.; Wang, S.; Wu, P.; Yu, G. Named entity recognition in aircraft design field based on deep learning. In Proceedings of the International Conference on Web Information Systems and Applications, Guangzhou, China, 23–25 September 2020; pp. 333–340.
16. Wang, H.; Zhu, H.; Lin, H. Research on Causality Extraction of Civil Aviation Accident. *Comput. Eng. Appl.* **2020**, *56*, 6.
17. Goldberg, Y.; Levy, O. word2vec Explained: Deriving Mikolov et al.'s negative-sampling word-embedding method. *arXiv* **2014**, arXiv:1402.3722. *preprint*.
18. Pennington, J.; Socher, R.; Manning, C.D. Glove: Global vectors for word representation. In Proceedings of the 2014 conference on empirical methods in natural language processing (EMNLP), Doha, Qatar, 25–29 October 2014; pp. 1532–1543.
19. Sarzynska-Wawer, J.; Wawer, A.; Pawlak, A.; Szymanowska, J.; Stefaniak, I.; Jarkiewicz, M.; Okruszek, L. Detecting formal thought disorder by deep contextualized word representations. *Psychiatry Res.* **2021**, *304*, 114135. [CrossRef]
20. Devlin, J.; Chang, M.-W.; Lee, K.; Toutanova, K. BERT: Pre-training of deep bidirectional transformers for language understanding. In Proceedings of the 2019 Conference of the North American Chapter of the Association for Computational Linguistics: Human Language Technologies, NAACL HLT 2019, Minneapolis, MN, USA, 2–7 June 2019; pp. 4171–4186.

21. Brown, T.; Mann, B.; Ryder, N.; Subbiah, M.; Kaplan, J.D.; Dhariwal, P.; Neelakantan, A.; Shyam, P.; Sastry, G.; Askell, A. Language models are few-shot learners. *Adv. Neural Inf. Process. Syst.* **2020**, *33*, 1877–1901.
22. Zeng, D.; Liu, K.; Lai, S.; Zhou, G.; Zhao, J. Relation classification via convolutional deep neural network. In Proceedings of the COLING 2014, the 25th international conference on computational linguistics: Technical papers, Dublin, Ireland, 23–29 August 2014; pp. 2335–2344.
23. Zeng, D.; Liu, K.; Chen, Y.; Zhao, J. Distant supervision for relation extraction via piecewise convolutional neural networks. In Proceedings of the 2015 conference on empirical methods in natural language processing, Lisbon, Portugal, 17–21 September 2015; pp. 1753–1762.
24. Xu, Y.; Mou, L.; Li, G.; Chen, Y.; Peng, H.; Jin, Z. Classifying relations via long short term memory networks along shortest dependency paths. In Proceedings of the 2015 conference on empirical methods in natural language processing, Lisbon, Portugal, 17–21 September 2015; pp. 1785–1794.
25. Zhang, N.; Deng, S.; Sun, Z.; Wang, G.; Chen, X.; Zhang, W.; Chen, H. Long-tail relation extraction via knowledge graph embeddings and graph convolution networks. In Proceedings of the 2019 Conference of the North American Chapter of the Association for Computational Linguistics: Human Language Technologies, NAACL HLT 2019, Minneapolis, MN, USA, 2–7 June 2019; pp. 3016–3025.
26. E, H.; Zhang, W.; Xiao, S.; Cheng, R.; Hu, Y.; Zhou, Y.; Niu, P. Survey of Entity Relationship Extraction Based on Deep Learning. *J. Sofrware* **2019**, *30*, 1793–1818.
27. Zhong, Z.; Chen, D. A Frustratingly Easy Approach for Entity and Relation Extraction. In Proceedings of the 2021 Conference of the North American Chapter of the Association for Computational Linguistics: Human Language Technologies, NAACL-HLT 2021, Virtual, Online, 6–11 June 2021; pp. 50–61.
28. Yu, B.; Zhang, Z.; Shu, X.; Liu, T.; Wang, Y.; Wang, B.; Li, S. Joint extraction of entities and relations based on a novel decomposition strategy. In Proceedings of the 24th European Conference on Artificial Intelligence, ECAI 2020, including 10th Conference on Prestigious Applications of Artificial Intelligence, PAIS 2020, Santiago de Compostela, Spain, 29 August–8 September 2020; pp. 2282–2289.
29. Wei, Z.; Su, J.; Wang, Y.; Tian, Y.; Chang, Y. A novel cascade binary tagging framework for relational triple extraction. In Proceedings of the 58th Annual Meeting of the Association for Computational Linguistics, ACL 2020, Virtual, Online, 5–10 July 2020; pp. 1476–1488.
30. Wang, Y.; Yu, B.; Zhang, Y.; Liu, T.; Zhu, H.; Sun, L. TPLinker: Single-stage Joint Extraction of Entities and Relations Through Token Pair Linking. In Proceedings of the 28th International Conference on Computational Linguistics, COLING 2020, Virtual, Online, 8–13 December 2020; pp. 1572–1582.
31. Zheng, H.; Wen, R.; Chen, X.; Yang, Y.; Zhang, Y.; Zhang, Z.; Zhang, N.; Qin, B.; Xu, M.; Zheng, Y. PRGC: Potential relation and global correspondence based joint relational triple extraction. In Proceedings of the Joint Conference of the 59th Annual Meeting of the Association for Computational Linguistics and the 11th International Joint Conference on Natural Language Processing, ACL-IJCNLP 2021, Virtual, Online, 1–6 August 2021; pp. 6225–6235.
32. Noy, N.F.; McGuinness, D.L. *Ontology Development 101: A Guide to Creating Your First Ontology*; Stanford University: Stanford, CA, USA, 2001.
33. Wu, Y.; Zhu, Y.; Li, J.; Zhang, C.; Gong, T.; Du, X.; Wu, T. Unmanned Aerial Vehicle Knowledge Graph Construction with SpERT. In Proceedings of the China Conference on Knowledge Graph and Semantic Computing, Shenyang, China, 24–27 August 2023; pp. 151–159.
34. Cui, Y.; Che, W.; Liu, T.; Qin, B.; Wang, S.; Hu, G. Revisiting pre-trained models for Chinese natural language processing. In Proceedings of the Findings of the Association for Computational Linguistics, ACL 2020: EMNLP 2020, Virtual, Online, 16–20 November 2020; pp. 657–668.
35. Eberts, M.; Ulges, A. Span-based joint entity and relation extraction with transformer pre-training. In Proceedings of the 24th European Conference on Artificial Intelligence, ECAI 2020, including 10th Conference on Prestigious Applications of Artificial Intelligence, PAIS 2020, Santiago de Compostela, Spain, 29 August–8 September 2020; pp. 2006–2013.
36. Yan, M.; Lou, X.; Chan, C.A.; Wang, Y.; Jiang, W. A semantic and emotion-based dual latent variable generation model for a dialogue system. *CAAI Trans. Intell. Technol.* **2023**, 1–12. [CrossRef]
37. Cui, Y.; Che, W.; Liu, T.; Qin, B.; Yang, Z. Pre-training with whole word masking for chinese bert. *IEEE/ACM Trans. Audio Speech Lang. Process.* **2021**, *29*, 3504–3514. [CrossRef]
38. Riedel, S.; Yao, L.; McCallum, A. Modeling relations and their mentions without labeled text. In Proceedings of the Machine Learning and Knowledge Discovery in Databases: European Conference, ECML PKDD 2010, Barcelona, Spain, 20–24 September 2010; Proceedings, Part III 21. pp. 148–163.

drones

MDPI

Article

UAV-Assisted Traffic Speed Prediction via Gray Relational Analysis and Deep Learning

Yanliu Zheng [1], Juan Luo [1,2,*], Ying Qiao [1] and Han Gao [1]

1 College of Computer Science and Electronic Engineering, Hunan University, No. 2 Lushan South Road, Changsha 410082, China; yanliu_zheng@hnu.edu.cn (Y.Z.); qy2020@hnu.edu.cn (Y.Q.); hangao1974@hnu.edu.cn (H.G.)

2 Research Institute of Hunan University, Chongqing 401120, China

* Correspondence: juanluo@hnu.edu.cn

Abstract: Accurate traffic prediction is crucial to alleviating traffic congestion in cities. Existing physical sensor-based traffic data acquisition methods have high transmission costs, serious traffic information redundancy, and large calculation volumes for spatiotemporal data processing, thus making it difficult to ensure accuracy and real-time traffic prediction. With the increasing resolution of UAV imagery, the use of unmanned aerial vehicles (UAV) imagery to obtain traffic information has become a hot spot. Still, analyzing and predicting traffic status after extracting traffic information is neglected. We develop a framework for traffic speed extraction and prediction based on UAV imagery processing, which consists of two parts: a traffic information extraction module based on UAV imagery recognition and a traffic speed prediction module based on deep learning. First, we use deep learning methods to automate the extraction of road information, implement vehicle recognition using convolutional neural networks and calculate the average speed of road sections based on panchromatic and multispectral image matching to construct a traffic prediction dataset. Then, we propose an attention-enhanced traffic speed prediction module that considers the spatiotemporal characteristics of traffic data and increases the weights of key roads by extracting important fine-grained spatiotemporal features twice to improve the prediction accuracy of the target roads. Finally, we validate the effectiveness of the proposed method on real data. Compared with the baseline algorithm, our algorithm achieves the best prediction performance regarding accuracy and stability.

Keywords: deep learning (DL); gray relational analysis (GRA); spatiotemporal prediction; traffic information extraction; UAV

Citation: Zheng, Y.; Luo, J.; Qiao, Y.; Gao, H. UAV-Assisted Traffic Speed Prediction via Gray Relational Analysis and Deep Learning. *Drones* **2023**, *7*, 372. https://doi.org/10.3390/drones7060372

Academic Editors: Bo Li, Daqing Chen, Ming Yan and Chunwei Tian

Received: 25 April 2023
Revised: 27 May 2023
Accepted: 29 May 2023
Published: 2 June 2023

Correction Statement: This article has been republished with a minor change. The change does not affect the scientific content of the article and further details are available within the backmatter of the website version of this article.

1. Introduction

1.1. Background

The maturity and implementation of 6G technology have accelerated the continuous evolution of telematics in the direction of efficiency, intelligence, and convenience. The development of low-altitude UAV traffic management systems provides data and decision support for ground air collaborative traffic state perception and management [1–3]. However, with the highly developed urban transportation system, traffic congestion has become a serious problem in people's daily lives. Dynamic analysis and the prediction of road traffic status is an important part of the intelligent transportation system, which can achieve the real-time and accurate prediction of short-term traffic status [4] and reduce urban traffic conditions. Among them, the vehicle speed is one of the main parameters reflecting the road operation status, which can affect the accuracy of short-term traffic prediction. Therefore, accurate vehicle speed prediction can improve the accuracy of short-term traffic prediction, help people plan their travel time and travel routes, and reduce the waste of time.

Existing intelligent transport systems collect data through the deployment of road infrastructure and in-vehicle mobile devices, such as loop detectors, remote traffic microwave sensors, cameras, and vehicle-equipped GPS sensors, but this method of collection is costly, difficult to deploy, complex to maintain, covers a small area and has serious traffic information redundancy. It is time consuming and computationally expensive, and it is difficult to balance real-time and forecast accuracy.

With the continuous development of UAV remote sensing technology, it provides a new method for the collection of traffic information in intelligent transportation systems, which is based on high-resolution UAV imagery data. The application scenario is shown in Figure 1. UAV imagery can visually describe traffic information at a certain time and show the relationship between the road and the number and distribution of vehicles on the road. UAV imagery at different moments can more intuitively describe changes in traffic status information. Government departments can save many sensing equipment and computing resources by collecting traffic information data through UAV imagery. Therefore, predicting traffic flow based on UAV imagery can improve real-time performance while ensuring accuracy.

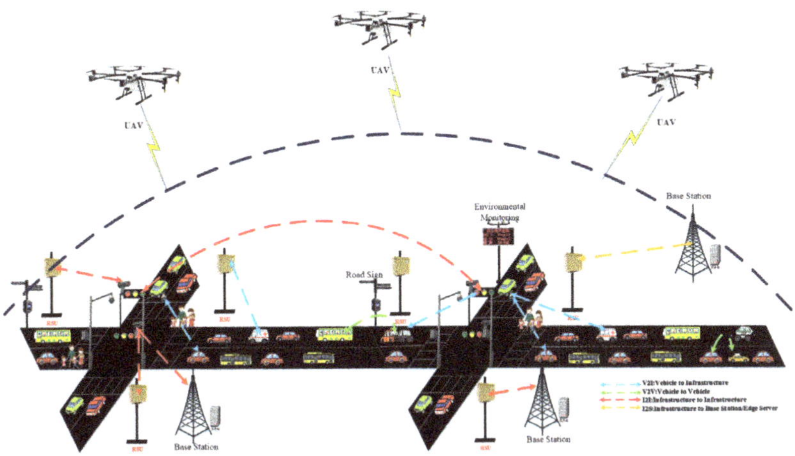

Figure 1. 6G-based intelligent transportation system.

The existing applications based on UAV remote sensing data in the field of transportation focus more on extracting traffic information [5], such as road and vehicle identification, analyzing vehicle density and traffic flow, judging road congestion, etc., ignoring the prediction of future traffic speed. UAV imagery data are periodic and sparse, which has a certain impact on the accuracy of prediction. Therefore, several problems must be solved in the current UAV image-processing-based methods for extracting traffic information and predicting traffic speed. Firstly, road extraction methods are less efficient and time consuming, and areas with similar features to roads are not easily distinguished. Secondly, traditional methods rely on the selection of the optimal threshold, and the accuracy of vehicle recognition algorithms is low for UAV imagery from different sources. Finally, vehicle speed estimation methods are not highly automated and require manual assistance during image processing, which is time consuming and labor intensive.

Therefore, we combine GPS data with UAV imagery to improve the extraction of information from a single UAV imagery and use the extracted results for traffic speed prediction. In this paper, we develop a framework for traffic speed extraction and prediction based on UAV imagery processing, which consists of a traffic information extraction module based on UAV imagery recognition and a traffic speed prediction module based on deep learning, making full use of the different effects of spatiotemporal relationships and spatiotemporal characteristics of the traffic information. We also use the Chengdu taxi GPS dataset to correct the road and vehicle speed information extracted from UAV imagery and enhance the data quality to improve the accuracy of the traffic speed prediction.

1.2. Contributions

We develop a cross-layer short-term traffic speed prediction architecture based on UAV imagery and GPS data, named GATLSTM, to improve the accuracy of the traffic speed prediction. The contributions of this paper are 3-fold:

- We propose a framework for traffic speed extraction and prediction based on UAV imagery and GPS data processing, including data collection, traffic information extraction and speed prediction layers, which intelligently realizes the deep fusion of remote sensing image and GPS data, and improves the prediction accuracy of the traffic speed.
- We designed a traffic information extraction method based on UAV imagery recognition. Firstly, we identify the road and vehicle information in UAV imagery to calculate the average speed of vehicles and roads. Then, we calculate the average speed of roads based on the GPS trajectory and order data collected by crowdsourcing and extract the spatial features using gray relational analysis (GRA) to fuse and correct these two types of data, which is to improve the completeness and accuracy of information extracted from a single UAV imagery. To the best of our knowledge, this is the first time to try to use GRA to extract spatial features.
- We design an attention-enhanced traffic speed spatiotemporal prediction algorithm, which combines the attention mechanism and LSTM, namely ATLSTM. LSTM is used to extract the time features, and the attention mechanism is to extract fine-grained spatiotemporal features again. ATLSTM fully fuses the spatiotemporal relations and accurately predicts traffic speed.

2. Related Works

Traffic prediction is the basis for intelligent traffic control and guidance, road network planning, and several prediction models with higher prediction accuracy widely used in intelligent traffic systems (ITS), mainly including Kalman filter, chaos theory, wavelet analysis, deep neural network, and other prediction models. At the beginning, traffic speeds were predicted using predictive models based on linear statistical theory. In 1960, Kalman proposed the Kalman filter algorithm, which can not only deal with the analysis of smooth data but also analyze and predict non-smooth data. Ref. [6] proposed a scalar adaptive traffic speed prediction model based on Kalman filter theory, which dynamically adjusts the noise in each iteration by minimizing the variance between the actual measured speed and its estimated speed. The above methods were simple to calculate and easy to update in real time; however, they cannot reflect the non-linear changes of traffic flow.

Then, prediction models based on nonlinear statistical theory were introduced into the field of traffic prediction [7,8]. To solve the problem of non-linear, stochastic, and highly non-smooth traffic flow, Ref. [9] explored the application of gray system theory and introduced a new gray system model to predict traffic parameters. However, those methods used unique variable time as the analysis factor and did not take into account the non-linearity characteristics of traffic data, which led to problems, such as large calculation errors, and dependence on historical data, and poor resistance to interference. Therefore, predictive models based on artificial intelligence are used to predict traffic speed. In 1993, Ref. [10] proposed using the artificial neural network (ANN) to predict the state of urban road traffic for the first time. As artificial intelligence techniques mature, deep learning-based methods are widely used to predict traffic speeds [9,11]. Ref. [12] proposed a traffic speed prediction method based on path selection. By analyzing GPS trajectories to select the key road, the Bi-LSTM neural network was introduced to model each critical path to achieve traffic speed prediction.

In recent years, graph-based methods have attracted researchers' attention, and are introduced into the traffic field for short-term traffic prediction [13–15]. Ref. [16] proposed an attention-based spatiotemporal graph convolution network (ASTGCN) model, which used the spatiotemporal attention mechanism and convolution mechanism to capture space and temporal information, and combined them with weights for traffic flow prediction. Ref. [17] took the traffic network as a graph and proposed a graph wavelet gated recurrent neural network and used it as the key component to extract spatial features. It used gated recursive structure to learn the time correlation in sequence data. This method solved the problem of a lack of flexibility in the local feature extraction process.

The preceding works used artificial intelligence to predict traffic flow and considered the spatiotemporal correlation. However, they require a large number of training samples and have a complicated training process, resulting in slow convergence speed and poor scalability. To overcome the shortcomings of a single method, many research works used multiple forecasting models to combine forecasts to obtain more ideal forecasting results [18,19]. Ref. [20] developed an attention-based Conv-LSTM module to extract spatial and short-term temporal features, and proposed a Bi-LSTM model to extract the periodic characteristics of traffic flow prediction. Combining GN and LSTM cells, Ref. [21] designed a new graph LSTM (GLSTM) framework to capture the spatiotemporal representation in road speed prediction. Combination model prediction methods can take advantage of different prediction methods and improve the prediction accuracy. However, if the combination method is improper, the prediction effect cannot be guaranteed.

In addition, traffic congestion prediction has also attracted the attention of scholars. Ref. [22] discussed and analyzed in detail the prediction of traffic congestion using artificial intelligence methods. However, these prediction methods are based on sensor data, with high data collection costs, small sensor coverage, and severe redundancy of traffic information. At the same time, it is computationally expensive and difficult to balance real-time and forecast accuracy. With the continuous development of UAV remote sensing technology and the increasing resolution of UAV imagery, people have expanded their traffic data acquisition from physical microsensors to UAV remote sensing data [23]. Ref. [24] proposed an automatic extraction method for vehicle trajectories. Ref. [25] presented a detailed methodological framework for automated UAV video processing to extract multi-vehicle trajectories for specific road segments. S. P. Hoogendoorn et al. [26] developed a new data collection system, tested with data collected from helicopters by digital cameras, for the determination of individual vehicle trajectories from digital aerial image sequences. In [27], a morphology-based vehicle detection method was developed on 3 m resolution planetary images using high temporal resolution multispectral images, first generating background images from multi-temporal images and then subtracting the background images to identify moving targets, generating traffic density trends for five cities and regions at an average temporal resolution of 7.1 days, and using low-resolution UAV imagery images to estimate small geographic scale traffic intensity. Ref. [28] developed an optimal air traffic assignment model, which is a one-dimensional convolutional neural network and encoder–decoder LSTM framework to compute and measure air traffic flow complexity in the neighborhood of a UAV at a given time. Ref. [29] presented a comparative study of air traffic flow management measures generated by a computational agent based on reinforcement learning, which established delays upon takeoff schedules of aircraft departing from certain terminal areas so as to avoid congestion or saturation in the air traffic control sectors due to a possible imbalance between demand and capacity.

Image-processing techniques are increasingly used in a wide range of applications, such as target recognition, image classification, and sentiment-semantic analysis [30]. Extracting traffic information from UAV imagery for traffic status analysis is also receiving increasing attention from scholars [31,32]. Ref. [33] combines high-resolution imagery, GIS, and deep convolutional neural networks to identify vehicles in movable windows of high-resolution imagery for the assessment of traffic congestion. Ref. [34] used the object-oriented classification method to establish the vehicle detection and processing process of high-resolution satellite images. Ref. [35] used a deep learning algorithm to detect vehicles in high-resolution UAV imagery. Ref. [36] proposed an image-processing-based vehicle speed measurement and vehicle tracking as well as license plate detection system.

These studies focus on extracting traffic information from UAV imagery, such as road identification, vehicle identification, vehicle trajectory extraction, etc., but few studies are used to predict traffic status. Secondly, the data source is singular, comprising only UAV imagery data. Traffic speed prediction is the key to induce traffic and relieve congestion in an intelligent transportation system, so it is crucial to integrate UAV imagery and GPS data for accurate traffic speed prediction.

3. Framework of Speed Prediction

3.1. Problem Definition

Traffic speed prediction is a typical spatiotemporal data prediction problem. ITSs generate massive data, such as video, image, text, etc,, which are changing dynamically over time. We use $\{V_t|t = 1, 2, \cdots, n\}$ to predict the traffic speed of the target road at some point in the future, where V_t is the historical variables of all roads in a given research area over a period of time.

3.2. System Model

We propose a framework for traffic speed extraction and prediction based on UAV imagery and GPS data processing, as shown in Figure 2, which consists of two parts: a traffic information extraction module based on UAV imagery recognition, and a traffic speed prediction module based on deep learning. We use this framework to automatically extract road and speed information from UAV imagery and construct traffic speed prediction datasets. With UAV imagery processing as the important module, we build a method for autonomous extraction of traffic information and prediction of road speed from three modules—road extraction, vehicle recognition, and traffic speed calculation and prediction—combining traditional image processing and deep learning methods.

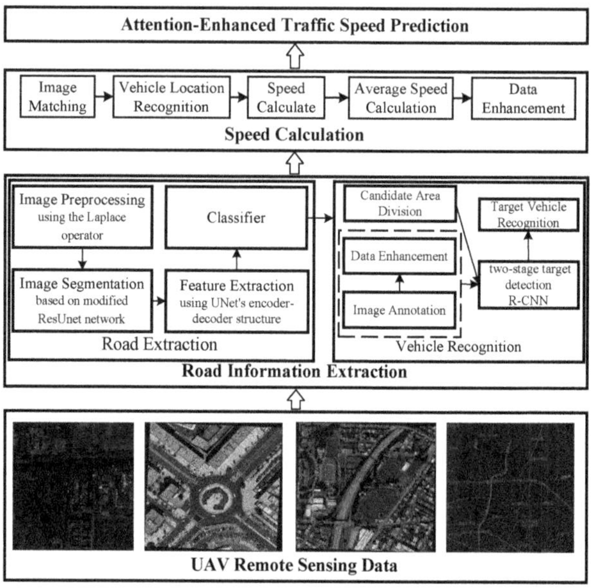

Figure 2. The cross-layer short-term traffic speed prediction architecture based on UAV imagery and GPS data.

In the 6G-based intelligent transportation system, the data collection layer collects urban road, vehicle GPS track and order data based on UAVs and physical sensors. The traffic information extraction layer first identifies the road and vehicle information in the UAV imagery and calculates the average speed of vehicles and road sections, then calculates the average speed of the road sections based on the GPS trajectories and orders collected by crowdsourcing. The two layers are fused and corrected to improve the accuracy of speed extraction and form the final speed prediction dataset, which is uploaded to the cloud. These processes are completed at intelligent edge nodes. A speed prediction model is built in the cloud and trained on the received traffic feature dataset to predict the short-term traffic speed at the next moment. The in-car navigation system plans the driving route for the user in real time based on the predicted values. Thus, our proposed cross-layer short-term traffic speed prediction model avoids congested roads, improves the efficiency of the road network, and can help traffic management authorities guide traffic.

3.3. Traffic Data

Traffic data acquisition techniques in intelligent transportation systems are mainly divided into fixed detector acquisition techniques and mobile traffic information acquisition techniques. In this paper, we select high-definition UAV imagery to extract road and speed information. As UAV imagery have a long shooting period and relatively sparse data, we use crowdsourced GPS data to improve and support the road and speed information extracted from aerial images and enhance the data quality. The UAV imagery data are of Jinniu District, Chengdu. The trajectory dataset was obtained from the DiDi "Gaia" data open project (accessed on: 8 March 2021) (https://gaia.didichuxing.com), i.e., the trajectory data of the ordered drivers on the DDT Express platform on the 65 square highway of the second ring road in Chengdu, with coordinates ranging in [104.043333, 30.727818], [104.129076, 30.726490], [104.129591, 30.655191], [104.042102, 30.652828], with the date range from 1 November to 30 November 2016, and the track point collection interval is 2–4 s. The track points are bound to the road to ensure that the data correspond to the road. The track points are tied to the road to ensure that the data correspond to the road.

Table 1 shows an example of the GPS track dataset used in this paper. Each record includes a driver ID, order ID, timestamp, and latitude and longitude information, which is corrected to match the OpenStreetMap community's open-source map data [37]. Each driver corresponds to one vehicle. In addition, the road segment information in this paper is obtained from the OpenStreetMap community's open-source map data.

Table 1. Dataset [1].

Drive ID	Order ID	Time Stamp	Longitude	Latitude
rJpebi8h4kuy xhrrbrwAlAo	mDxe8figiv DxikA0hvplI	1477969147	104.07513	30.72724
rJpebi8h4kuy xhrrbrwAlAo	mDxe8figiv DxikA0hvplI	1477969150	104.07513	30.72702
jurg2ccjbgxy rdtpmzvlCy	nIBn2lekm pJtdlx2juovJ	1477997455	104.0785	30.72702
...	...	...	...	...

[1] Data are from: https://gaia.didichuxing.com (accessed on: 8 March 2021).

4. Traffic Information Extraction Based on UAV Imagery and GPS Data

4.1. Road Extraction

UAV imagery is usually large in scope and rich in information, containing other scenes such as buildings and roads. We aim to obtain traffic information from UAV imagery, so extracting road areas is a key point.

Firstly, the UAV imagery is pre-processed. During the formation and transmission of UAV imagery, the weather, light, system noise, and shadows of features may influence the image quality. We use the Laplace operator to sharpen the images and highlight the detailed features of the images. Secondly, we use a modified ResUnet network to extract features from the images, using UNet's encoder–decoder structure to capture more contextual information and improve the accuracy and efficiency of the road area extraction.

4.2. Vehicle Identification

We use the R-CNN two-stage target detection algorithm to identify vehicles in the UAV imagery. Candidate regions are first generated using image segmentation, and each region is scaled to a uniform size to extract features. Regions with similar features, such as color and texture, are combined with filtering out possible vehicle locations. Then, the candidate regions are fed into a convolutional neural network for spatial feature extraction. Finally, the extracted features are fed into an SVM for sample classification to identify the vehicle. The two-stage target detection algorithm can identify objects in the figure more accurately and detect the location of the objects.

4.3. Vehicle Speed Calculation

By combining panchromatic and multispectral images, we can estimate the vehicle speed by calculating the vehicle's displacement. Due to the different sensor positions, there is a small time difference between the panchromatic and multispectral images. The vehicle position cannot be extracted directly from the multispectral image due to the low resolution of the multispectral image. Therefore, based on the vehicle position extracted in the high-resolution image, the panchromatic image is matched with the multispectral image to indirectly obtain the vehicle position in the multispectral image, which calculates the vehicle displacement and then estimates the vehicle speed based on the difference in imaging time between the two. The steps are first to use the principal component analysis method to perform multi-band multispectral image conversion to obtain a multispectral image. Secondly, the panchromatic and multispectral images are adjusted to the same resolution to obtain the vehicle template and the target image. Finally, the position of the target vehicle in the multispectral image is obtained by the similarity matrix calculation.

As there is a limit to the speed at which vehicles can travel in the road area, the vehicle displacement is not large for a short period and is reflected in the image by about 20–30 pixels. The size of the vehicle template in the panchromatic image is $N \times M$, and the size of the target image is $N' \times M'$, and then its length and width correspond to

$$N' = N + \frac{2v\Delta t}{p}, \tag{1}$$

$$M' = M + \frac{2v\Delta t}{p}, \tag{2}$$

where v is the maximum vehicle speed, Δt is the imaging time difference between the multispectral and panchromatic images, and p is the image resolution.

In the panchromatic and multispectral image stages, the vehicle template is matched for similarity with each possible position of the vehicle in the target image to obtain a similarity matrix, the value of which corresponds to the degree of matching at this position, and the maximum value in the similarity matrix indicates the position of the vehicle template in the target image. We use the spatial cosine theorem to calculate the similarity between the vehicle template and the target image:

$$p_{ij} = \frac{\sum_{m=1}\sum_{n=1}\left(A_{m,n} \times B_{i+m,j+m}\right)}{\sqrt{\sum_{m=1}\sum_{n=1} A_{m,n}^2}\sqrt{\sum_{m=1}\sum_{n=1} B_{i+m,j+n}^2}}, \tag{3}$$

where A is the vehicle template, B is the target image, and i and j denote the position in the target image. By selecting the maximum value of the similarity matrix, the vehicle's position in the multispectral image can be located to calculate the vehicle's displacement at different time phases, and the vehicle speed information can be estimated.

The average speed of a road is an important parameter of traffic state in this paper, which can better reflect the operation state of traffic in a specific section and time period. The calculation method is shown in Equation (4). We select the average speed of the road v_l as the characteristic parameter of traffic state for analysis and prediction:

$$v_l = \frac{1}{N}\sum_{i=1}^{N} v_i. \quad (4)$$

4.4. GPS Data Enhancement

UAV remote sensing data are periodic and sparse, which has a certain impact on the accuracy of the prediction. Given that the data may be lost or abnormal in the process of data collection and transmission, the GPS data are initially preprocessed, and feature extraction is then performed. The process of trajectory data preprocessing and feature extraction is shown in Figure 3.

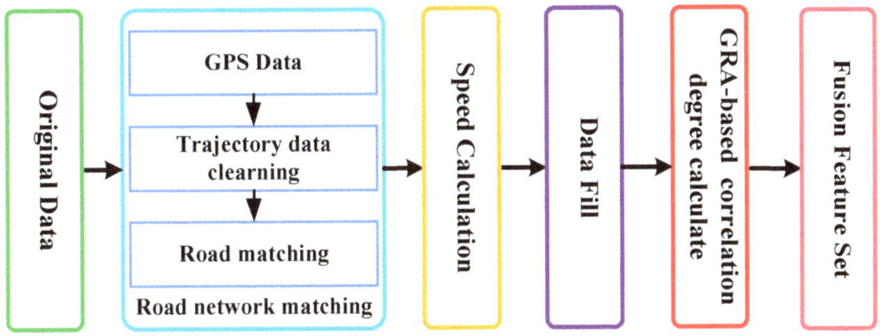

Figure 3. GRA-based trajectory data process.

In this paper, we take the Chengdu taxi trajectory data as original data that include the road name, road number, the longitude and latitude of the starting point of the road, and the length of the road. The final data of the traffic characteristic parameters are obtained by cleaning the original GPS data, matching road, calculating average speed, processing abnormal and missing value, fusing external feature data, and extracting spatial feature.

We select the target road network data from the original data. The target road network consists of nodes and roads; long roads are divided into short roads based on the traffic lights. We match the selected trajectory data with roads by the longitude and latitude range of the start point and end point of the road, completing the road network matching.

We calculate the average speed of the matched roads and complete the preliminary traffic state feature extraction based on GPS data. We set the observation period as 5 min, and extract 288 sets of traffic state features for each road section in one day. According to Equation (4) and the trajectory dataset, the average speed of each vehicle passing through the target roads in the observation period is calculated. The set of vehicles that pass through road i is denoted as $C = \{C_{i,1}, C_{i,2}, \cdots, C_{i,N}\}, (i = 1, 2, \cdots, S)$, S is the number of roads, and the position latitude and longitude of vehicle $C_{i,k}$ at time t are denoted as $(W_{i,k,t}, J_{i,k,t}), (k = 1, 2, \cdots, N)$. According to the shortest spherical distance [38], the distance (unit: km) traveled by the vehicle $C_{i,k}$ from time $t-1$ to time t can be calculated as follows:

$$d_{i,k,t} = 2R \times \arcsin\left(\sqrt{\sin^2\left(\frac{w_{i,k,t} - w_{i,k,t-1}}{2}\right) + \cos w_{i,k,t} \times \cos w_{i,k,t-1} \times \sin^2\left(\frac{j_{i,k,t} - j_{i,k,t-1}}{2}\right)}\right), \quad (5)$$

where R is the radius of the earth.

The average speed (unit: km/h) of the vehicle $C_{i,k}$ in the t-th time interval is defined and calculated by

$$\bar{v}_{i,k,t} = d_{i,k,t} \div (\frac{5}{60}). \quad (6)$$

The average speed of N vehicles on road section i in the t-th time interval is calculated to obtain the average speed $v_{i,t}$ (unit: km/h) of the road section in the t-th time interval, shown as follows:

$$V_{i,t} = \frac{1}{N}\sum_{k=1}^{N} \bar{v}_{i,k,t}. \quad (7)$$

The missing data are mainly for the period of less traffic flow, which may lead to the missing average speed data due to no vehicles passing through during the observation time. The average speed threshold in this paper is set as

$$0 < V_{i,t} \leq \phi_V V_f, \quad (8)$$

where V_f is the limited speed specified in the observation section, ϕ_V is the correction factor, and the range usually takes [1.3, 1.5].

For the missing and abnormal data, we use a time-series average method to interpolate such values. The time-series average method takes the current time data as the center and takes the arithmetic average value of the average speed of several adjacent periods in the current time period as the traffic state characteristics at the current time, calculated by Equation (9):

$$\widehat{V}_{i,t} = \frac{1}{2n}(V_{i,t-n} + V_{i,t-1} + V_{i,t+1} + V_{i,t+n} + 3L). \quad (9)$$

In Equation (9), $\widehat{V}_{i,t}$ is the estimated value of the average speed of the road with a missing or abnormal interval in the t-th format interval, and $2n$ is the group number of adjacent periods. This method is suitable for those with fewer missing values. To meet the real-time requirements of traffic forecast and the demand of intelligent traffic guidance, n should not be too large here, usually $n = 1$.

When the data of a certain period of time are continuously missing, the later experimental error will be relatively large by using the time-series average method to fill the data. Therefore, we use the least squares polynomial fitting method to fill the data, calculated by Equation (10):

$$f(v) = a_0 + a_1 v + a_2 v^2 + \cdots + a_{n-1} v^{n-1} + a_n v^n. \quad (10)$$

On the basis of the above analysis, we process the track data of vehicles to obtain the final set of traffic characteristic parameters. The trajectory data of vehicles are extracted for preprocessing and traffic state feature extraction, which travel from north to south in 1 to 2 November 2016 at Julidi Middel Road. A total of 576 groups of characteristic parameters are obtained, as shown in Figure 4. From the figure, the average speed of the road section, which is a characteristic parameter of traffic state, can express the traffic operation trend and state after preprocessing the original track data. By processing the missing and abnormal data, the curve of the average speed of the road section becomes complete and smooth.

4.5. Data Correlation Analysis Based on GRA

Traffic prediction can not only provide a scientific basis for traffic managers, perceive traffic congestion in advance, and restrict vehicle driving but also provide a safe road choice and guarantee the improvement of travel efficiency for urban passengers. However, traffic forecasting has always been an arduous task due to its complex spatiotemporal dependence.

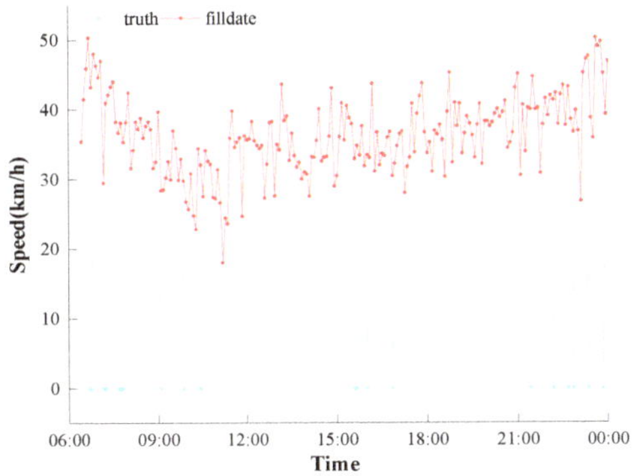

Figure 4. The result of speed data filling.

It is well known that traffic data have strong spatiotemporal correlation with each other, which poses a challenge for traffic speed prediction. The spatial change of traffic speed is mainly determined by the topological structure of an urban road network. The traffic condition of the upstream road affects the traffic condition of the downstream road through the transmission effect, and the traffic condition of the downstream road affects the traffic condition of the upstream through the feedback effect. As shown in Figure 5, the average speed of a certain road section is not only affected by the historical traffic state of a single section but is also closely related to the traffic conditions of adjacent sections. The dynamic change of the road speed with time is mainly reflected in the periodicity and trend. As shown in Figure 6, the traffic speed on road 1 shows a similar daily variation trend from Monday to Sunday.

We use GRA to solve the complex spatiotemporal dependence problem. GRA examines the correlation degree of various factors in a system by comparing the similarity between the geometric shapes of each factor curve [39]. It has the advantages of a small amount of data, convenient calculation, and no need to calculate statistical characteristics, and its analysis results are basically consistent with the results of qualitative analysis. To find the adjacent roads that are related to the traffic speed of the target section, this paper chooses to use the gray correlation analysis to measure the correlation degree between the adjacent sections in the peak period.

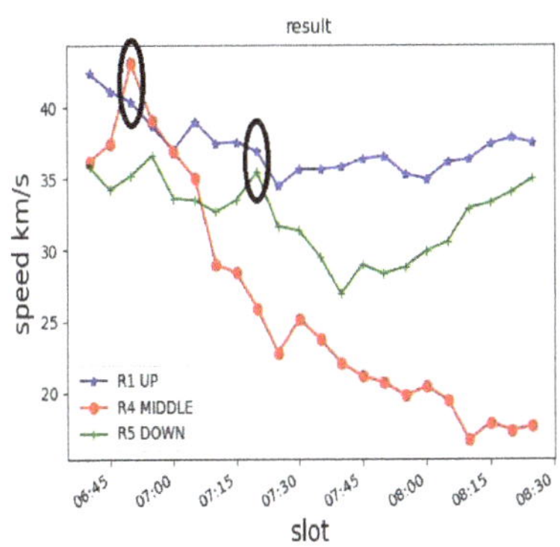

Figure 5. Spatial correlation of traffic speed.

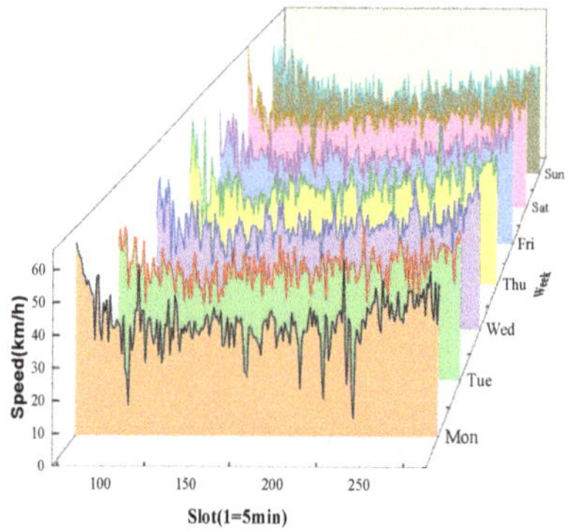

Figure 6. Time correlation of traffic speed.

Assuming that the target road has l adjacent roads, we determine the influence degree of adjacent roads on the target road by using GRA. We use $v_0(t)$ to denote the traffic state feature sequence of the target road $v_i(t)(i = 1, 2, \cdots, l)$ to denote the traffic state feature sequence of adjacent roads. The computation steps are as follows.

Step 1: The difference between the target road sequence and the adjacent road sequence at each time point is calculated by

$$\Delta_i(t) = |v_0(t) - v_i(t)|, \tag{11}$$

and

$$\Delta_i = (\Delta_i(1), \Delta_i(2), K, \Delta_i(l)), \tag{12}$$

where n is the length of the sequence, $t = 1, 2, \cdots, n$.

Step 2: Equation (13) calculates the gray correlation coefficient $\gamma_i(t)$ between the target link sequence $v_0(t)$ and the adjacent link sequence $v_i(t)$ at each time point. The correlation coefficient reflects the closeness of the two compared sequences at each time point $0 < \gamma \leq 1$:

$$\gamma_i(t) = \frac{\Delta_i min + \Delta_i max}{v_i(t) + \Delta_i max}, \tag{13}$$

where $\Delta_i max$ and $\Delta_i min$ are the maximum and minimum differences between the feature sequence of road i and the feature sequence of the target road section, respectively.

Step 3: Equation (14) calculates the gray correlation degree between the traffic state characteristics of the target road section and the adjacent road section:

$$\gamma_{0,i} = \frac{1}{n} \sum_{i=1}^{n} \gamma_i(t). \tag{14}$$

The gray correlation degree of each road section in the neighboring area is sorted with the target road section. The greater the correlation degree, the greater the influence of the adjacent section on the traffic state of the target section, which can be used as the basis for extracting the spatial characteristics of traffic speed.

5. Vehicle Speed Prediction Based on Deep Learning

5.1. LSTM Network

The LSTM network is a special type of recurrent neural network (RNN). It can learn long-term dependent information and overcome the shortcomings of traditional RNN. In the dynamic gate structure of LSTM, the forgetting gate controls whether the current moment memory module retains the hidden cell state of the upper layer with a certain probability, which reads as h_{t-1}, x_t and C_{t-1}. h_t is the output of time t of the LSTM memory unit, and C is the state of the memory unit. f_t outputs a value between 0 and 1, with 0 for complete discard and 1 for full reservation.

Usually, when using LSTM to perform natural language processing, a softmax layer is added at the end of the LSTM layer as the final output. The traffic speed prediction in this paper belongs to a regression problem. Therefore, we add the linear transformation layer after LSTM as the final predicted value output and use the gradient descent method to train the model. Moreover, the objective function in the training process is the sum of mean square errors (MSEs) between the predicted and actual values.

To minimize the training error, the Adam algorithm is used as the optimization algorithm, which is an adaptive learning rate algorithm. Adam calculates the first and second moments of each parameter according to the loss function. The learning rate of each parameter is dynamically adjusted to make the parameter change stable during the training process. The convergence speed of Adam is fast, and the learning effect in the training process is good. In addition, the dropout mechanism [40] is introduced into the network to prevent the model from overfitting. The input nodes and recursive connections are inactivated with a certain probability during the training process, which reduces the interaction between nodes and strengthens the generalization ability of the model.

5.2. Principle of Attention Mechanism

Although LSTM introduces a memory unit to determine the information to be retained or removed through its "gate" structure, obtaining the final reasonable vector representation with a long time series is still difficult. In addition, when the time span is large, the calculation will be heavy and the time consumption will be extremely long.

The attention mechanism is a method for solving problems by imitating human attention. It can quickly screen high-value information from a large number of information, which is mainly used to overcome the limitations of the LSTM feature extraction. The attention mechanism selectively learns these inputs when it trains a model, and associates the output sequence when it outputs a model, which achieves information filtering.

In our model, we combine an attention mechanism with gated convolution to adaptively adjust the key features importance, where the importance of value can be calculated as

$$E = V_e \cdot \sigma(ReLU(X \cdot a) + b_e), \tag{15}$$

and

$$E'_{i,j} = \frac{exp(E_{i,j})}{\sum_{i=1}^{T_r-1} exp(E_{i,j})}, \tag{16}$$

where $a = softmax(X)$, V_e and b_e are learnable parameters, E is the correlation matrix, $E_{i,j}$ is the strength of dependence between i and j, and $E'_{i,j}$ is the attention value.

5.3. ATLSTM Model Construction

In view of the spatial dynamic instability and long-term time dependence of the road traffic speed, we introduce the attention mechanism into LSTM. By combining the unique unit structure of LSTM and attention in mining temporal and spatial series data rules, ATLSTM weighs all of the input features one by one, pays attention to the specific space and channel, extracts significant fine-grained feature information of the space–time series, and then predict the traffic speed. The structure of the model is shown in Figure 7.

We aim to predict the average speed of the target section in the $t+1$ time interval according to the historical traffic speed of the target road in $[t-p, t]$ time period and the traffic speed of the four adjacent roads of the target road. In the data preprocessing stage, the traffic data and external influence data in the historical data are initially extracted and time correlated to form the fusion feature set. Then, through the GRA, the data of four adjacent roads are extracted to form the final feature set, which are more closely related to the target road space.

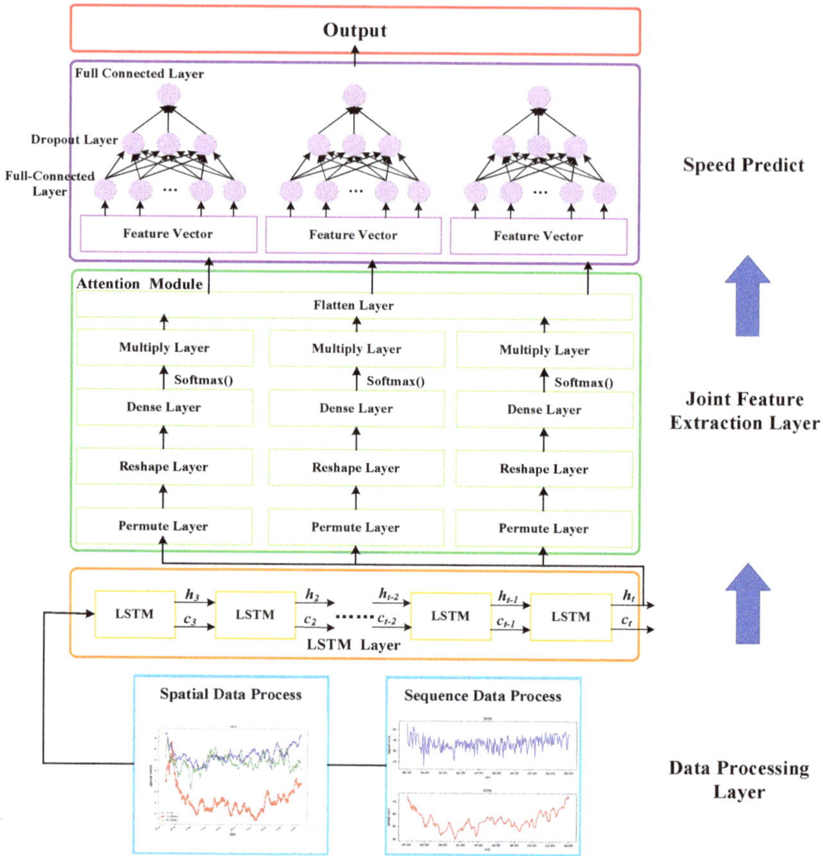

Figure 7. ATLSTM model.

In the stage of traffic speed prediction, the training set data are initially inputted into the LSTM network for training. The LSTM network trains the set data and then extracts the characteristics of short-term time series. The data trained by the LSTM network are inputted into the attention mechanism, which consists of five layers, i.e., permute, reshape, dense, permute and multiply layers. The first permute layer converts the received LSTM data into a specific format. The reshape layer transforms the data into a specific shape. The dense layer activation function selects softmax to calculate the weight of each feature, which is the density layer in the attention structure. In this paper, the attention weight of multi-dimensional features in attention is not shared.

Given that the second structure of attention is multiplication, it needs to multiply the corresponding elements. Thus, the second permute layer is used to transform the dimension again. In the multiply layer, the weight is multiplied by the input, and the attention layer is finished. After the flatten layer, the multi-dimensional data are tiled into one dimension and connected with the output layer. Finally, through the full link layer, the average speed of after 5 min is predicted.

6. Experiment Analysis

6.1. Experiment Settings

In this section, the Jiulidi Middle Road in Chengdu is selected as the target section, and the coordinates range from [104.057548, 30.698718] to [104.057468, 30.696901], and the vehicle trajectory data [41] from north to south are collected as traffic data. The road section information is from the Chengdu Traffic Administration Bureau. The data related to the date and time, such as holidays (weekends), are generated by the actual situation.

The above data are preprocessed and feature extracted by the methods described in Sections 5.1 and 5.2. The data from 1 to 21 November are selected as train samples, and the data from 22 to 23 November are selected as test samples. The prediction model is sensitive to the numerical range of the training input data. Thus, we use the min-max normalization method to linearly transform the traffic state characteristics and map the data to the interval [0, 1], which does not change the overall the trend of the original data. In this manner, the network can converge quickly and eliminate the influence of different data orders on the training effect of the model.

In the existing prediction model, no good method is available to determine the parameters, which are usually obtained by repeated experiments based on relevant experience. We establish a prediction model with two-layer LSTM and an attention mechanism. There is a dense layer behind the two-layer LSTM. The activation function is linear. The dropout function is used in the hidden layer of the LSTM model, and 20% nodes are selected to discard each round of weight update. The loss function is MSE, and the optimization algorithm is Adam. In the attention mechanism, the softmax function calculates the weight of each feature, and the attention weight of the multi-dimensional features is not shared. Multi-attention calculates the attention value. Through the fully connected layer, the average speed of after 5 min is predicted. The six types of comparison algorithms selected in this paper are LSTM, GRU, SAEs, CNNLSTM, GCN, and TGCN.

6.2. Evaluation Metrics

We adapt five performance metrics to evaluate the effectiveness of the proposed algorithm,which are described in detail as follows.

Mean absolute error (MAE):

$$MAE = \frac{1}{n}\sum_{t=1}^{n} |y_t - \widehat{y}_t|, \tag{17}$$

where y_t is the average speed of the actual road vehicle in the t time interval, $\widehat{y}_t$ represents the predicted value of the t time interval, and n is the number of samples.

Mean squared error (MSE):

$$MSE = \frac{1}{n}\sum_{t=1}^{n} (y_t - \widehat{y}_t)^2. \tag{18}$$

Mean absolute percentage error (MAPE):

$$MAPE = \frac{1}{n}\sum_{t=1}^{n} \frac{|y_t - \widehat{y}_t|}{y_t}. \tag{19}$$

Fitting degree of the model $\boldsymbol{R^2}$: When evaluating the designed prediction model and comparing the prediction results with the actual situation, the larger the fitting degree, the better the fitting degree will be. The method is defined as Equation (20):

$$\boldsymbol{R^2} = \boldsymbol{1} - \frac{SS_{residual}}{SS_{total}} = \frac{\sum_{i=1}^{n}(y_i - \widehat{y}_i)^2}{\sum_{i=1}^{n}(y_i - \sum_{i=1}^{n} y_i)^2}. \tag{20}$$

Accuracy (ACC): the accuracy of the model prediction is calculated as follows:

$$Accuracy = 1 - \sum_{n}^{i=1} \frac{y_i - \widehat{y}_i}{y_i}. \tag{21}$$

6.3. Spatial Correlation Analysis of Traffic Data

We analyze seven roads in the 3rd North Section of the First Ring Road in Chengdu. We find that the traffic speed trends of adjacent road sections have different degrees of similarity. Therefore, when predicting the traffic speed of a certain road, it is necessary to consider the influence of the adjacent area roads.

Table 2 shows the correlation degree between each road in the selected area and target road. The correlation degree of the adjacent roads to the target road is different. The bigger the correlation degree, the greater the influence on the traffic speed of the target road will be.

When too few adjacent roads are used for the prediction, it cannot make full use of the spatial correlation of traffic speed, while too many adjacent roads may produce noise data and affect the prediction accuracy. So we select four adjacent roads, whose average correlation degree is greater than 0.7 in a week, to help predict speed. As we can see in Table 2, the adjacent road 2, road 3, road 4, and road 7 are selected to predict the target road speed.

Table 2. Correlation coefficient between road 1 and adjacent roads.

Week	Road 2	Road 3	Road 4	Road 5	Road 6	Road 7
Monday	0.798	0.819	0.786	0.735	0.722	0.840
Tuesday	0.797	0.796	0.771	0.683	0.728	0.833
Wednesday	0.790	0.819	0.824	0.708	0.695	0.808
Thursday	0.794	0.787	0.789	0.673	0.670	0.814
Friday	0.772	0.855	0.797	0.656	0.665	0.805
Saturday	0.821	0.729	0.792	0.740	0.669	0.786
Sunday	0.812	0.808	0.788	0.709	0.654	0.777
AveRelate	0.791	0.802	0.792	0.700	0.686	0.809

6.4. Speed Prediction Results Analysis of GATLSTM

To reduce the input data dimension and improve the efficiency of the algorithm, for the traffic state prediction at time t, we select the first half hour, that is, the traffic speed in the time period $[t-5, t]$ and the speed of the four adjacent road sections with a higher degree of correlation as the traffic feature sequence input of the prediction algorithm.

To control the complexity of the model and the accuracy of model prediction, we change the number of neurons and the number of iterations in conducting the experiments. As shown in Figure 8, the prediction accuracy of the model increases with the number of neurons. When the number of neurons reaches 128, the prediction accuracy of the algorithm decreases instead. Thus, we select 128 as the number of neurons in each layer. As shown in Figure 9, the prediction accuracy of the model and the time for model training increases with the number of iterations. To balance the prediction accuracy and computational efficiency, we set the number of iterations as 400.

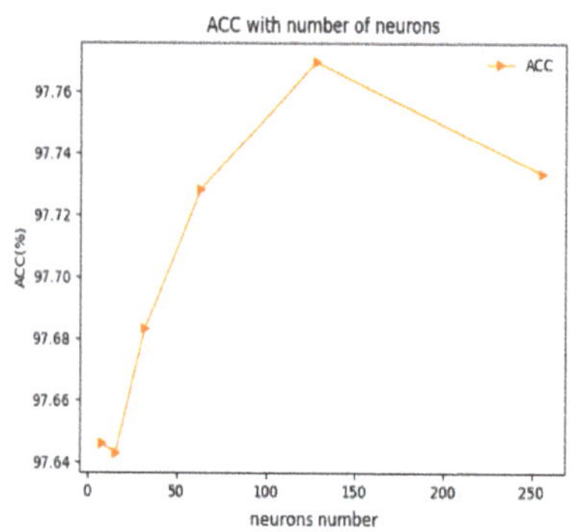

Figure 8. ACC with different number of neurons.

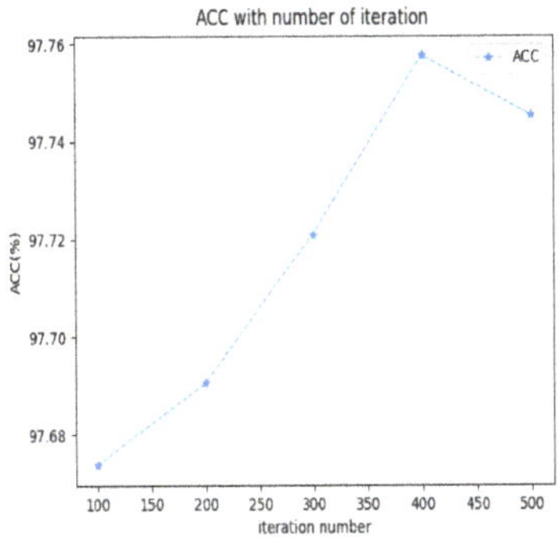

Figure 9. ACC with different number of iterations.

After the experimental parameters are determined, we predict the traffic speed of the target road section on weekdays and holidays (weekends). Figures 10 and 11 show the prediction results of the GATLSTM. As shown from the figures, the morning rush hour is more obvious on workdays, the traffic speed fluctuates greatly throughout the day during holidays (weekends), and the evening rush hour is longer. Moreover, the traffic speed prediction curve in this paper can better fit the measured speed curve in both cases.

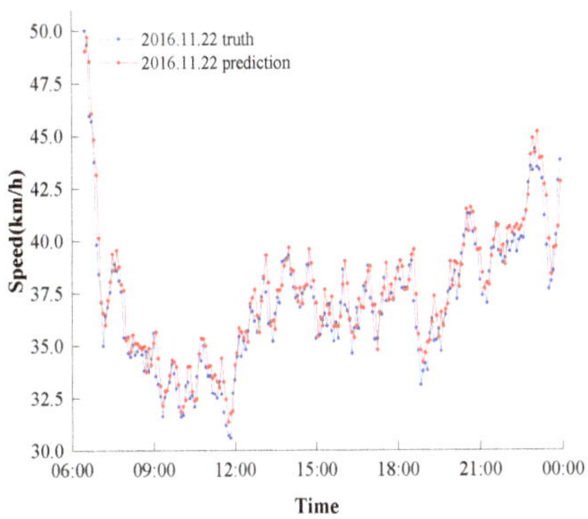

Figure 10. GATLSTM predict result (WorkDay).

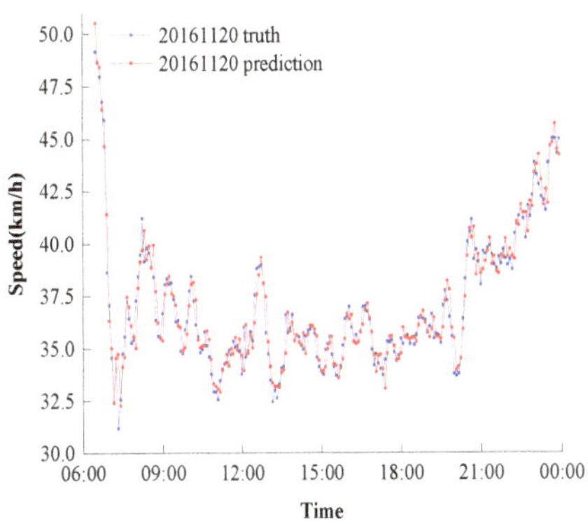

Figure 11. GATLSTM predict result (Holiday).

6.5. Comparative Experiments

To validate the effectiveness of the GATLSTM algorithm, selecting LSTM, GRU, SAEs, CNNLSTM, GCN, and TGCN models as baseline methods, we compare the proposed method with the baseline method on a real dataset.

Figures 12 and 13 show the results of different methods, and Table 3 shows the performance metrics of different methods. In Figures 12 and 13, the blue line is the real speed value of the target section, the red line is the speed value predicted by the GATLSTM model, and the other color lines are the predicted values from the baseline algorithm. The closer the distance to the blue line, the closer the prediction result is to the true value, and we can easily see that our prediction result is closest to the blue line, indicating that our proposed model can better learn the potential features between the traffic speed spatiotemporal sequences. As shown in Table 3, the MAE of the ATLSTM is the smallest, and its accuracy is the highest. Therefore, GATLSTM has the best performance compared to the baseline algorithm for this prediction problem. We also found that the prediction results of GCN and TGCN models are less accurate, which indicates that the traffic data are highly spatiotemporal correlated and the best prediction results cannot be obtained by time-series models and spatial-prediction models alone. Moreover, we conclude that GCN and TGCN may be more suitable for large cities with a complex road network, but GATLSTM is more suitable for small- and medium-sized cities. We further find that the prediction is accurate and efficient when the data pre-processing process is simple.

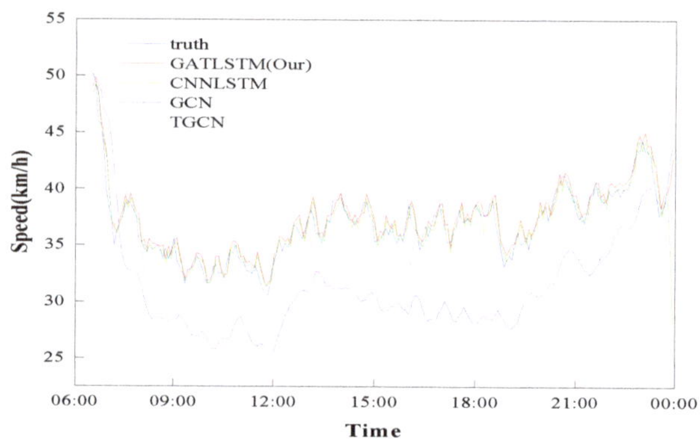

Figure 12. Spatiotemporal prediction.

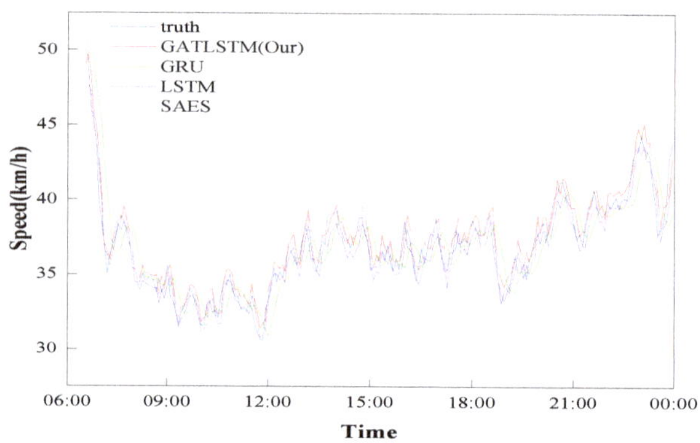

Figure 13. Time-series prediction.

Table 3. Performance measurement.

Algorithm	ACC	MAE	MAPE	RMSE	R^2
GATLSTM	0.979	0.565	2.08	0.768	0.944
CNNLSTM	0.965	0.708	2.20	0.889	0.928
GCN	0.777	5.507	16.24	6.767	0.627
TGCN	0.866	3.052	6.48	4.076	0.865
LSTM	0.967	0.969	2.14	1.209	0.925
GRU	0.913	1.407	3.55	2.647	0.943
SAES	0.962	0.858	2.68	1.408	0.809

Finally, to compare the prediction effects of the models more intuitively, we compared the performance of each prediction model under the condition of large traffic flow fluctuation at 12:00–14:30. The results are shown in Figure 14. In comparison with other algorithms, our method still has the best prediction results during peak periods.

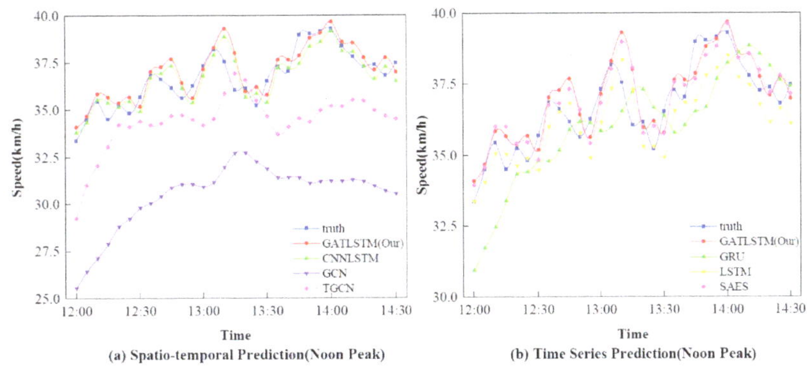

Figure 14. The prediction results of noon peak.

7. Conclusions

Considering that the existing physical sensor-based traffic data collection methods have high transmission costs, serious traffic information redundancy, large calculation volumes of spatiotemporal data processing, and sparse UAV imagery data, which make it difficult to ensure the real-time accuracy of traffic prediction, we use a novel end-to-end deep learning architecture to effectively improve the prediction accuracy. The architecture consists of a traffic information extraction module based on UAV imagery recognition and an attention-enhanced spatiotemporal prediction (ATLSTM) module. The traffic information extraction module deeply fuses the UAV imagery and GPS data to extract traffic information. ATLSTM extracts important fine-grained spatiotemporal features again to accurately predict short-term traffic speed. Finally, we evaluate the performance of the algorithm on a real-world dataset. Our method achieves the best prediction performance in terms of accuracy and stability, is more suitable for small- and medium-scale road networks, and significantly reduces graph-based computational complexity.

Author Contributions: Y.Z. and J.L. proposed the prediction architecture design, Y.Z. and Y.Q. performed the experiments and analyzed the results. Y.Z. and H.G. wrote the paper. J.L. and Y.Q. revised the paper and provided valuable advises. All authors have read and agreed to the published version of the manuscript.

Funding: This work was supported in part by the National Natural Science Foundation of China under Grant 61972140, Natural Science Foundation of Chongqing under Grant cstc2021jcyj-msxmX0460, Key scientific and technological research and development plan of Hunan Province under Grant 2022NK2046.

Acknowledgments: The authors would also like to thank the anonymous referees for their valuable comments and helpful suggestions.

Conflicts of Interest: The authors declare no conflict of interest.

References

1. Shrestha, R.; Oh, I.; Kim, S. A Survey on Operation Concept, Advancements, and Challenging Issues of Urban Air Traffic Management. *Front. Future Transp.* **2021**, *2*, 1–19. [CrossRef]
2. Davies, L.; Vagapov, Y.; Grout, V.; Cunningham, S.; Anuchin, A. Review of Air Traffic Management Systems for UAV Integration into Urban Airspace. In Proceedings of the 2021 28th International Workshop on Electric Drives: Improving Reliability of Electric Drives (IWED), Moscow, Russia, 27–29 January 2021; pp. 1–6. [CrossRef]
3. Kopardekar, P.; Rios, J.; Prevot, T.; Johnson, M.; Jung, J.; Robinson, J.E. Unmanned aircraft system traffic management (UTM) concept of operations. In Proceedings of the AIAA Aviation and Aeronautics Forum (Aviation 2016), Washington, DC, USA, 13–17 June 2016; pp. 1–16.
4. Li, Z.; Li, Y.; Li, L. A Comparison of Detrending Models and Multi-Regime Models for Traffic Flow Prediction. *IEEE Intell. Transp. Syst. Mag.* **2014**, *6*, 34–44.
5. Macioszek, E.; Kurek, A. Extracting Road Traffic Volume in the City before and during COVID-19 through Video Remote Sensing. *Remote Sens.* **2021**, 13, 2329. [CrossRef]
6. Mir, Z.; Filali, F. An adaptive Kalman filter based traffic prediction algorithm for urban road network. In Proceedings of the 2016 12th International Conference on Innovations in Information Technology (IIT), Al Ain, United Arab Emirates, 28–30 November 2016; pp. 1–6. [CrossRef]
7. Zhong, J.T.; Ling, S. Key Factors of K-Nearest Neighbours Nonparametric Regression in Short-Time Traffic Flow Forecasting. In Proceedings of the 21st International Conference on Industrial Engineering and Engineering Management 2014; Atlantis Press: Paris, Frace, 2015; pp. 9–12.
8. Sun, P.; Aljeri, N.; Boukerche, A. A Fast Vehicular Traffic Flow Prediction Scheme Based on Fourier and Wavelet Analysis. In Proceedings of the 2018 IEEE Global Communications Conference (GLOBECOM), Abu Dhabi, United Arab Emirates, 9–13 December 2018. [CrossRef]
9. Comert, G.; Begashaw, N.; Huynh, N. Improved grey system models for predicting traffic parameters. *Expert Syst. Appl.* **2021**, *177*, 114972. [CrossRef]
10. Vythoulkas, P. Alternative approaches to short term traffic forecasting for use in driver information systems. *Transp. Traffic Theory* **1993**, *12*, 485–506.
11. Yisheng, L.; Duan, Y.; Kang, W.; Li, Z. Traffic Flow Prediction With Big Data: A Deep Learning Approach. *IEEE Trans. Intell. Transp. Syst.* **2014**, *16*, 865–873. [CrossRef]
12. Wang, J.; Chen, R.; He, Z. Traffic speed prediction for urban transportation network: A path based deep learning approach. *Transp. Res. Part C Emerg. Technol.* **2019**, *100*, 372–385. [CrossRef]
13. Guo, G.; Yuan, W. Short-term traffic speed forecasting based on graph attention temporal convolutional networks. *Neurocomputing* **2020**, *410*, 387–393. [CrossRef]
14. Zhang, Z.; Li, Y.; Song, H.; Dong, H. Multiple dynamic graph based traffic speed prediction method. *Neurocomputing* **2021**, *461*, 109–117. [CrossRef]
15. Li, Y.; Yu, R.; Shahabi, C.; Liu, Y. Graph Convolutional Recurrent Neural Network: Data-Driven Traffic Forecasting. *arXiv* **2017**. [CrossRef]
16. Guo, S.; Lin, Y.; Feng, N.; Song, C.; Wan, H. Attention Based Spatial-Temporal Graph Convolutional Networks for Traffic Flow Forecasting. In Proceedings of the AAAI Conference on Artificial Intelligence, Honolulu, HI, USA, 27 January–1 February 2019; Volume 33, pp. 922–929. [CrossRef]
17. Cui, Z.; Ke, R.; Pu, Z.; Ma, X.; Wang, Y. Learning traffic as a graph: A gated graph wavelet recurrent neural network for network-scale traffic prediction. *Transp. Res. Part C Emerg. Technol.* **2020**, *115*, 102620. [CrossRef]
18. Luo, W.; Dong, B.; Wang, Z. Short-term Traffic Flow Prediction Based on CNN-SVR Hybrid Deep Learning Model. *J. Transp. Syst. Eng. Inf. Technol.* **2017**, *17*, 68–74.
19. Yu, H.; Wu, Z.; Wang, S.; Wang, Y.; Ma, X. Spatiotemporal recurrent convolutional networks for traffic prediction in transportation networks. *Sensors* **2017**, *27*, 1501. [CrossRef] [PubMed]
20. Zheng, H.; Lin, F.; Feng, X.; Chen, Y. A Hybrid Deep Learning Model with Attention-Based Conv-LSTM Networks for Short-Term Traffic Flow Prediction. *IEEE Trans. Intell. Transp. Syst.* **2020**, *22*, 6910–6920. [CrossRef]
21. Lu, Z.; Lv, W.; Cao, Y.; Xie, Z.; Peng, H.; Du, B. LSTM variants meet graph neural networks for road speed prediction. *Neurocomputing* **2020**, *400*, 34–45. [CrossRef]
22. Akhtar, M.; Moridpour, S. A review of traffic congestion prediction using artificial intelligence. *J. Adv. Transp.* **2021**, *2021*, 1–18. [CrossRef]
23. Zhang, C.; Tang, Z.; Zhang, M.; Wang, B.; Hou, L. Developing a More Reliable Aerial Photography-Based Method for Acquiring Freeway Traffic Data. *Remote Sens.* **2022**, *14*, 2202. [CrossRef]
24. Azevedo, C.L.; Cardoso, J.L.; Ben-Akiva, M.; Costeira, J.P.; Marques, M. Automatic Vehicle Trajectory Extraction by Aerial Remote Sensing. *Procedia—Soc. Behav. Sci.* 2014, *111*, 849–858. [CrossRef]

25. Khan, M.A.; Ectors, W.; Bellemans, T.; Janssens, D.; Wets, G. Unmanned Aerial Vehicle-Based Traffic Analysis: A Case Study for Shockwave Identification and Flow Parameters Estimation at Signalized Intersections. *Remote Sens.* **2018**, *10*, 458. [CrossRef]
26. Hoogendoorn, S.P.; Zuylen, H.J.V.; Schreuder, M.; Gorte, B.; Vosselman, G. Microscopic Traffic Data Collection by Remote Sensing. *Transp. Res. Rec.* **2003**, *1855*, 121–128. [CrossRef]
27. Chen, Y.; Qin, R.; Zhang, G.; Albanwan, H. Spatial Temporal Analysis of Traffic Patterns during the COVID-19 Epidemic by Vehicle Detection Using Planet Remote-Sensing Satellite Images. *Remote Sens.* **2021**, *13*, 208. [CrossRef]
28. Alharbi, A.; Petrunin, I.; Panagiotakopoulos, D. Deep Learning Architecture for UAV Traffic-Density Prediction. *Drones* **2023**, *7*, 78. [CrossRef]
29. Crespo, A.M.F.; Weigang, L.; Barros, A.G.D. Reinforcement learning agents to tactical air traffic flow management. *Int. J. Aviat. Manag.* **2017**, *1*, 145–161. [CrossRef]
30. Yan, M.; Lou, X.; Chan, C.A.; Wang, Y.; Jiang, W. A semantic and emotion-based dual latent variable generation model for a dialogue system. *CAAI Trans. Intell. Technol.* **2023**. [CrossRef]
31. Guo, Y.; Wu, C.; Du, B.; Zhang, L. Density Map-based vehicle counting in remote sensing images with limited resolution. *ISPRS J. Photogramm. Remote Sens.* **2022**, *189*, 201–217. [CrossRef]
32. Yin, Z.; Tang, Y. Analysis of Traffic Flow in Urban Area for Satellite Video. In Proceedings of the IGARSS 2020—2020 IEEE International Geoscience and Remote Sensing Symposium, Waikoloa, HI, USA, 26 September–2 October 2020; pp. 2898–2901. [CrossRef]
33. Chakraborty, D.; Mohan, S.; Dutta, D.; Jha, C.S. Assessment of traffic congestion with high-resolution remote sensing data and deep convolution neural network. *Geocarto Int.* **2022**, *37*, 6808–6825. [CrossRef]
34. Guo, D.; Zhu, S.; Wei, J. Research on Vehicle Identification Based on High Resolution Satellite Remote Sensing Image. In Proceedings of the 2019 International Conference on Intelligent Transportation, Big Data and Smart City (ICITBS), Changsha, China, 12–13 January 2019; pp. 62–65. [CrossRef]
35. Tan, Q.; Ling, J.; Hu, J.; Qin, X.; Hu, J. Vehicle Detection in High Resolution Satellite Remote Sensing Images Based on Deep Learning. *IEEE Access* **2020**, *8*, 153394–153402. [CrossRef]
36. Kusetogullari, H.; Demirel, H.; Celik, T.; Bayindir, S. Real time detection and tracking of vehicles for speed measurement and licence plate detection. In Proceedings of the The Seventh IASTED International Conference on Visualization, Imaging and Image Processing, Palma de Mallorca, Spain, 29–31 August 2007; pp. 53–58.
37. Vargas-Munoz, J.E.; Srivastava, S.; Tuia, D.; Falcão, A.X. OpenStreetMap: Challenges and Opportunities in Machine Learning and Remote Sensing. *IEEE Geosci. Remote Sens. Mag.* **2021**, *9*, 184–199. [CrossRef]
38. Jiménez-Meza, A.; Arámburo-Lizárraga, J.; de la Fuente, E. Framework for Estimating Travel Time, Distance, Speed, and Street Segment Level of Service (LOS), based on GPS Data. *Procedia Technol.* **2013**, *7*, 61–70. [CrossRef]
39. Tseng, F.; Yu, H.C.; Tzeng, G. Applied Hybrid Grey Model to Forecast Seasonal Time Series. *Technol. Forecast. Soc. Chang.* **2001**, *67*, 291–302. [CrossRef]
40. Zaremba, W.; Sutskever, I.; Vinyals, O. Recurrent Neural Network Regularization. *arXiv* **2014**.
41. Data Source. [Didi Chuxing GAIA Initiative]. Available online: https://gaia.didichuxing.com (accessed on 8 March 2021).

drones

Article

PFFNET: A Fast Progressive Feature Fusion Network for Detecting Drones in Infrared Images

Ziqiang Han [1], Cong Zhang [1,*], Hengzhen Feng [2], Mingkai Yue [1] and Kangnan Quan [1]

[1] School of Equipment Engineering, Shenyang Ligong University, Shenyang 110159, China; hanzq@sylu.edu.cn (Z.H.); ymkig@sylu.edu.cn (M.Y.); quankn@stu.sylu.edu.cn (K.Q.)
[2] Science and Technology on Electromechanical Dynamic Control Laboratory, Beijing Institute of Technology, Beijing 100081, China; 7520210110@bit.edu.cn
* Correspondence: zcbxl@sylu.edu.cn

Abstract: The rampant misuse of drones poses a serious threat to national security and human life. Currently, CNN (Convolutional Neural Networks) are widely used to detect drones. However, small drone targets often reduced amplitude or even lost features in infrared images which traditional CNN cannot overcome. This paper proposes a Progressive Feature Fusion Network (PFFNET) and designs a Pooling Pyramid Fusion (PFM) to provide more effective global contextual priors for the highest downsampling output. Then, the Feature Selection Model (FSM) is designed to improve the use of the output coding graph and enhance the feature representation of the target in the network. Finally, a lightweight segmentation head is designed to achieve progressive feature fusion with multi-layer outputs. Experimental results show that the proposed algorithm has good real-time performance and high accuracy in drone target detection. On the public dataset, the intersection over union (IOU) is improved by 2.5% and the detection time is reduced by 81%.

Keywords: background clutter; counter-drones; progressive fusion; lightweight network

Citation: Han, Z.; Zhang, C.; Feng, H.; Yue, M.; Quan, K. PFFNET: A Fast Progressive Feature Fusion Network for Detecting Drones in Infrared Images. *Drones* **2023**, *7*, 424. https://doi.org/10.3390/drones7070424

Academic Editor: Giordano Teza

Received: 1 May 2023
Revised: 19 June 2023
Accepted: 22 June 2023
Published: 26 June 2023

1. Introduction

Various types of small unmanned aerial vehicles threaten infrastructure, hardware and people seriously [1,2]. At the same time, the accurate detection of drone targets in low-resolution, visually blurred infrared images is a challenging task. There are two main problems:

(1) The influence of the target itself: due to the flight altitude being usually below 500 m, drone targets often have a few to several tens of pixels in an infrared image. In addition, drones usually have a low signal-to-noise ratio (SCR) and are easily submerged in strong noise and cluttered backgrounds [3–5]. Therefore, the radiation intensity of the target is lower and it lacks significant morphological features, making target detection in infrared images difficult [6,7].

(2) The contradiction between the target and the detection algorithm: compared to RGB images, detecting drones in infrared images presents more problems, such as the lack of shape and texture features. After filtering and convolution calculations, it is easy to weaken or even lose the representative features of drones (such as wings) [8,9]. Besides, although building shallow networks can improve performance in deep learning algorithms, the contradiction between advanced semantic features and high resolution still cannot be resolved [10].

Overall, there are too many negative samples in the image due to the large variation in target size and the extremely low percentage of pixels in infrared images, resulting in the loss of most of the available information during algorithm operation [11]. In addition, most negative samples are easily classified, which makes it difficult for the algorithm to optimize in the expected direction [12]. Therefore, the nets designed for normal objects are hardly used to detect drones in infrared images.

To detect drone targets in infrared images, researchers have proposed many traditional methods over the past few years. The traditional detection method involves implementing SIRST (Single-frame InfraRed Small Target) detection by calculating the non-coherence between the target and background. Typical methods include filter-based methods [13–15], which can only suppress uniform and smooth background noise. They are also unable to adapt to complex backgrounds and may have a higher false alarm rate. The HVS method [16–19] uses the ratio of gray values between each pixel position and its neighboring region as an enhancement factor. It can effectively enhance the real target. However, it cannot effectively suppress the background noise. The low-rank representation [20–22] can adapt to infrared images with low SCR ratios. However, in complex backgrounds, there is still a high false alarm rate for small and shape-varying targets. Most traditional methods heavily rely on manual features. These methods are simple calculations without training or learning. However, designing hand-crafted features and tuning hyperparameters require expert knowledge and a significant amount of engineering effort.

With the development of CNN methods, more data-driven methods are being applied to infrared small target detection [23–26]. Data-driven methods are suitable for more complex real scenarios and are less affected by target size, shape, and background changes. These methods require a large amount of data to demonstrate strong model fitting ability and have achieved better detection performance than traditional methods. Based on data-driven methods, the convolutional segmentation network can simultaneously produce pixel-level classification and location output [27]. The first segmentation-based SIRST detection method ACM was proposed by [28], which designed a semantic segmentation network using an asymmetric context module. On this basis, Dai [29] further introduced expanded local contrast to improve their model. They use bottom-up local attention modulation modules to embed subtle low-level details at higher levels by combining traditional methods with deep learning methods. Zhang [30] used an attention mechanism to guide the pyramid context network to detect targets. The feature map is divided into blocks to compute local correlations. Then global contextual attention is used to compute the correlation between semantics. Finally, the decoding maps of different scales are fused to improve detection performance.

As well as using segmentation networks to solve the problem of drone target detection, researchers also offer some other ideas. In [31], a balance between missed detection (MD) and false alarms (FA) was achieved. The cGAN networks were used to separately build models for miss detection (MD) and false alarm (FA) as two subtasks as generators. Next, a discriminator for image classification is used to distinguish the outputs of the two generators and ground-truth images. Chen [32] uses visible light images to achieve drone detection. By lightweight improvement of the backbone network and using the multi-scale fusion method to enhance the use of shallow features. A new non-maximum suppression method is designed to solve the problem of drone loss in multi-scale detection, ultimately achieving real-time detection. However, the above methods still have many shortcomings. First, the problem of small target feature loss in the deep layers of the network still exists, and the contradiction between high-level semantic features and high resolution cannot be resolved. Second, the coding maps generated by each downsampling layer cannot be well used. These problems will make the drone detection algorithm less robust to scene changes (such as cluttered backgrounds, and targets with different SCR, shapes and sizes).

To solve these problems, we propose a data-driven progressive feature fusion detection method (PFFNet) from the perspective of infrared unmanned aerial vehicle target detection. First, global features were extracted from the input infrared image. Then, it passes the encoding maps output by downsampling to the FSM and PFM modules. The deep features that include high-level semantic information, the shallow features that contain rich image contour and the position information can be fully fused, thereby improving the utilization of the output encoding maps of the downsampling layer. In addition, the output feature maps are cross-scale fused to enhance the response amplitude of infrared unmanned aerial vehicle targets in the deep network and solve the problem of feature loss in small targets in the deep layers of the network. The high-level semantic information and shallow semantic information are superimposed and output through dimensional cascading. The confidence map is obtained through threshold segmentation to output the final detection result. Finally, to verify the effectiveness of PFFNet, we conducted extensive ablation studies on FSM and PFM and comparative experiments with existing methods on the SIRST Aug and IRSTD datasets. The experimental results show that the various modules of PFFNet have improved the detection of infrared unmanned aerial vehicle targets. Our algorithm has stronger robustness, better detection performance, and faster target detection time.

2. Methods

Given an input image I, we aim to classify each pixel by end-to-end convolving a neural network to determine whether it is a drone target. Finally, we output a segmentation result that is the same size as I. The PFFNet detection algorithm is divided into two parts: the global feature extractor and the progressive feature fusion network. The global feature extractor extracts the basic features of the input infrared image I by looking at the entire image. The redundant information in the image can effectively reduce by obtaining these basic features.

The progressive fusion network is divided into two modules: the Neck and the Head. The Neck includes the Pool Pyramid Fusion Model (PFM) and the Feature Selection Model (FSM). The former is used to enhance the feature response amplitude in the deep network of the infrared drone target. The latter acts as a bridge for information interaction between high and low layers, increasing the utilization rate of the downsampling output encoding map. The Head implements the progressive fusion of feature maps of different scales and generates a segmentation mask.

As shown in Figure 1, the input image I is encoded into different dimensions and resolutions by the backbone to generate encoding maps b_a ($a = 2, 3, 4$). The low-level spatial position information of the target's salient features is obtained from b_a (a = 2, 3) by the FSM. Locating the high-frequency response area to reduce the influence of redundant signals on the target position information and output the feature maps f_a ($a = 2, 3$). The b_4 is used as the input of the PFM to output the decoded image p. The PFM is composed of four different pooling structures in parallel to form a pyramid network. The high-frequency response amplitude of deep target features is enhanced and then passes to the FSM after upsampling. The FSM and PFM extract local features of targets and use the progressive fusion method to calculate the phase output feature maps y_a ($a = 1, 2, 3$). After being processed by the Ghost Model [33], y_a is doubled in size and element-wise added. This process greatly simplifies the task of small target detection by sharing the same weight for all convolution blocks and reduces the parameters of the P algorithm by using element-wise addition while reducing the network inference time.

Then, the fused output is upsampled and dimensionally cascaded through convolution calculation. We proposed a multi-scale fusion strategy to progressively fuse feature maps of different sizes. Furthermore, the confidence map O is obtained by performing the final threshold segmentation on the fused feather map. The backbone is mainly used to expand the receptive field and extract deep semantic features. Upsampling helps to restore the size of the feature map. The progressive multi-scale feature fusion is achieved by upsampling and downsampling. The FSM and PFM modules are used to ensure the feature representation of small targets in the network.

To achieve good context data modeling ability, the simplest way is to repeat and stack the network depth. The more layers the network has, the richer the semantic information and the larger the receptive field [34–37]. However, infrared small targets have significant differences in size and a very low pixel ratio. If the network depth is blindly increased, the problem of feature disappearance may occur after the drone target undergoes multiple downsampling operations. Therefore, we should design special modules to extract high-level features while ensuring the representation of small targets with a very small pixel ratio in the deep network.

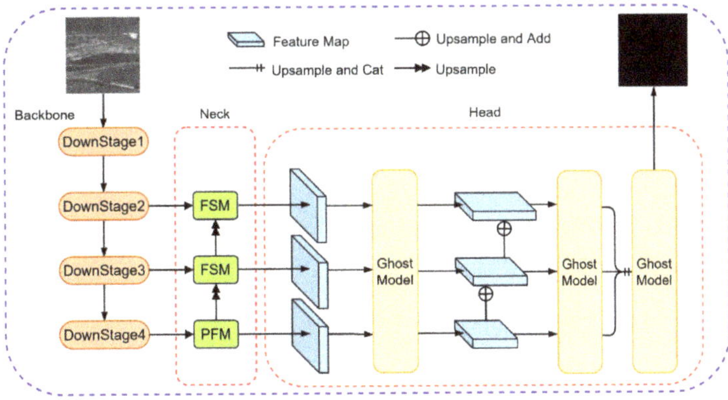

Figure 1. Structure and composition of PFFNET.

2.1. Feature Selection Module

We hope to improve the ability of channel information to interact between different downsampling layers based on a feature fusion perspective. Inspired by CBAM [38] and ShuffleNet [35], a new module called Feature Selection Module (FSM) is proposed to aggregate low-level and deep semantic information. The FSM is mainly divided into two parts: Location Selection Model (LSM) and Channel Selection Model (CSM). Experiments have shown that high-level semantic features contain detailed semantic information about the target, while low-level semantic features contain precise information about the target's location. As shown in Figure 2, FSM processes the different categories of information separately and uses the semantic information of each dimension to achieve information interaction between different coding graphs.

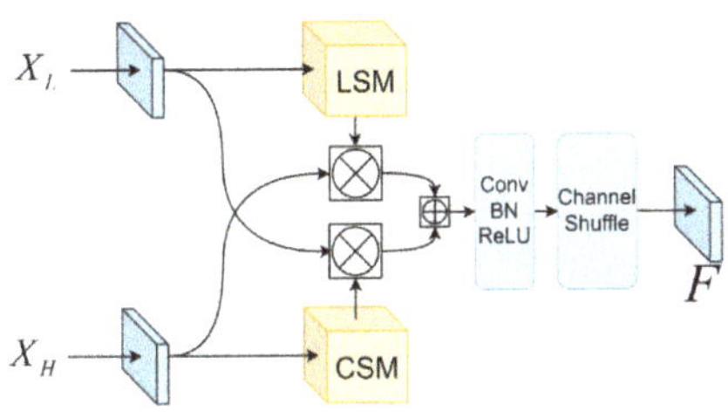

Figure 2. Structure and composition of Feature Selection Module.

To preserve the feature of drone targets in deep networks and encode the spatial details of targets' positions, this paper combines LSM and CSM. In order to cover more parts of the target to be recognized with features, this module first aggregates the channels via CSM. LSM was used to obtain more contour features and accurate location information. Then, the output of the current branch is multiplied by the input of another branch to enhance the high-low level channel features and facilitate multi-channel information interaction. Through 5×5 convolution we obtain a larger receptive field. Finally, the information exchange between different channels is established based on the Channel Shuffle idea. The output of the feature selection module $F = \mathbb{R}^{C \times H \times W}$ is represented as:

$$F\prime = \mu(C(X_H) \otimes X_L \oplus L(X_L) \otimes X_H) \quad (1)$$

$$F = \sigma(\varepsilon_{5\times5}(F\prime)) \quad (2)$$

where X_H is the deep feature that includes high-level semantic information, X_L is the shallow feature that contains rich image contour information and position information, $\otimes$ and $\oplus$ represents element-wise multiplication and addition of vectors, C and L represent the CSM and LSM modules, respectively. E is the convolution calculation. Σ represents the activation function of the Rectified Linear Unit. M is used to enhance feather representation and is a positive integer.

2.2. Channel Selection Model

To solve the problem of losing or weakening the target area response value during up-sampling of drone targets, the CSM is used to enhance the target area response amplitude.

As shown in Figure 3a, the channel features at each spatial position are individually aggregated. The high-frequency response channel weights of small targets are directionally enhanced to highlight the subtle details of deep targets. In Equation (3), this module performs average pooling and max pooling operations on the input feature map X_H to generate different 3D tensors x_{hi}. Coupling the global information of the feature map X in its internal channel. Then, a 1×1 convolution is used to evaluate the importance of each channel and calculate the corresponding weight. The aggregated output $H(X) \in \mathbb{R}^{C\times H\times W}$ can be represented as:

$$x_{h1} = \frac{1}{H \times W}\sum\nolimits_{w=h=1}^{H,W} X_H[:, w, h] \quad (3)$$

$$x_{h2} = \max(X_H[:, W, H]) \quad (4)$$

where x_{h1} and x_{h2} are the feature vector calculated by average pooling and maximum pooling. W and h represent the width and height of the feature map, respectively. The output of CSM as $C(X_H) \in \mathbb{R}^{C\times H\times W}$ is

$$C(X_H) = \sigma \sum_{i=1,2} \varepsilon_{1\times1}(\delta(\varepsilon_{1\times1}(x_{hi}))) \quad (5)$$

where δ represent the Sigmoid function. The output through $\varepsilon_{1\times1}$ are (c, c/r_f, 1, 1) and (c/r_f, c, 1, 1), respectively. R_f is the channel descent ratio.

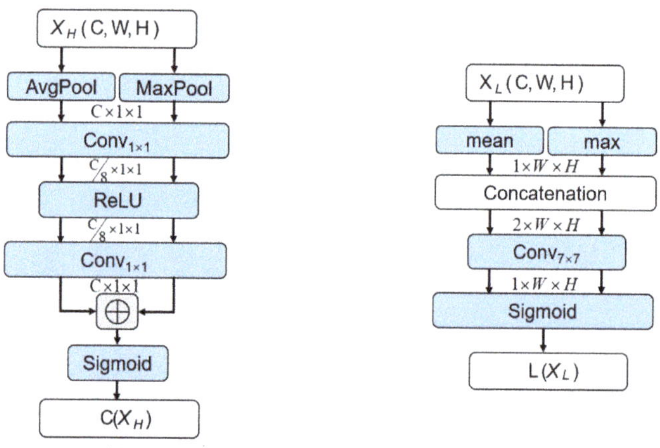

Figure 3. Structure and composition of CSM and LSM, (**a**) Channel selection model; (**b**) Location selection model.

2.3. Location Selection Model

The SCR of drone targets in infrared images is extremely low, which easily take interference signals into the process of feature extraction. The LSM could quickly locate visual salient regions. As shown in Figure 3b, this module calculates the maximum and mean values of the input feature map X_L, respectively.

$$x_{l1} = \frac{1}{C}\sum_{1}^{C} X_L[c,:,;] \tag{6}$$

$$x_{l2} = \max(X_L[c,:,;]) \tag{7}$$

where x_{l1} and x_{l2} represent the mean and maximum calculation of channel dimension. Perform cascading operations in the channel direction before performing convolution operations. Here, a 7×7 convolution can further expand the receptive field of the convolution kernel. It can also capture areas with higher local response amplitudes from the lower-level network. In addition, the accurate position of the drone target in the feature map is ensured. The output $L(X) \in \mathbb{R}^{C \times H \times W}$ can be calculated using the following equation:

$$L(X_L) = \delta(\varepsilon_{7\times7}(\mathbb{C}(x_{l1}, x_{l2}))) \tag{8}$$

where C represents the dimension cascade operation. The final output size of the feature map of this module is (1, W, H).

2.4. Pooling Pyramid Fusion Module

Deeper neural networks can obtain more detailed semantic information about the target, but this method is not suitable for smaller targets. As the number of downsamplings increases, the feature of drone targets (such as propellers and arms) weakens or even disappears. To solve this problem, this paper proposes a Pooling Pyramid Fusion Module (PFM) for drone target detection. Affected by [39], the PFM is only used to process the encoding map of the highest downsampling layer. Therefore, it can provide a more effective global context prior to pixel-level scene parsing. The PFM compresses spatial dimensions through different global adaptive pooling layer structures. Besides, the corresponding dimension mean value can be extracted to enhance the feature representation of small targets in deep networks.

As shown in Figure 4, the input feature map $I \in \mathbb{R}^{C \times H \times W}$ is parallelly input into the pyramid pooling module for decoding, generating four encoding structures of different sizes 1×1, 2×2, 3×3, and 6×6. Then, 1×1 convolution is used to reduce the feature dimension to $1/4C$. r_p is the channel descent ratio. The four feature maps of different sizes are upsampled by bilinear interpolation. Then concatenating with the input feature map in the channel dimension. Finally, a 3×3 convolution is performed to output the feature map $O \in \mathbb{R}^{C \times H \times W}$, and form a contextual pyramid through five feature maps of the same dimension but different scales.

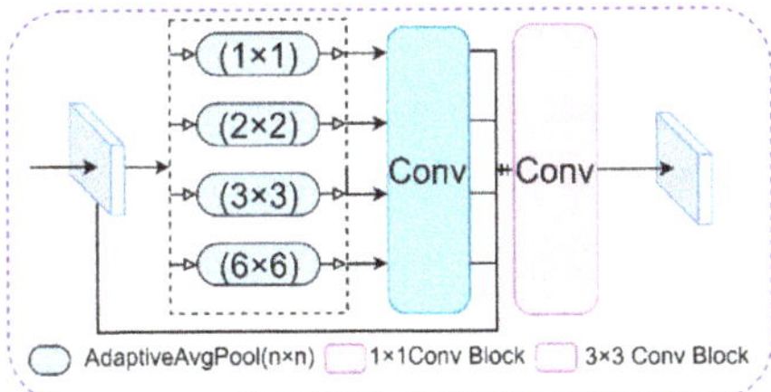

Figure 4. Structure and composition of PFM.

2.5. Segmentation Head

After multiple downsampling and convolution calculations, the targets' feature response in the deepest layer of the convolutional network will weaken. In response to this problem, we proposed a progressive feature fusion structure that is better suited for drones. This segmentation head is designed and improved based on FPN [40]. It can fuse different sizes of feature maps and enable the stacking of information between high and low layers to enhance the high-frequency response amplitude of the target.

As shown in Figure 5, the input I with different sizes proceeds through the Ghost Model. Due to the small proportion of drone targets in the image, ordinary convolution calculations can generate a large number of feature maps with the same texture information or even without drone target information. However, the Ghost Model could generate encoding maps with the same number and texture information through simple linear calculations. This step reduces the convolution parameter volume and improves training and inference efficiency.

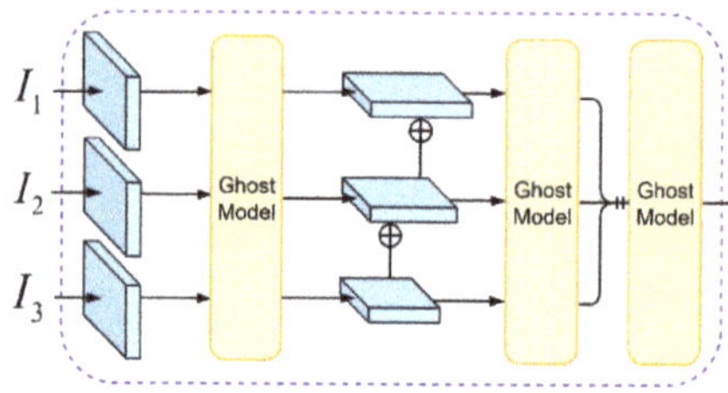

Figure 5. Structure and composition of segmentation head.

Finally, we train the entire network by *SoftIoULoss* and *CELoss* in PFFNet and optimize the weighted loss between the predicted and segmented images. They can be expressed by the following equations:

$$CELoss = -\sum_{cls} T\log(P) \tag{9}$$

$$SoftIoULoss_{smooth} = \frac{\sum_{pixels} TP + smooth}{\sum_{pixels} (T + P - TP) + smooth} \tag{10}$$

$$Loss = \alpha(1 - IOU) + \beta(1 - CELoss) \tag{11}$$

where T and P represent the pixel values of the real target and the output prediction, respectively. Based on the initial loss value during training, $\alpha = 3$ and $\beta = 1$ were set to balance the individual loss with the total loss to optimize the algorithm in the expected direction. To ensure stability in the calculation, this paper sets $smooth = 1$. Different weight balances may affect performance indicators [10].

3. Experiments

This section mainly introduces the implementation details and evaluation metrics of the algorithm and compares it with other methods on two different datasets. In order to verify the effectiveness of the data-driven model PFFNet, comparative experiments and ablation experiments were conducted, respectively.

3.1. Datasets

Performance-based on data-driven methods is highly affected by the quality, quantity, and diversity of the data. Infrared image datasets have fewer images compared to visible datasets. Most methods are trained and evaluated on their own private datasets. Wang [31] established an open infrared small target dataset that includes 10,000 images. However, many of the targets in the dataset are too large and the annotations are not accurate which affects the training effectiveness. Dai [28] released a dataset with high-quality semantic segmentation masks. However, the dataset is small, which easily leads to unstable model training, overfitting, and model convergence problems. To verify the reliability and robustness of the algorithm, experiments were conducted on two publicly available datasets with different image sizes (SIRST Aug [30] and IRSTD 1k [41]). The dataset [30] includes 8525 images in the training set and 545 images in the test set with an image size of 256 × 256, which is sufficient to satisfy the training requirements of data-driven models. The image size in the dataset [41] is 512 × 512 and includes different types of small targets such as drones, organisms, ships, and vehicles. This dataset also covers many different scenes, including seawater, fields, mountains, cities, and clouds, with a cluttered background and severe noise. Therefore, it is sufficient to verify the proposed detection method.

3.2. Experimental Preparation and Evaluation Method

PFFNet will conduct ablation experiments and multi-algorithm comparison experiments on the SIRST Aug and IRSTD 1k open datasets.

This paper uses classic semantic segmentation evaluation indicators such as F1-score, receiver operating characteristic curve (ROC), and Intersection over Union (IoU). To measure the connection between precision and recall, *F1-score* is introduced. Meaning that the network must be able to detect targets and ensure as few false alarms as possible. ROC is a qualitative indicator that reflects the connection between the target detection rate (P_d) and the false alarm rate (P_f). *Precision*, *recall*, target P_d, and P_f are defined as follows:

$$Precision = \frac{T_P}{T_P + F_P}, P_d = Recall = \frac{T_P}{T_P + F_N}, P_F = \frac{F_P}{N} \tag{12}$$

where T_P represents the target pixels that are correctly matched with the true label by the predicted pixels. F_P represents the background label pixels that are incorrectly predicted as targets. F_N represents the number of target pixels that are incorrectly classified as background. N represents the total number of pixels in the image. *F1-score* and IoU can be defined as:

$$F1 - score = \frac{2 * Precision * Recall}{Precision + Recall}, \text{IoU} = \frac{T_P}{T + P - T_P} \tag{13}$$

PFFNet is implemented based on Pytorch. The optimizer uses stochastic gradient descent (SGD), with momentum and weight decay coefficients set to 0.9 and 0.0001, respectively. The initial learning rate is 0.05, and a poly decay strategy is used. In SIRST Aug, the batch size is set to 32, and 30 epochs are trained. In IRSTD 1k, the batch size is set to 8, and 150 epochs are trained. In terms of hardware, we use a Tesla P100 GPU for training and a 3060 GPU for inference.

3.3. Comparative Experiments

We compared PFFNet with four classic methods of different types. Results in [28,30] show that data-driven methods are superior to model-driven methods. Therefore, only data-driven methods are compared in our experiments. In the data-driven scheme, we select AGPCNet, ACM, MDFA, ALC and PFFNet for comparison. PFFNet-S and PFFNet-R represent the selection of Swin Transformer V1 [42] and ResNet-18 [43] as the global feature extractors, respectively. The purpose is to verify whether the algorithm is compatible with different feature extraction structures. The hyperparameters of other algorithms are not changed and remain at their default settings.

As shown in Table 1, the maximum and secondary values of each column are highlighted in bold and underlined, respectively. PFFNet achieved the best detection performance on both different datasets, with an IoU of up to 73.7% on the SIRST AUG dataset. PFFNet-R and AGPCNet were selected for comparison because they both chose ResNet-18 as their global feature extractor. Obviously, compared to AGPCNet, PFFNet-R increased IoU by 8.9% (73.7 vs. 67.6) and 5.5% (66.1 vs. 62.6) on both datasets, respectively. This method reduces detection time by 55.8% (0.01 vs. 0.05) while maintaining detection accuracy. Compared to ALC, PFFNet-S increased IoU by 2.5% (73.7 vs. 71.9) and 3.2% (64.2 vs. 62.2), respectively, and reduced the detection time by 81% (0.01 vs. 0.06). Furthermore, we calculated the average time of different methods on 1000 infrared images, among which PFFNet-S is 6 ms slower than ACM, but PFFNet-S has a better detection performance.

Table 1. Comparisons with other methods on SIRST Aug and IRSTD-1k.

Methods	SIRST Aug (%)				IRSTD 1k (%)				Time on GPU/s
	Precision	Recall	IoU	F1-Score	Precision	Recall	IoU	F1-Score	
ACM	87.2	70.7	64.1	78.1	76.6	74.8	60.9	75.7	**0.005**
ALC	88	74.9	71.9	81	80.3	73.4	62.2	76.7	0.058
MDFA	81.1	65.4	56.7	72.4	66.5	70	51.7	68.2	0.064
AGPCNet	87.7	74.7	67.6	80.7	76.1	78	62.6	77	0.052
PFFNet-S	81	**89**	73.7	84.8	78.4	**77.9**	64.2	78.2	0.011
PFFNet-R	**88.7**	81.3	**73.7**	**84.9**	**81.8**	77.4	**66.1**	**79.6**	0.023

Besides, it can also be seen that each algorithm performs significantly better on the SIRST Aug dataset compared to IRSTD 1k. Just like the IoU values of PFFNet-S on different datasets (73.7% on SIRST Aug dataset and 64.2% on IRSTD 1k dataset). There are two reasons for this result:

(1) Number of data used for model training

The results of detection based on data-driven methods are affected by the number of data. The SIRST Aug dataset is used to train models with more data than IRSTD 1k (8525:800). Due to the higher resolution of the images contained in IRSTD 1k (512 × 512), good detection results were also achieved even with a small amount of data. The higher the image resolution, the better the detection result, which has been proven [44].

(2) The complexity of the data

The IRSTD 1k dataset contains more challenging scenarios for detection. For example, drone targets with different shapes, low contrast, low signal-to-noise ratio, as well as more complex backgrounds and more noise interference. The complexity of its data is much higher than SIRST Aug.

Furthermore, to visually compare the AUC, the ROC curves of these methods on two different datasets are shown in Figure 6. Experimental results demonstrate that PFFNet can greatly suppress the background. This method fully learns highly discriminative semantic features from diverse training data to achieve highly robust object detection results. In addition, it can segment targets more accurately than other state-of-the-art methods.

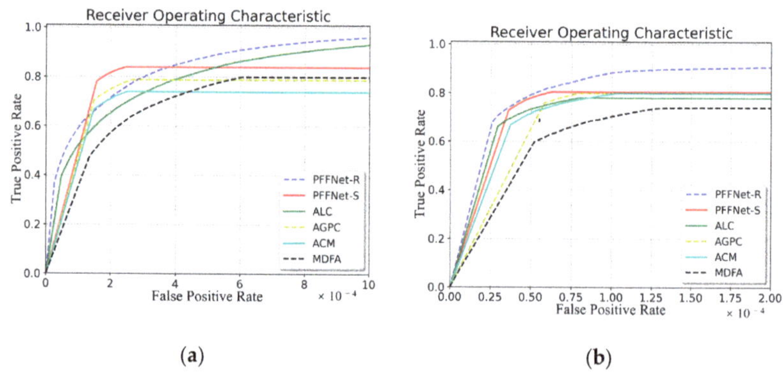

Figure 6. The ROC curve compared with other methods on different data sets, (**a**) SIRST Aug dataset; (**b**) IRSTD 1k dataset.

This paper also selects four types of data for visual evaluation of the above algorithms. As shown in Figure 7, Introducing infrared images and GT (Ground Truth) for four different scenarios. The 'ori' represents the original images and the 'after' represents the enhanced ones. The local contrast between the target and background is extremely low in Figure 7a,d. Figure 7b,c have a high local contrast with a complex background. Meanwhile, Figure 7a,c have multiple drone targets.

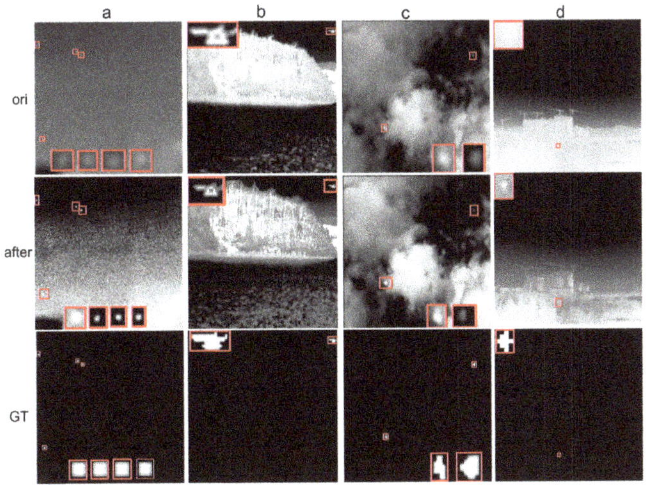

Figure 7. Four scenarios with drones and their GT.

Figure 8 shows the output of the GT map with five different methods. Although the enhanced after-a has stronger local contrast, ACM and MDFA still cannot detect all targets. When there are few targets (after-b) or the background is relatively simple (after-c), all algorithms have good performance. For after-d, although the image features have been enhanced, false alarms (ACM, ALC) or missed detections (MDFA) still occur. In addition, both AGPCNet and PFFNet can detect all targets, although PFFNet has a shorter detection time. Obviously, PFFNet is more suitable for drone target detection.

3.4. Ablation Experiments

To investigate the effectiveness of each module in PFFNet, we conducted multiple ablation studies using SIRST Aug and IRSTD 1k. Although ResNet-18 has better performance, it has a longer training and inference time. Therefore, we choose Swin Transformer v1, which performed well and had a shorter detection time as the backbone for ablation.

In Table 2, ablation experiments were conducted on Segmentation Head, FSM, and PFM, respectively. First, for Swin Transformer v1, after adding the Segmentation Head module, the IoU increased by 49.1% (71.3 vs. 47.8) and 49.3% (60.3 vs. 40.4). It can prove the effectiveness of using different downsampling output encoding and multi-scale fusion when segmenting targets. To verify the effectiveness of other modules, we use the Backbone and Segmentation Head modules as baselines to fine-tune the model. After adding the FSM module, the IoU increased by 3% (73.4 vs. 71.3) and 5% (63.4 vs. 60.3). Therefore, the detection accuracy could be improved by enhancing the information exchange between high-level and low-level features. After adding the PFM module, the IoU increased by 0.52% (71.6 vs. 71.3) and 1.8% (61.3 vs. 60.3). Although the texture information of the top-level target is more abundant, there is a lack of effective global contextual priors. Using PFM modules to enhance global feature representation can achieve better performance.

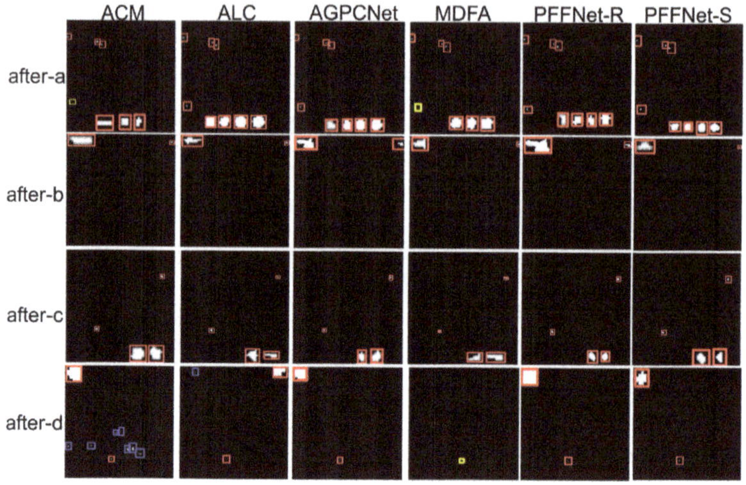

Figure 8. Visual results of different methods. The close-up attempts to display the zoomed-in targets or detection results. Boxes in red, yellow and blue represent correctly detected targets, miss detected targets and false detected targets, respectively. After-(a,b,c,d) indicates the image (a,b,c,d) after enhancement.

Table 2. Ablation of the Segmentation Head, FSM and PFM.

Segmentation Head	FSM	PFM	SIRST Aug(%)		IRSTD 1k(%)	
			IoU	F1-Score	IoU	F1-Score
			47.5	55	40.4	48.7
√			71.3	82.4	60.3	75.3
√	√		**73.4**	**84.7**	**63.4**	**77.6**
√		√	71.6	83.1	61.3	76

For FSM, as shown in Table 3. CSM is used to directly enhance the response weight of targets of the upsampling layer. Although good results can be achieved (71.3 vs. 72.5, 62 vs. 60.3), effective use of lower layer feature maps can achieve better results (72.8 vs. 71.3, 62.6 vs. 60.3). This is because the lower sampling output contains more information of target location and drone contour. Therefore, using different strategies for different sampling output results is beneficial for drone target detection.

Table 3. Ablation of the CSM and LSM in FSM.

CSM	LSM	SIRST Aug (%)				IRSTD 1k (%)			
		IoU	F1-Score	Precison	Recall	IoU	F1-Score	Precison	Recall
		71.3	82.4	77.1	88.6	60.3	75.3	69.8	81.7
√		72.5	83.6	78.6	90.3	62	76.5	72.4	81.2
	√	72.8	84.3	78	91.7	62.6	77.3	75.5	79.1

Dimensionality reduction can reduce redundant information and greatly accelerate the training speed in the network. However, this may result in the loss of useful information. In PFFNet, two dimensionality reductions were performed in FSM and PFM separately, and the reduction ratios were denoted as (r_f, r_p). We choose different dimensionality reduction ratios to explore the best way to segment small infrared targets. According to [38], we set $r_f = 8$. As shown in Table 4, the best result was IoU (73.7%,64.2%) when $r_f = 8$ and $r_p = 4$.

Table 4. Ablation on Dimensionality Reduction Ratio.

Reduction Ratios	SIRST Aug (%)				IRSTD 1k (%)			
	IoU	F1-Score	Precision	Recall	IoU	F1-Score	Precision	Recall
(8,1)	71.8	83.6	77.8	90.3	63.8	77.9	76.6	79.2
(8,2)	73.1	84.4	80.1	89.3	64	78.1	75.7	80.6
(8,4)	73.7	84.8	81	89	64.2	78.2	78.4	77.9
(8,8)	70.6	82.8	74.8	92.7	62.8	77.1	74.2	80.3

With the network deepening and the number of channels increasing, more complex features are learned. In order to enhance the network's expressive ability, it is necessary to cover as many key features as possible. However, it does not apply to all kinds of target detection, such as small drone targets, which have small size and low signal-to-noise ratio. After multiple downsampling, there may be repeated features or even feature loss. Setting a smaller ratio of dimensionality reduction becomes meaningless, such as $r_p = 1$ or 2. Alternatively, choosing a larger ratio of dimensionality reduction may result in the model being unable to learn more complex features, such as $r_p = 8$, which is clearly not what we expected. Therefore, an appropriate proportion with different kinds of targets should be chosen to enhance the expressive ability of the model.

4. Discussion

From the above experiments, MDFA divides the generator into miss detection and false alarm and uses GAN to detect infrared targets. However, the results of the downsampling cannot be used effectively. Additionally, MDFA can easily lead to false alarms and miss-detection when the local contrast is low. ACM was the first to use segmentation methods to detect infrared targets, providing a theoretical basis and experimental data for researchers. However, it has many limitations, such as the global information could not be used effectively. ALC combines data-driven and model-driven methods to achieve high precision. However, it ignored the effect of the target's surrounding environment because of only enlarged the scale and enhanced the local contrast. AGPCNet used attention modules and self-attention-guided algorithms to learn more contextual information. However, it can easily lead to miss detection when the target is in a complex environment. Overall, the above methods do not balance the detection accuracy and time cost. Table 5 shows the comparison of all methods.

Table 5. Advantages and disadvantages of all methods (H-high, L-low).

Methods	Advantage	Disadvantage
ACM	L-time	H-FA&MD, L-accuracy
ALC	H-accuracy	H-time, L-recall
MDFA	H-robustness	H-FA&MD, H-time
AGPCNet	H-recall	H-time, L-accuracy
Ours	H-robustness and accuracy	L-FA

In our method, ablation was used to verify the role of each module. For drone targets, the segmentation head was designed to achieve progressive multi-scale fusion. In addition, the problem of feature disappearance was solved. Then, considering the influence of global context prior on the top layer output of the network, the PFM was used to enhance drone feature representation in networks. Finally, FSM was designed to promote information exchange between layers to improve detection accuracy.

5. Conclusions

This paper proposes a fast detection method for infrared small targets: Progressive Feature Fusion Network (PFFNet). The FSM was used to promote the use of shallow features and enhance upsampling output amplitude of the target. Then, the high-low level encoding outputs are fused to improve the accuracy of the model. The PFM was used to process the output coding diagram of the highest downsampling layer in the backbone and provide more effective global contextual priors for pixel-level scene parsing, guiding algorithms to learn more target features. Then, from the perspective of multi-scale fusion, we design a lightweight segmentation head to adapt to drone targets based on FPN. Enhance the ability to detect small targets by progressively fusing low-level and high-level semantics. Comparative and Ablation experiments demonstrate that PFFNet has the ability to accurately detect drones in complex scenes. Additionally, it can balance the detection accuracy and detection time. This study could promote drone target detection and even expand drone remote sensing segmentation.

However, there are still some problems with the algorithm that need further research, such as dealing with network overfitting, utilizing more efficient contextual information, etc. In future work, RGB images and fusion images will continue to be explored for their application in infrared drone detection.

Author Contributions: Conceptualization, C.Z. and Z.H.; methodology, Z.H.; software, Z.H.; validation, Z.H., K.Q. and M.Y.; formal analysis, M.Y.; investigation, H.F.; resources, C.Z.; data curation, C.Z.; writing—original draft preparation, Z.H.; writing—review and editing, C.Z.; visualization, Z.H.; supervision, C.Z.; project administration, C.Z.; funding acquisition, C.Z. and M.Y. All authors have read and agreed to the published version of the manuscript.

Funding: This research was funded by Liaoning Provincial Department of Education, grant number LJKMZ20220605.

Institutional Review Board Statement: Not applicable.

Informed Consent Statement: Not applicable.

Data Availability Statement: Not applicable.

Conflicts of Interest: The authors declare no conflict of interest.

References

1. Kapoulas, I.K.; Hatziefremidis, A.; Baldoukas, A.K.; Valamontes, E.S.; Statharas, J.C. Small Fixed-Wing UAV Radar Cross-Section Signature Investigation and Detection and Classification of Distance Estimation Using Realistic Parameters of a Commercial Anti-Drone System. *Drones* **2023**, *7*, 39. [CrossRef]
2. Li, B.; Song, C.; Bai, S.; Huang, J.; Ma, R.; Wan, K.; Neretin, E. Multi-UAV Trajectory Planning during Cooperative Tracking Based on a Fusion Algorithm Integrating MPC and Standoff. *Drones* **2023**, *7*, 196. [CrossRef]
3. Zhao, M.; Cheng, L.; Yang, X.; Feng, P.; Liu, L.; Wu, N. TBC-Net: A Real-Time Detector for Infrared Small Target Detection Using Semantic Constraint. *arXiv* **2019**, arXiv:2001.05852.
4. Hou, Q.; Wang, Z.; Tan, F.; Zhao, Y.; Zheng, H.; Zhang, W. RISTDnet: Robust Infrared Small Target Detection Network. *IEEE Geosci. Remote Sens. Lett.* **2022**, *19*, 1–5. [CrossRef]
5. Liao, K.-C.; Wu, H.-Y.; Wen, H.-T. Using Drones for Thermal Imaging Photography and Building 3D Images to Analyze the Defects of Solar Modules. *Inventions* **2022**, *7*, 67. [CrossRef]
6. Li, B.; Yang, Z.-P.; Chen, D.-Q.; Liang, S.-Y.; Ma, H. Maneuvering target tracking of UAV based on MN-DDPG and transfer learning. *Def. Technol.* **2021**, *17*, 457–466. [CrossRef]
7. Fernández, A.; Usamentiaga, R.; de Arquer, P.; Fernández, M.; Fernández, D.; Carús, J.L.; Fernández, M. Robust Detection, Classification and Localization of Defects in Large Photovoltaic Plants Based on Unmanned Aerial Vehicles and Infrared Thermography. *Appl. Sci.* **2020**, *10*, 5948. [CrossRef]
8. Chen, F.; Gao, C.; Liu, F.; Zhao, Y.; Zhou, Y.; Meng, D.; Zuo, W. Local Patch Network with Global Attention for Infrared Small Target Detection. *IEEE Trans. Aerosp. Electron. Syst.* **2022**, *58*, 3979–3991. [CrossRef]
9. Ying, X.; Wang, Y.; Wang, L.; Sheng, W.; Liu, L.; Lin, Z.; Zhou, S. Local Motion and Contrast Priors Driven Deep Network for Infrared Small Target Superresolution. *IEEE J. Sel. Top. Appl. Earth Obs. Remote Sens.* **2022**, *15*, 5480–5495. [CrossRef]
10. Chen, Y.; Li, L.; Liu, X.; Su, X. A Multi-Task Framework for Infrared Small Target Detection and Segmentation. *IEEE Trans. Geosci. Remote Sens.* **2022**, *60*, 1–9. [CrossRef]
11. Wang, C.; Meng, L.; Gao, Q.; Wang, J.; Wang, T.; Liu, X.; Du, F.; Wang, L.; Wang, E. A Lightweight Uav Swarm Detection Method Integrated Attention Mechanism. *Drones* **2022**, *7*, 13. [CrossRef]
12. Li, B.; Xiao, C.; Wang, L.; Wang, Y.; Lin, Z.; Li, M.; An, W.; Guo, Y. Dense Nested Attention Network for Infrared Small Target Detection. *IEEE Trans. Image Process.* **2022**, *32*, 1745–1758. [CrossRef]

13. Bai, X.; Zhou, F. Analysis of new top-hat transformation and the application for infrared dim small target detection. *Pattern Recognit.* **2010**, *43*, 2145–2156. [CrossRef]
14. Chang, B.; Meng, L.; Haber, E.; Ruthotto, L.; Begert, D.; Holtham, E. Reversible Architectures for Arbitrarily Deep Residual Neural Networks. *Proc. Conf. AAAI Artif. Intell.* **2018**, *32*, 2811–2818. [CrossRef]
15. Rivest, J.; Fortin, R. Detection of dim targets in digital infrared imagery by morphological image processing. *Opt. Eng.* **1996**, *35*, 1886–1893. [CrossRef]
16. Chen, C.L.P.; Li, H.; Wei, Y.; Xia, T.; Tang, Y.Y. A Local Contrast Method for Small Infrared Target Detection. *IEEE Trans. Geosci. Remote Sens.* **2014**, *52*, 574–581. [CrossRef]
17. Wei, Y.; You, X.; Li, H. Multiscale patch-based contrast measure for small infrared target detection. *Pattern Recognit.* **2016**, *58*, 216–226. [CrossRef]
18. Han, J.; Ma, Y.; Zhou, B.; Fan, F.; Liang, K.; Fang, Y. A Robust Infrared Small Target Detection Algorithm Based on Human Visual System. *IEEE Geosci. Remote Sens. Lett.* **2014**, *11*, 2168–2172. [CrossRef]
19. Han, J.; Moradi, S.; Faramarzi, I.; Liu, C.; Zhang, H.; Zhao, Q. A Local Contrast Method for Infrared Small-Target Detection Utilizing a Tri-Layer Window. *IEEE Geosci. Remote Sens. Lett.* **2019**, *17*, 1822–1826. [CrossRef]
20. Zhang, L.; Peng, Z. Infrared Small Target Detection Based on Partial Sum of the Tensor Nuclear Norm. *Remote Sens.* **2019**, *11*, 382. [CrossRef]
21. Zhu, H.; Liu, S.; Deng, L.; Li, Y.; Xiao, F. Infrared Small Target Detection via Low-Rank Tensor Completion with Top-Hat Regularization. *IEEE Trans. Geosci. Remote Sens.* **2020**, *58*, 1004–1016. [CrossRef]
22. Dai, Y.; Wu, Y.; Song, Y.; Guo, J. Non-negative infrared patch-image model: Robust target-background separation via partial sum minimization of singular values. *Infrared Phys. Technol.* **2017**, *81*, 182–194. [CrossRef]
23. Fu, J.; Liu, J.; Tian, H.; Li, Y.; Bao, Y.; Fang, Z.; Lu, H. Dual Attention Network for Scene Segmentation. *arXiv* **2019**, arXiv:1809.02983.
24. Zheng, Z.; Wang, P.; Liu, W.; Li, J.; Ye, R.; Ren, D. Distance-IoU Loss: Faster and Better Learning for Bounding Box Regression. *arXiv* **2019**, arXiv:1911.08287. [CrossRef]
25. Liu, W.; Anguelov, D.; Erhan, D.; Szegedy, C.; Reed, S.; Fu, C.-Y.; Berg, A.C. SSD: Single Shot MultiBox Detector. *Eur. Conf. Comput. Vis.* **2016**, *9905*, 21–37. [CrossRef]
26. Wang, C.; Shi, Z.; Meng, L.; Wang, J.; Wang, T.; Gao, Q.; Wang, E. Anti-Occlusion UAV Tracking Algorithm with a Low-Altitude Complex Background by Integrating Attention Mechanism. *Drones* **2022**, *6*, 149. [CrossRef]
27. Liu, S.; Qi, L.; Qin, H.; Shi, J.; Jia, J. Path Aggregation Network for Instance Segmentation. *arXiv* **2018**, arXiv:1803.01534.
28. Dai, Y.; Wu, Y.; Zhou, F.; Barnard, K. Asymmetric Contextual Modulation for Infrared Small Target Detection. *arXiv* **2020**, arXiv:2009.14530.
29. Dai, Y.; Wu, Y.; Zhou, F.; Barnard, K. Attentional Local Contrast Networks for Infrared Small Target Detection. *IEEE Trans. Geosci. Remote Sens.* **2021**, *59*, 9813–9824. [CrossRef]
30. Zhang, T.; Cao, S.; Pu, T.; Peng, Z. AGPCNet: Attention-Guided Pyramid Context Networks for Infrared Small Target Detection. *arXiv* **2021**, arXiv:2111.03580.
31. Wang, H.; Zhou, L.; Wang, L. Miss Detection vs. False Alarm: Adversarial Learning for Small Object Segmentation in Infrared Images. In Proceedings of the 2019 IEEE/CVF International Conference on Computer Vision (ICCV), Seoul, Republic of Korea, 27 October–2 November 2019; IEEE: Piscataway, NJ, USA, 2019; pp. 8508–8517. [CrossRef]
32. Cheng, Q.; Wang, H.; Zhu, B.; Shi, Y.; Xie, B. A Real-Time UAV Target Detection Algorithm Based on Edge Computing. *Drones* **2023**, *7*, 95. [CrossRef]
33. Han, K.; Wang, Y.; Tian, Q.; Guo, J.; Xu, C.; Xu, C. GhostNet: More Features from Cheap Operations. *arXiv* **2020**, arXiv:1911.11907.
34. Howard, A.; Sandler, M.; Chu, G.; Chen, L.C.; Chen, B.; Tan, M.; Wang, W.; Zhu, Y.; Pang, R.; Vasudevan, V.; et al. Searching for MobileNetV3. In Proceedings of the 2019 IEEE/CVF International Conference on Computer Vision (ICCV), Seoul, Republic of Korea, 27 October–2 November 2019; pp. 1314–1324.
35. Ma, N.; Zhang, X.; Zheng, H.-T.; Sun, J. ShuffleNet V2: Practical Guidelines for Efficient CNN Architecture Design. *arXiv* **2018**, arXiv:1807.11164. [CrossRef]
36. Xiong, Y.; Liu, H.; Gupta, S.; Akin, B.; Bender, G.; Wang, Y.; Kindermans, P.J.; Tan, M.; Singh, V.; Chen, B. MobileDets: Searching for Object Detection Architectures for Mobile Accelerators. In Proceedings of the 2021 IEEE/CVF Conference on Computer Vision and Pattern Recognition (CVPR), Nashville, TN, USA, 20–25 June 2021; pp. 3824–3833.
37. Huang, G.; Liu, Z.; van der Maaten, L.; Weinberger, K.Q. Densely Connected Convolutional Networks. *arXiv* **2018**, arXiv:1608.06993.
38. Woo, S.; Park, J.; Lee, J.-Y.; Kweon, I.S. CBAM: Convolutional Block Attention Module. *arXiv* **2018**, arXiv:1807.06521.
39. Zhao, H.; Shi, J.; Qi, X.; Wang, X.; Jia, J. Pyramid Scene Parsing Network. In Proceedings of the 2017 IEEE Conference on Computer Vision and Pattern Recognition (CVPR), Honolulu, HI, USA, 21–26 July 2017; IEEE: Piscataway, NJ, USA, 2017; pp. 6230–6239. [CrossRef]
40. Lin, T.-Y.; Dollar, P.; Girshick, R.; He, K.; Hariharan, B.; Belongie, S. Feature Pyramid Networks for Object Detection. In Proceedings of the 2017 IEEE Conference on Computer Vision and Pattern Recognition (CVPR), Honolulu, HI, USA, 21–26 July 2017; IEEE: Piscataway, NJ, USA, 2017; pp. 936–944. [CrossRef]

41. Zhang, M.; Zhang, R.; Yang, Y.; Bai, H.; Zhang, J.; Guo, J. ISNet: Shape Matters for Infrared Small Target Detection. In Proceedings of the 2022 IEEE/CVF Conference on Computer Vision and Pattern Recognition (CVPR), New Orleans, LA, USA, 18–24 June 2022; IEEE: Piscataway, NJ, USA, 2022; pp. 867–876. [CrossRef]
42. Liu, Z.; Lin, Y.; Cao, Y.; Hu, H.; Wei, Y.; Zhang, Z.; Lin, S.; Guo, B. Swin Transformer: Hierarchical Vision Transformer Using Shifted Windows. *arXiv* **2021**, arXiv:2103.14030.
43. He, K.; Zhang, X.; Ren, S.; Sun, J. Deep Residual Learning for Image Recognition. *arXiv* **2015**, arXiv:1512.03385.
44. Chen, L.-C.; Papandreou, G.; Schroff, F.; Adam, H. Rethinking Atrous Convolution for Semantic Image Segmentation. *arXiv* **2017**, arXiv:1706.05587.

drones

MDPI

Article

Hierarchical Maneuver Decision Method Based on PG-Option for UAV Pursuit-Evasion Game

Bo Li [1], Haohui Zhang [1], Pingkuan He [1], Geng Wang [1,*], Kaiqiang Yue [1] and Evgeny Neretin [2]

1 School of Electronics and Information, Northwestern Polytechnical University, Xi'an 710072, China; libo803@nwpu.edu.cn (B.L.); zhanghaohui@mail.nwpu.edu.cn (H.Z.); npuhpk@163.com (P.H.); ykq15929955434@163.com (K.Y.)

2 School of Robotic and Intelligent Systems, Moscow Aviation Institute, 125993 Moscow, Russia; e.s.neretin@mai.ru

* Correspondence: wanggeng@nwpu.edu.cn; Tel.: +86-133-8922-3600

Abstract: Aiming at the autonomous decision-making problem in an Unmanned aerial vehicle (UAV) pursuit-evasion game, this paper proposes a hierarchical maneuver decision method based on the PG-option. Firstly, considering various situations of the relationship of both sides comprehensively, this paper designs four maneuver decision options: advantage game, quick escape, situation change and quick pursuit, and the four options are trained by Soft Actor-Critic (SAC) to obtain the corresponding meta-policy. In addition, to avoid high dimensions in the state space in the hierarchical model, this paper combines the policy gradient (PG) algorithm with the traditional hierarchical reinforcement learning algorithm based on the option. The PG algorithm is used to train the policy selector as the top-level strategy. Finally, to solve the problem of frequent switching of meta-policies, this paper sets the delay selection of the policy selector and introduces the expert experience to design the termination function of the meta-policies, which improves the flexibility of switching policies. Simulation experiments show that the PG-option algorithm has a good effect on UAV pursuit-evasion game and adapts to various environments by switching corresponding meta-policies according to current situation.

Keywords: UAV pursuit-evasion game; hierarchical reinforcement learning; meta-policy; policy gradient

Citation: Li, B.; Zhang, H.; He, P.; Wang, G.; Yue, K.; Neretin, E. Hierarchical Maneuver Decision Method Based on PG-Option for UAV Pursuit-Evasion Game. *Drones* **2023**, 7, 449. https://doi.org/10.3390/drones7070449

Academic Editor: Diego Gonzalez-Aguilera

Received: 23 April 2023
Revised: 30 June 2023
Accepted: 4 July 2023
Published: 6 July 2023

1. Introduction

Unmanned aerial vehicles (UAVs) [1–7] are used in many fields, such as intelligent confrontation [8], target rounding [9] and intelligent transportation [10], because of their characteristics of being unmanned, having good concealment and having no casualties. UAV pursuit-evasion [11] involves a game between two UAVs with competing interests. In the process of UAV pursuit-evasion, being able to make effective maneuvering decisions [12] to destroy the other side and capture the other side is the key to victory. Among these, the real-time intelligent maneuvering decision-making ability of UAV is the core of problem solving. The maneuvering decision-making mechanism reflects the intelligence level of a UAV in the pursuit-evasion game. Therefore, it is necessary to design an effective maneuvering policy in the process of the UAV pursuit-evasion game.

At present, decision algorithms in UAV pursuit-evasion mainly include differential game theory [13], influence graph method [14], heuristic search algorithm [15], etc. F. Yu et al. [13] take into account the impact of environmental impediments in the pursuit-evasion game between UAVs and UGVs, qualitatively assess the pursuit problem of the difference game and use the differential game in the pursuit-evasion game. Q. Pan et al. [14] propose a cooperative maneuver decision method for multiple unmanned aerial vehicles based on the influence graph theory. A state predicted influence diagram model is used to analyze elements, and an unscented Kalman filter model is used for belief state updating. Mikhail et al. [15] propose schemes to solve the pursuit-evasion problem using Apollonius circles and UAVs in a non-deterministic environment. However, the disadvantages of traditional UAV pursuit-evasion decision methods are that the rule base is complicated and difficult to cover all pursuit-evasion situations, and it has poor flexibility.

Since the 21st century, the development of artificial intelligence [16] has reached its climax. Artificial intelligence methods such as deep learning [17] and reinforcement learning [18] have been well applied in many high-tech fields. The application of artificial intelligence to UAV pursuit-evasion has also achieved good results. Researchers regard UAV as an agent in reinforcement learning, enabling them to gradually acquire optimized polices in complex environments after a certain stage of trial-and-error learning in the environment [19]. Deep reinforcement learning combines the perceptual ability of deep learning with the decision-making ability of reinforcement learning in a general form, providing a way to solve the problem of UAV pursuit-evasion [20–24]. Zhang et al. [20] design a guiding reward function based on the Deep Deterministic Policy Gradient (DDPG) algorithm for the UAV pursuit-evasion task and introduce a soft update strategy based on a sliding average. A UAV swarm can successfully carry out pursuit missions. However, the game between multi-UAVs and one UAV results in a significant advantage gap between the opposing sides. Additionally, the two-dimensional simulation environment differs greatly from actual pursuit-evasion tasks. Zhang et al. [21] construct a multi-agent coronal bidirectional coordinated target prediction network (CBC-TP network) by vectorial extension of the multi-agent depth Deterministic strategy gradient formula, so as to realize the process of UAV pursuit-evasion. However, the pursuit-evasion game environment is singular, and this paper does not comprehensively consider various situations. Fu et al. [22] propose IL-DDPG, which combines the DDPG algorithm and imitation learning algorithm. At the same time, the proportional guidance law is introduced as a guidance strategy. Compared with the DDPG algorithm, it improves the search efficiency and realizes the fast tracking of a pursuing UAV to avoid a UAV. However, the two-dimensional simulation environment differs significantly from the real pursuit-evasion environment. Additionally, the trained UAV can only perform the task of pursuit, resulting in limited functionality. Sun et al. [23] study the pursuit-evasion game problem of multiple UAVs with multiple static obstacles in a two-dimensional bounded environment and propose a multi-agent deep deterministic policy gradient based on attention algorithm to solve the pursuit-evasion problem of multi-UAVs. However, the paper does not take into account the threat posed by an opponent UAV, and there is a lack of a game process between the opponent UAV and our UAV. Vlahov et al. [24] discuss a framework for the development of reactive strategies that can learn to exploit behaviors, apply the A3C algorithm to UAV pursuit decision-making and verify the effectiveness of learning strategies through Monte Carlo experiments. However, it does not consider the generalization of the algorithm. When the environment changes, the performance advantage of the algorithm is not manifested. Additionally, it also does not comprehensively consider multiple situational scenarios.

Although reinforcement learning has a good performance in the UAV pursuit-evasion game, it also faces a problem: in the training process, due to the large state space, it may lead to dimensional disaster, which in turn affects the convergence speed of training. In addition, the pursuit-evasion environment changes in real time. The above algorithms do not comprehensively consider the overall situation, and only one strategy is trained so that the trained UAV can only complete the task of pursuit or evasion, and its performance is relatively simple, which cannot realize the function of turning the defeat into victory.

At present, hierarchical reinforcement learning algorithms are widely used in many fields [25–28]. Hierarchical reinforcement learning is for the performance of multiple sub-tasks in a hierarchical manner, which improves the decision-making efficiency of the model. Based on the above motivations, this paper considers applying the hierarchical reinforcement learning algorithm to the intelligent decision-making of UAV pursuit-evasion. Based on the framework of hierarchical reinforcement learning based on option [29], this paper chooses to use a policy gradient (PG) algorithm to train the top-level policy selector and proposes a UAV hierarchical maneuver decision-making method based on the PG-option. Four options of the advantage game (AG), quick escape (QE), situation change (SC) and quick pursuit (QP) are designed by considering various situations in the UAV pursuit-evasion game comprehensively, and the corresponding meta-policy is trained by the SAC algorithm. The meta-policy termination function is designed to improve the flexibility of meta-policy switching. The experimental results show that the UAV trained by the PG option can flexibly switch meta-policies in the pursuit-evasion game and can use different meta-policy combinations to deal with different complex scenarios, reflecting the superiority and robustness of hierarchical maneuver decision-making method based on the PG option in the future.

The main contributions of this paper are summarized as follows:

(1) Comprehensively considering the constraint information in the process of UAV pursuit-evasion, this paper conducts a flight control model for a UAV and introduces the concept of threat zone, which makes the simulation more realistic;
(2) Considering various situations in the process of UAV pursuit-evasion comprehensively, four meta-policies are designed for UAV flight decision-making, which not only enrich the maneuver library but also effectively improve the effectiveness of the algorithm;
(3) A hierarchical maneuvering decision-making method for the UAV is designed to ensure that the UAV can flexibly switch meta-polices under different situations and achieve victory.

2. Problem Formulation and Preliminaries

In this section, to simplify the UAV pursuit-evasion scenario and subtract some unnecessary influencing factors, some assumptions are made for the scenario. The UAV flight control model and threat zone model are established for subsequent experiments.

2.1. Scenario Description and Assumptions

In this paper, we focus on the problem of UAV maneuvering decisions in a one-to-one UAV pursuit-evasion game. In the process of pursuit-evasion, many factors may affect the flight of UAV, which can cause the model to be more complex. However, it is not necessary to take all factors into account. Therefore, to pay more attention to our research and simplify the complexity of scenario, this paper proposes the following assumptions:

(a) The UAV is assumed to be a rigid body, and the gravity acceleration is a unified value;
(b) This paper ignores the influence of earth curvature, earth rotation and earth revolution on the UAV flight;
(c) To simplify the complexity of the UAV flight, this paper only considers the kinematics model.

The scenario is defined as a three-dimensional pursuit-evasion scenario. Therefore, a three-dimensional kinematics model of a UAV is established, and the UAV is described by physical quantities such as position, speed and attitude. The OXYZ northeast coordinate system is established. The coordinate origin is O, which indicates the center point of the scenario. The X-axis points to the north direction; the Z-axis points to the east direction; and the Y-axis points to the vertical direction. The situation diagram of both sides in scenario is shown in Figure 1.

The situation information of the red UAV is $\vec{R}_m = (X_m, Y_m, Z_m)$ and $\vec{V} = (v_{xm}, v_{ym}, v_{zm})$. The situation information of the blue UAV is $\vec{R}_t = (X_t, Y_t, Z_t)$ and $\vec{V}_t = (v_{xt}, v_{yt}, v_{zt})$.

The relative position vector of red UAV and the blue UAV is $\vec{D}$, and the distance scalar is d. The azimuth angle q represents the angle between the red UAV's velocity $\vec{V}$ and $\vec{D}$. The formula is as follows:

$$\vec{D} = (X_t - X_m, Y_t - Y_m, Z_t - Z_m) \tag{1}$$

$$q = \text{acos}(\frac{\vec{D} \cdot \vec{V}}{\left\|\vec{D}\right\| \cdot \left\|\vec{V}\right\|}) = \text{acos}(\frac{(X_t - X_m)v_{xm} + (Y_t - Y_m)v_{ym} + (Z_t - Z_m)v_{zm}}{d * (\sqrt{v_{xm}^2 + v_{ym}^2 + v_{zm}^2})}) \quad (2)$$

$$d = \left\|\vec{D}\right\| \quad (3)$$

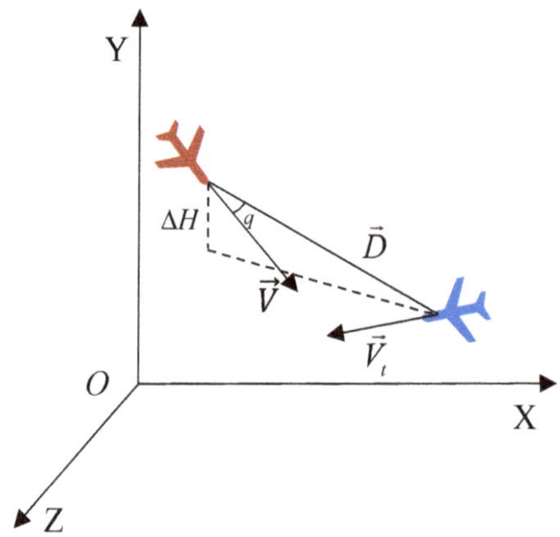

Figure 1. The situation in UAV pursuit-evasion.

2.2. UAV Model

It is assumed that our UAV can obtain the position and speed of the opponent through the ground command center. Our UAV needs to evaluate the current situation and make maneuvering decisions to effectively capture the opponent.

In the process of UAV kinematics modeling, it is considered a rigid body and regarded as a mass point. The kinematics equation of a three-degree-of-freedom UAV is as follows:

$$\begin{cases} X_{t+dT} = X_t + V_{t+dT} * \cos(\theta_{t+dT}) * \cos(\varphi_{t+dT}) * dT \\ Y_{t+dT} = Y_t + V_{t+dT} * \sin(\theta_{t+dT}) * dT \\ Z_{t+dT} = Z_t + V_{t+dT} * \cos(\theta_{t+dT}) * \sin(\varphi_{t+dT}) * dT \end{cases} \quad (4)$$

where φ denotes the heading angle; θ denotes the pitch angle; V is the speed of the UAV; $[X, Y, Z]$ denotes the component of the UAV in coordinate axes.

To make the model more real, this paper sets the threat range with the UAV as the center. The threat range is a conical area centered on the UAV and formed by the deviation of the UAV's speed direction from the $q_{\max}$ angle, which is called the threat zone in this paper. The threat zone contains three important indicators: maximum threat distance $D_{\max}$, minimum threat distance $D_{\min}$ and maximum threat angle $q_{\max}$. The two-dimensional diagram of the threat zone is shown in Figure 2.

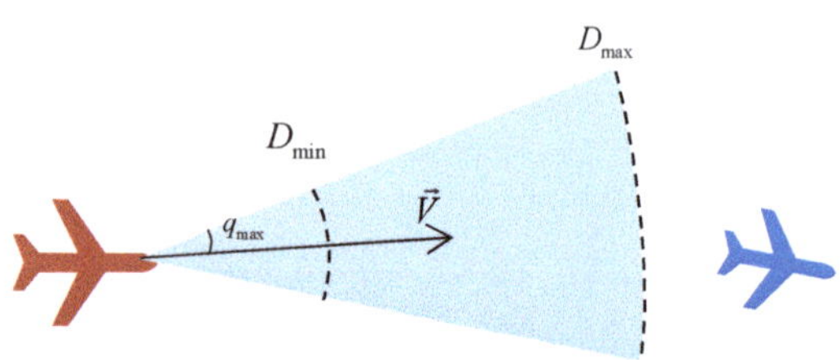

Figure 2. Threat zone in 2D.

When the distance between the two sides is less than the minimum threat distance $D_{\min}$, it is considered that the two sides collide and crash, and our UAV failed. When the continuous flight time of blue UAV in our threat zone that is defined as t_{in} exceeds the maximum time threshold $t_{\max}$, it is considered that the red UAV successfully captured the blue UAV. The formula is as follows:

$$\begin{cases} t_{in} > t_{\max} \\ q < q_{\max} \\ D_{\min} < \left|\vec{D}\right| < D_{\max} \end{cases} \tag{5}$$

3. Hierarchical Maneuver Decision Method for UAV Pursuit-Evasion Game

In this section, a hierarchical maneuver decision-making method for the UAV pursuit-evasion game based on the PG option is proposed. Considering the situation of both sides comprehensively, UAV maneuver decision-making be divided into the four meta-policies of advantage game (AG), quick escape (QE), situation change (SC) and quick pursuit (QP). Because different meta-polices perform different maneuvering characteristics, the corresponding reward function is designed and trained by the SAC algorithm. Then, the PG algorithm is integrated into the traditional hierarchical reinforcement learning to train the upper policy selector, and the expert experience is introduced to design meta-policy termination function. The structure of hierarchical decision-making is shown in Figure 3. The hierarchical decision-making model designed in this paper consists of two policy layers. There are four meta-polices at the bottom of the hierarchical decision model. The meta-policies are pre-trained so that they can perform corresponding actions according to the current pursuit-evasion situation when they are selected. There is a policy selector at the top of the hierarchical decision model, which is used to select the corresponding meta-policy according to the current situation.

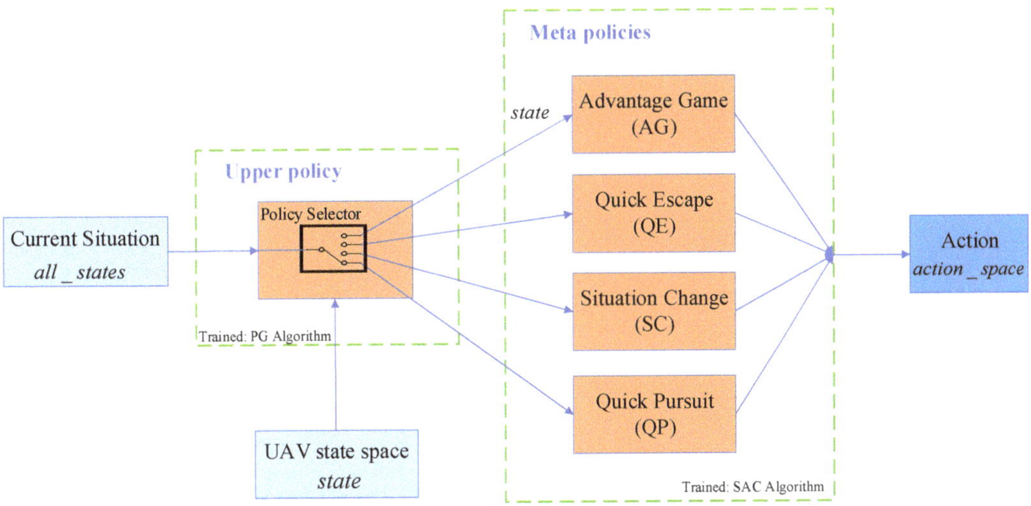

Figure 3. The structure of hierarchical decision.

3.1. Meta-Policy Model Training Algorithm Based on SAC

For the four meta-policies training of AG, QE, SC and QP, the paper uses the maximum entropy Soft Actor-Critic (SAC) algorithm to train them. SAC is a classical reinforcement learning algorithm. The SAC [30–33] algorithm is an algorithm based on the Actor-Critic framework, which can randomize the policy. At the same time, entropy is introduced to represent the randomness of the policy. After being trained, the agent will achieve a balance between the reward value and the entropy, so that the agent can increase the exploration of the state space while maximizing the reward and achieving the purpose of accelerating the learning speed.

The four meta-policies all use the same state space, which is represented by nine tuples, as shown in Formula (6).

$$state = [X_m, Y_m, Z_m, v, \theta, \varphi, d, q_m, q_t] \tag{6}$$

where X_m, Y_m, Z_m represents the projection of UAV 's position on three coordinate axes; v is the speed of our UAV; θ denotes the pitch angle of our UAV; φ is the heading angle of the UAV; d denotes the distance between the opponent and our UAV; q_m is the relative azimuth of the opponent UAV related to our UAV; q_t is the relative azimuth of our UAV related to opponent UAV. q_t and q_m are shown in Figure 4, where the blue side is the opponent UAV and the red side is our UAV.

Figure 4. Relative azimuth.

Maneuvering speed change rate, pitch angle change rate and heading angle change the rate control of our UAV. The action space of each meta-policy can be represented by triples in the form of Formula (7).

$$action\_space = [\dot{v}, \dot{\theta}, \dot{\varphi}] \tag{7}$$

where $\dot{v}$ denotes the acceleration; $\dot{\theta}$ denotes the change rate of pitch angle; $\dot{\varphi}$ denotes the change rate of heading angle.

When using a reinforcement learning algorithm to train agents, the design of the reward function is often very important for training results. A good reward function is often more intuitive and simple, which can make the agent move in the optimal direction. Different reward functions need to be set for different meta-policy to explore the corresponding optimal policy.

To facilitate the setting of different reward functions for the four meta-policy trainings, the following reward and penalty items are set:

- R_{mq1} is the continuous angle penalty value for our UAV, which will continuously penalize the agent in a round. The equation R_{mq1} is: $R_{mq1} = -q_m/180$;
- R_{mq2} is the sparse angle reward value for our UAV, which will reward the agent under certain conditions in a round. The equation R_{mq2} is as follows: $R_{mq2} = 1, if\ q_m < q_{\max}$;
- R_{tq1} is the continuous angle penalty value for opponent UAV, which will continuously penalize the agent in a round. The equation R_{tq1} is as follows: $R_{tq1} = -q_t/180$;
- R_{tq2} is the sparse angle reward value for opponent UAV, which will reward the agent under certain conditions in a round. The equation R_{tq2} is as follows: $R_{tq2} = 1, if\ q_t < q_{\max}$;
- R_{d1} will reward the agent when the opponent UAV is in the threat zone of our UAV. The equation R_{d1} is as follows: $R_{d1} = 1\ , if\ D_{\min} < d < D_{\max}$;
- R_{d2} will penalize the agent when the opponent UAV is out of the threat zone of our UAV. The equation R_{d2} is as follows: $R_{d2} = -3,\ if\ d \geq D_{\max}\ or\ d \leq D_{\min}$;
- R_{d3} denotes the ratio of the distance between the two sides and maximum threat distance. The equation R_{d3} is as follows: $R_{d3} = d/D_{\max}$.

Then, we describe each meta-policy and set different reward function combinations for different meta-policies:

- Advantage Game (AG): AG refers to two UAVs in the same scenario, and they are in each other's threat zone. In this situation, our UAV needs to escape from the opponent's threat zone as soon as possible and keep the distance between two sides within a small range to facilitate our UAV to capture the opponent. Therefore, for the AG, the total reward is R_{AG}. The equation R_{AG} is as follows: $R_{AG} = w_1 * (R_{mq1} + R_{mq2}) - w_2 * (R_{tq1} + R_{tq2}) + w_3 * (R_{d1} + R_{d2} + R_{d3}/5)$. The paper holds that the reward of both sides is equally important to the task of AG, so $w_1 = w_2 = 0.4$. Furthermore, the primary task of AG is to ensure that UAV escapes from the opponent's threat zone quickly. Therefore, the weight of the distance reward is small; that is $w_3 = 0.2$.
- Quick Escape (QE): QE means that in the same scenario, our UAV is located in the opponent threat zone, while the opponent UAV is not in our threat zone, and the opponent's advantage is greater than our advantage. The primary task of our UAV is to maneuver quickly to escape the opponent's threat zone. For the QE task, the total reward is R_{QE}. The equation R_{QE} is as follows: $R_{QE} = w_1 * (R_{tq1} + R_{tq2}) - w_2 * (R_{d1} + R_{d3}/5)$. In the QE task, angle reward and distance reward are equally important; that is $w_1 = w_2 = 0.5$.
- Situation Change (SC): SC means that in the same scenario, both sides are not in the other's threat zone, but the opponent's advantage is greater than our advantage. Our primary task is to maneuver quickly to change the situation so that our advantage is greater than the opponent's advantage. For the SC task, the total rewards are R_{SC}. The equation R_{SC} is as follows: $R_{SC} = w_1 * (R_{mq1} + R_{mq2}) - w_2 * (R_{tq1} + R_{tq2}) + w_3 * R_{d3}$. This paper believes that our angle reward and distance reward are equally important to the SC task; that is $w_1 = w_3 = 0.4$, $w_2 = 0.2$.
- Quick Pursuit (QP): QP means that in the same scenario, both the opponent and our UAV are not in the opponent's threat zone, and our UAV's advantage is greater than the opponent UAV's advantage. The primary task of our UAV is to capture the opponent UAV. Our UAV needs to maneuver quickly to make the opponent UAV fall into our threat zone. The total rewards are R_{QP}. The equation R_{QP} is $R_{QP} = w_1 * (R_{mq1} + R_{mq2}) + w_2 * (R_{d1} + R_{d2}/5)$. In the QP task, angle reward and distance reward are equally important; that is $w_1 = w_2 = 0.5$.

The meta-policy uses the maximum entropy SAC algorithm for training. The network structure set in this paper is shown in Figure 5.

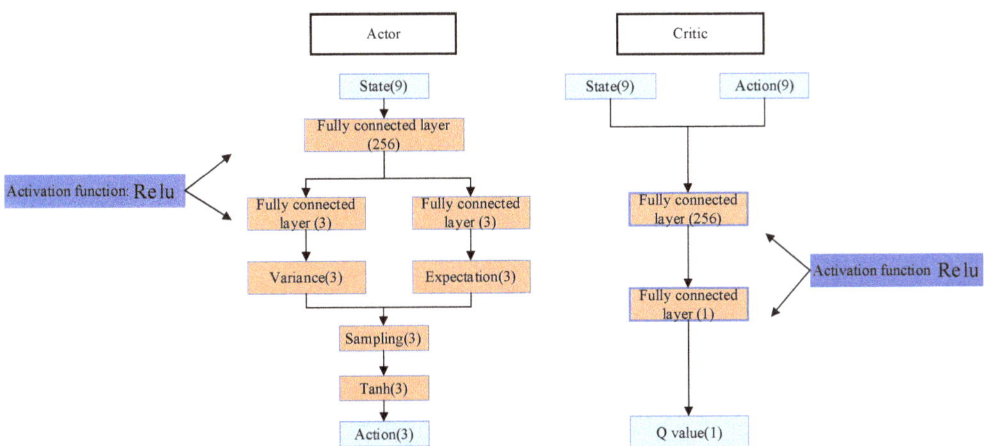

Figure 5. The structure of the Actor-Critic network.

The input of the Actor neural network is the state value of both sides, and the output is the speed change rate, pitch angle change rate and heading angle change rate of our UAV. The input of the Critic neural network is the state value and action value, and the output is the state action value (Q value). Experiments have shown that a wide and shallow fully connected neural network with the same number of neurons performs better than a narrow and deep network [34]. Therefore, both Actor and Critic networks are fully connected neural networks with two hidden layers, and the number of hidden layer neurons is 256. The activation function is Relu. In addition, the Actor neural network has two output layers, which are used to output the mean μ and variance σ, respectively, and the final action value can be obtained by sampling for agent execution.

Meta-policy training pseudo-code is shown in Algorithm 1.

Algorithm 1: Meta-Policy Training Algorithm

1. Randomly generated parameters: $\theta, \varphi_1, \varphi_2$
2. Initialize the policy network π_θ and two Soft-Q networks $Q_{\varphi_1}, Q_{\varphi_2}$
3. Initialize the target network of the Soft-Q network, and let $\varphi_1{\prime} \leftarrow \varphi_1, \varphi_2{\prime} \leftarrow \varphi_2$
4.   **FOR** *t* = *0*, *T*:
5.   Get state *s*
6.   **IF** *t* < *start_size*:
7. $a = random()$
8.   **ELSE**:
9. $\mu, \log\sigma = \pi_\theta(s), a = \tanh(\mu + \tau * \exp(\sigma))$
10.   The agent performs actions *a* and gets a reward *r* and the next state $s\prime$
11.   Store the array $< s, a, r, s\prime >$ in the experience pool
12.   **IF** Experience storage in the experience pool > *batch_size*:
13.   Sampling *batch_size* group data $< s, a, r, s\prime >$ from the experience pool
14.   Update Q function network parameters: $\theta_i \leftarrow \theta_i - \lambda_Q \hat{\nabla}_{\theta_i} J_Q(\theta_i)$
15.   Update policy network's weights: $\phi \leftarrow \phi - \lambda_\pi \hat{\nabla}_\phi J_\pi(\phi)$
16.   Adjust the parameters of temperature: $\alpha \leftarrow \alpha - \lambda_\pi \hat{\nabla}_\alpha J(\alpha)$
17.   Update the parameters of the target network: $\varphi_1{}' \leftarrow \varphi + (1-\tau)\varphi_1{}'$ $\varphi_2{}' \leftarrow \varphi + (1-\tau)\varphi_2{}'$
18.   **END IF**
19.   Jump to step 6, let $s \leftarrow s\prime$
20.   **END IF**
21. **END FOR**

3.2. Hierarchical Decision Method Based on PG-Option

The traditional hierarchical reinforcement learning algorithm based on option divides the target task into multiple options. Each option has its policy, which is called the meta-policy. The meta-policy is selected by the top-level policy selector. In the traditional hierarchical reinforcement learning algorithm based on this option, the policy selector uses the ε-greedy algorithm to select the meta-policy with the highest reward return in the current state as the decision. The problem with ε-greedy algorithm is that it generally does not consider other possible situations from the whole. Each choice is only a local optimal solution, and it takes a long time to capture the opponent. However, in a complex scene, our UAV needs to remove danger as soon as possible to achieve victory.

To better solve the problem of the complex UAV pursuit-evasion game, the paper designs a hierarchical decision model based on the traditional hierarchical reinforcement learning algorithm framework and the policy gradient (PG) algorithm. Instead of the ε-greedy policy, the neural network trained by the PG algorithm is used as the policy selector. Compared with the ε-greedy algorithm, PG is easier to converge the training results and can handle high-dimensional continuous state data.

Since the policy selector evaluates the overall situation, the state space of the upper policy selector adds the position $[X_t, Y_t, Z_t]$ and speed v_t of the opponent UAV based on the meta-policy state space, which is shown in Formula (8).

$$all\_states = [X_m, Y_m, Z_m, X_t, Y_t, Z_t, v, v_t, \theta, \varphi, d, q_m, q_t] \quad (8)$$

The action space outputs the probability of four meta-policies. If a meta-policy is selected, the final reward value will increase, which will increase the probability of this probability, and vice versa.

When the policy selector model is trained, the reward function is shown in Formula (9).

$$R = \begin{cases} r(\tau)/total\_step & if \text{ Success} \\ -3 & if \text{ Failed} \end{cases} \quad (9)$$

where $r(\tau)$ is the sum of the rewards at the end of each round; $total\_step$ is the all simulation step size in each round.

The policy selector can only be rewarded at the end of that moment. The success mark is to capture the opponent UAV, and the failure mark is that the opponent UAV captures our UAV or our UAV fails to capture opponent UAV within the maximum number of rounds.

In the process of training, the upper policy selector model is optimized by maximizing the reward value, and the network parameter of the whole hierarchical model is set as $\theta = (\phi, \varphi)$. ϕ denotes the network parameters of the policy selector, and φ denotes the network parameters of the meta-policy. The expected reward of the pursuit-evasion process is shown in Formula (10).

$$J(\theta) = E_{\tau \sim \pi_\theta(\tau)}[r(\tau)] = \int_{\tau \sim \pi_\theta(\tau)} \pi_\theta(\tau) r(\tau) \quad (10)$$

Next, we take the derivative of Formula (10) which is shown in Formula (11):

$$\nabla_\theta J(\theta) = E_{\tau \sim \pi_\theta(\tau)}[\nabla_\theta \log \pi_\theta(\tau) r(\tau)] \quad (11)$$

where $\pi_\theta(\tau)$ is as follows:

$$\pi_\theta(\tau) = p(s_{0,0}) \prod_{i=0}^{p-1} \pi_\phi(g_i|s_{i,0}) \prod_{j=0}^{T_i} \pi_\varphi(a_{i,j}|s_{i,g}, g_i) p(s_{i,j+1}|s_{i,j}, a_{i.j}) \quad (12)$$

Each round is divided into p stages. $s_{0,0}$ denotes the initial state of each round in UAV pursuit-evasion game, and $s_{i,0}$ denotes the initial state of the UAV in the first stage. At each stage, the UAV performs T_i steps, and $\pi_\phi(g_i|s_{i,0})$ denotes the probability that the meta-strategy g_i is selected by the policy selector. $\pi_\varphi(a_{i,j}|s_{i,j}, g_i)$ denotes the probability of the current action. $p(s_{i,j+1}|s_{i,j}, a_{i,j})$ denotes the state transition probability. In Section 2.1, the meta-policy is trained to converge, so the state transition probability is 1. Formula (12) can be rewritten:

$$\pi_\theta(\tau) = p(s_{0,0}) \prod_{i=0}^{p-1} \pi_\phi(g_i|s_{i,0}) \quad (13)$$

Then, we take the derivative of Formula (13) which is shown in Formula (14):

$$\nabla_\theta \log \pi_\theta(\tau) = \sum_{i=0}^{p-1} \nabla_\phi \log \pi_\phi(g_i|s_{i,0}) \quad (14)$$

Therefore, Formula (11) can be written as:

$$\nabla_\theta J(\theta) = E_{\tau \sim \pi_\theta(\tau)}[\sum_{i=0}^{p-1} \nabla_\phi \log \pi_\phi(g_i|s_{i,0}) r(\tau)] \quad (15)$$

In the process of model training, the optimization goal is to maximize the gradient of Formula (15), to achieve the convergence of the policy selector model.

In addition, in the traditional hierarchical reinforcement learning algorithm based on this option, for each time the agent performs an action and obtains a new state, the policy will reselect the meta-policy according to the current state. However, in this way, when the situation is complex, the switching of meta-policy is too frequent, which makes it difficult for the UAV to meet such high-frequency switching, thus making it difficult to complete the task.

To solve the above problems, the policy selector performs periodic state evaluation at a frequency of 10 Hz and introduces expert experience to design meta-policies' termination functions to obtain the signs of meta-policy success and failure which are shown in Table 1. After the end, the next meta-policy is selected. The framework of the model training is shown in Figure 6.

Table 1. The signs of meta-policy success and failure.

Meta-Policy	Success	Failure
AG	Out of the opponent UAV's threat zone 1000 m	Failed to break away from the opponent threat zone within 500 steps
QE	Out of the opponent UAV's threat zone 2000 m	Failed to break away from the opponent threat zone within 500 steps
SC	The relative azimuth of the opponent is less than 25°	Our relative azimuth is not bigger than the opponent's relative azimuth within 500 steps
QP	Successfully capturing the opponent	The opponent failed to exist our threat zone within 500 steps

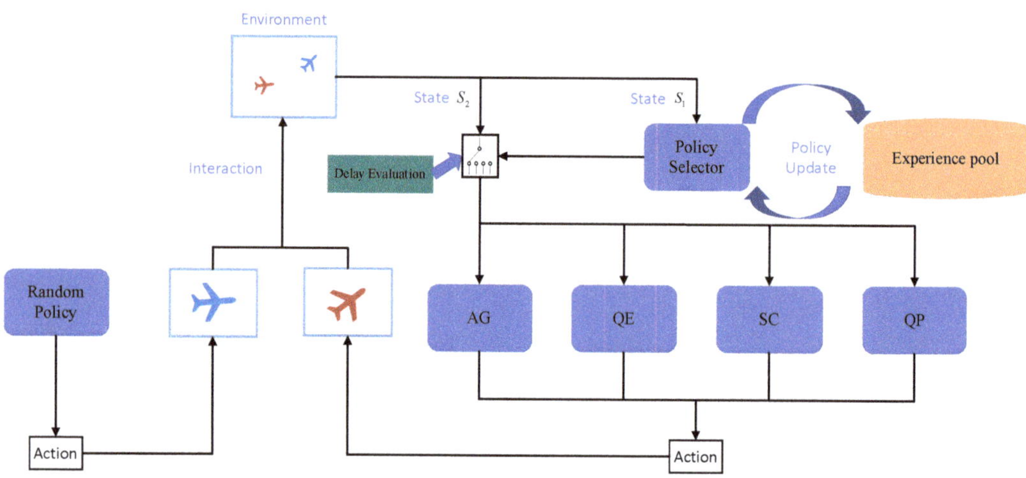

Figure 6. Training framework of hierarchical decision model.

The policy selector training pseudo-code is shown in Algorithm 2.

Algorithm 2: Hierarchical Maneuver Decision Algorithm Based on PG-Option

1. Randomly generate the policy selector network's parameters ϕ; Initialize the experience pool D
2. Initialize state variables $s_{Drevious}$, Initialize the initial state of the agent s_0, $s_0 = s_{Drevious}$
3. **FOR** $m = 1, M$:
4. Initialization counting stage: $p = 0$

Periodic selection of meta-policy g_{π} according to 10 Hz frequency
FOR t = 1, T:
    Generate actions a_t and execute actions; Get reward value r_t and new state s_{0+t}
    IF the current meta-policy execution ends:
    Save $(s_{Drevious}, g_{\pi}, r_t)$ into the experience pool
    $s_{Drevious} = s_{0+t}, p = p + l$
    END IF
    IF the end of the current round:
        IF capturing the opponent UAV:
            Update the reward value $r_t = r(\tau)/t$
        ELSE:
            Update the reward value $r_t = -3$
        END IF
        Update the network parameters of the policy selector ϕ, Clear the experience pool D
        Reinitialize the environment to obtain the initial state of the agent
    END IF
END FOR
END FOR

4. Experiments and Results

In this section, the proposed hierarchical maneuver decision method will be numerically simulated in many scenes. Firstly, the training parameters and hardware conditions of the simulation experiment are given. After the convergence of the four meta-policy trainings, the policy selector is trained and tested in the experimental scenario, which shows the feasibility of the method. Finally, the algorithm is tested by multiple UAV pursuit-evasion situations, and the simulation analysis is carried out to illustrate the generalization of the method.

4.1. Parameters and Hardware

In this paper, to avoid the instability of simulation results due to the large performance advantage gap between two UAVs, it is assumed that parameters of two UAVs are the same, as shown in Table 2. The heading angle variation is the range of the maximum allowable charge of the heading angle in the single-step execution of the UAV. The pitch angle variation is the range of the maximum allowable change of the pitch angle in the single-step execution of the UAV. The maximum speed of the UAV is the upper limit of the speed of the UAV. The maximum time threshold when the UAV falls into the other's threat zone is 2 s. Each step takes 0.1 s.

Table 2. The parameters of UAVs.

Parameter	Value
Our UAV pitch angle range	$[0°, 4°]$
Our UAV heading angle range	$[0°, 10°]$
Our UAV change range in speed	[0 m/s, 10 m/s]
Our UAV threat distance range	[1 km, 3 km]
Our UAV maximum speed	200 (m/s)
Our UAV Maximum threat angle	20 (°)
Our UAV Maximum time threshold	2 (s)
Opponent UAV pitch angle range	$[0°, 4°]$
Opponent UAV heading angle range	$[0°, 10°]$
Opponent UAV change range in speed	[0 m/s, 10 m/s]
Opponent UAV threat distance range	[1 km, 3 km]
Opponent UAV maximum speed	200 (m/s)
Opponent UAV Maximum threat angle	20 (°)
Opponent UAV Maximum time threshold	2 (s)

The four meta-policies are trained using the SAC algorithm, and the hyperparameters during training are as follows in Table 3. Batch_size is the number of samples extracted by the UAV from the experience pool during each training. The discount rate denotes the future reward discount when calculating the loss value. The optimization algorithm is the Adam algorithm. The regularization coefficient of entropy is initialized to 1. The target entropy is −3.

Table 3. The hyperparameters of SAC.

Hyperparameter	Value	Hyperparameter	Value
Actor-network learning rate	3×10^{-4}	Total number of rounds	10,000
Critic network learning rate	3×10^{-4}	Max steps in one round	500
Experience pool size	100,000	Soft update parameters	0.005
Batch_size	64	Discount rate	0.99
Target entropy value	−3	Regularization coefficient of entropy	1

When training the model, the Pytorch (v1.2.1) framework is used to design the neural network and optimizer. The CPU is an Intel i5-10400F, and the GPU is an Nvidia RTX 3060Ti which can accelerate the training process of the deep neural network. The GPU memory size is 12 G, and the computer memory is 32 G.

4.2. Results of Train and Simulation

In this paper, to train the PG-option hierarchical decision algorithm, the four meta-policies must firstly be trained to converge. According to Figure 7, it can be seen that at the beginning of training, our UAV adopts a random policy. After a period of training, the reward value convergence tends to be stable, and our UAV has been able to complete the corresponding tasks stably.

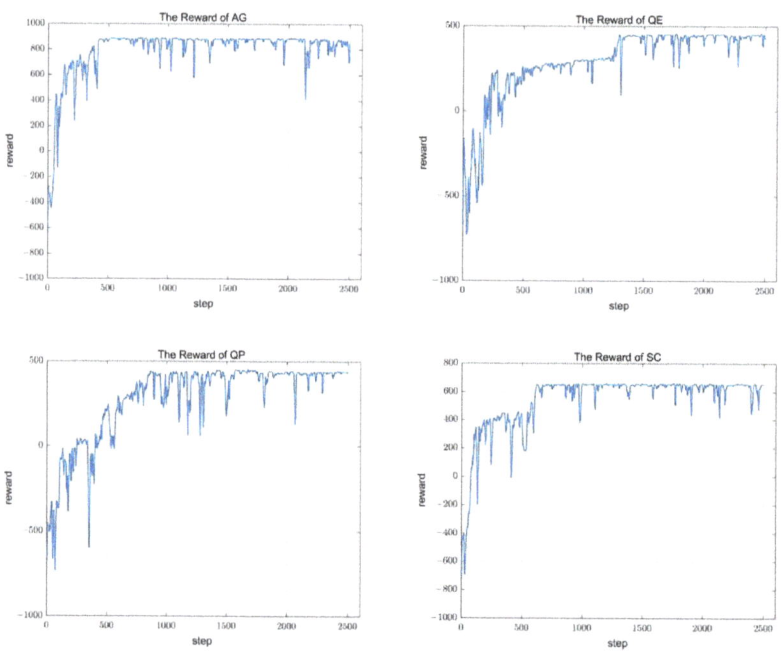

Figure 7. The reward function of meta-policies.

In this paper, to train the hierarchical decision algorithm based on the PG option, in the initial environment, the positions of both sides are randomly initialized within a certain range, and the opponent UAV is added to the disturbance in the original maneuvering mode.

Our UAV maneuvers use hierarchical decisions based on the PG-option algorithm. The opponent UAV maneuvers according to the predetermined trajectory and will threaten our UAV.

In this paper, the success rate is used to judge whether the policy selector network converges or not. The specific method is to count the success rate of the UAV in the training round. If the final success rate can converge within a certain range, it shows that the network of policy selectors has been trained and converged.

Training is performed in the initial environment using Table 4. The random value range is taken $[-1000, 1000]$, and the success rate is counted every 50 rounds.

Table 4. Initialization information in the training scene.

	X (km)	Y (km)	Z (km)	Pitch (°)	Yaw (°)	Speed (m/s)	Distance (km)	Azimuth (°)
Our UAV	2	3.5	−3	2	50	70	7.457	97.3
Opponent UAV	−3.5	3	2	1	−40	75		

It can be seen from Figure 8 that the final success rate is stable at about 82% under the above training, indicating that the policy selector training has converged after about 7200 rounds.

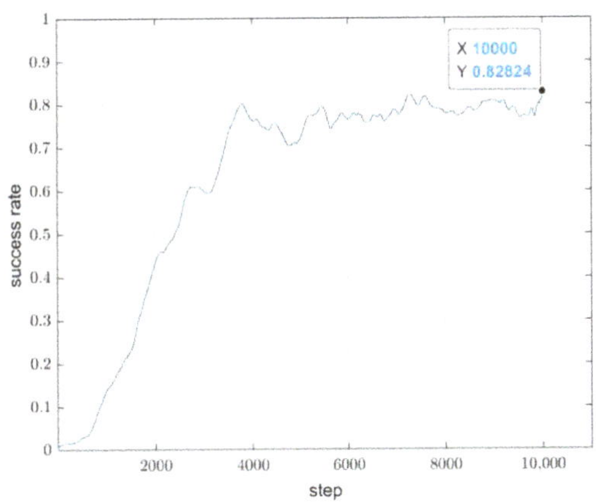

Figure 8. The curve of success rate in the training scene.

The trained model is tested in the training scene. Figure 9 shows the flight trajectory of the opponent and our UAV in the training scene.

At the initial time, the two UAVs are far away. It can be seen from the XOZ view that the height of the opponent UAV is lower than that of our UAV when the simulation begins. After that, our UAV changes the pitch angle, raises the height and shortens the distance between the two sides. From the XOY view, our UAV can reduce the opponent's azimuth angle against us, fly to the opponent's tail and fix the opponent UAV in the threat zone. Finally, the relative distance between the two sides is gradually shortened to meet the conditions of capturing.

From Figure 10a, it can be seen that before 80 steps, our UAV reduces the relative azimuth of the opponent UAV by rapidly rotating the heading angle and the pitch angle. After that, the relative azimuth of our UAV has always been kept in a small range. From

Figure 10b, it can be seen that our UAV can approach the target faster. After meeting the distance conditions of the threat zone, our UAV can converge stably at about 2000 m.

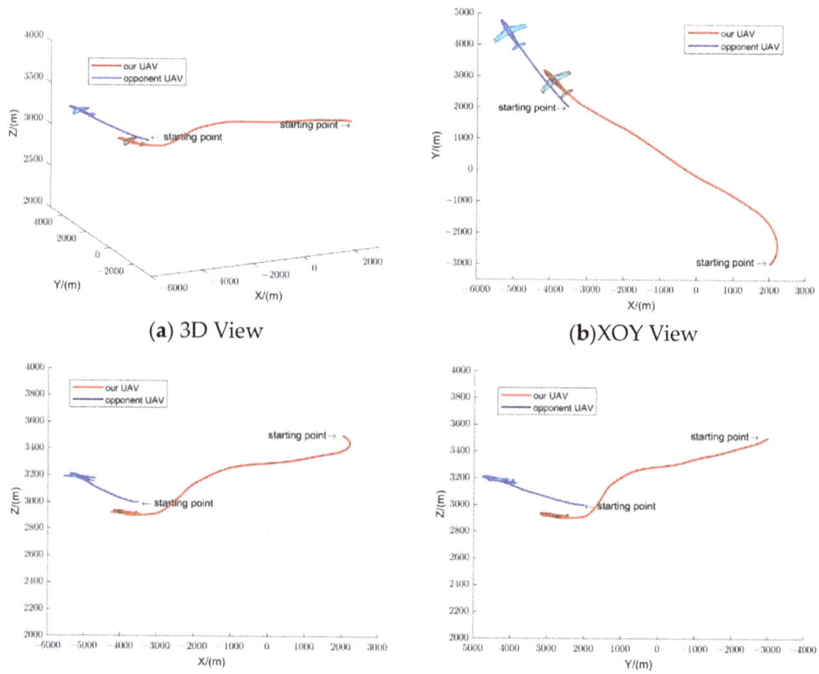

(**a**) 3D View (**b**)XOY View

(**c**) XOZ View (**d**) YOZ View

Figure 9. The flight trajectory in the training scene.

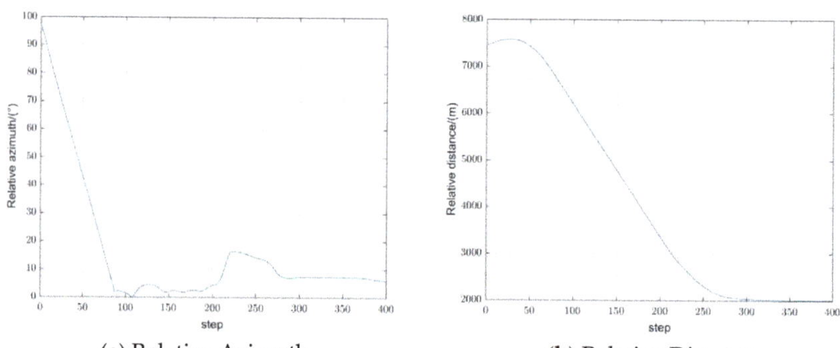

(**a**) Relative Azimuth (**b**) Relative Distance

Figure 10. Relative situation between opponent and our UAV in the training scene.

The selection of meta-policy is shown in Figure 11. At the beginning, the UAV chooses SC for situation conversion. After 220 steps, our UAV chooses the QP method to approach the opponent to capture it, and the selected meta-policy is more in line with the current situation.

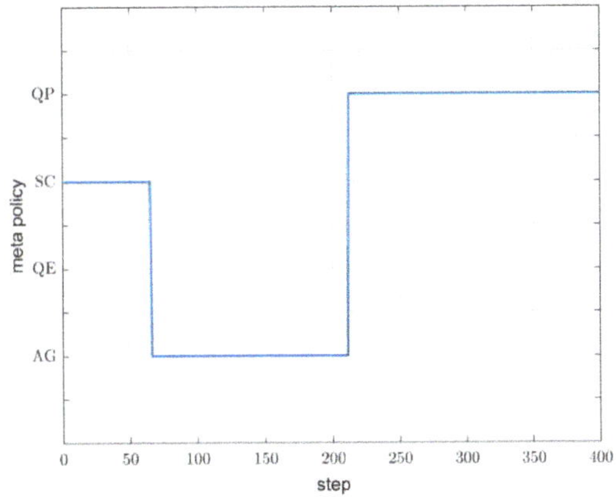

Figure 11. Meta-policy selection in the training scene.

4.3. Generalization Simulation

To test whether the PG-option hierarchical decision algorithm can realize the intelligent decision-making task under different initial situations, our UAV is placed in the pursued situation and the active capture situation, respectively, and its flight trajectory is analyzed.

4.3.1. Active Capture Situation

The active capture situation means that the relative azimuth angle of both sides is small at the beginning of the operation, and the overall situation is that the opponent UAV and our UAV are maneuvering in opposite directions. A simulation will be carried out in this situation. The initial state is shown in Table 5.

Table 5. Initialization information in active capture situation.

	X (km)	Y (km)	Z (km)	Pitch (°)	Yaw (°)	Speed (m/s)	Distance (km)	Azimuth (°)
Our UAV	2	2	3	2	50	70	7.991	29.6
Opponent UAV	−2	9	3.5	1	−40	80		

The flight trajectory is shown in Figure 12. From Figure 12a, it can be seen that at the beginning of the simulation, the height of the opponent UAV is higher than that of our UAV. Our UAV adjusts its pitch angle and shortens the distance between the opponent and our UAV. From Figure 12b, it can be seen that at the beginning of the simulation, both sides of the opponent UAV and our UAV are maneuvering toward each other. Our UAV continues to maneuver and threat around the tail of the opponent UAV.

It can be seen from Figure 13a that after the simulation, our UAV maneuvers towards the opponent and reduces the relative azimuth of the opponent. The relative azimuth meets the requirements of the threat zone from beginning to end. After our UAV escapes the opponent threat area, our UAV keeps the opponent within its threat area.

From Figure 13b, it can be seen that after the simulation begins, the two sides maneuver in opposite directions, and the distance between the opponent UAV and our UAV decreases rapidly. When the distance between the opponent UAV and our UAV meets the requirements of the threat zone, our UAV continues to maneuver and keeps the distance between the opponent and our UAV in this range.

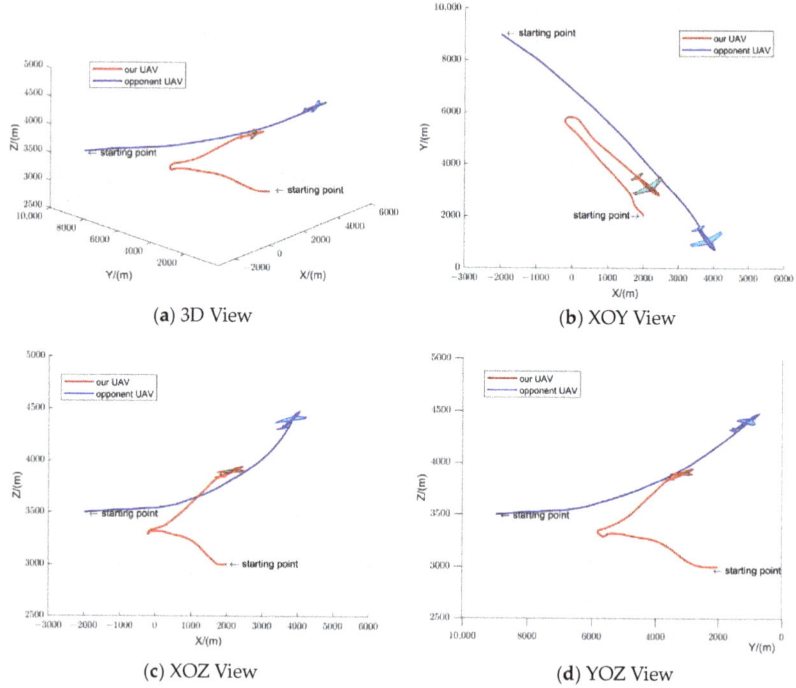

(a) 3D View (b) XOY View

(c) XOZ View (d) YOZ View

Figure 12. The flight trajectory in active capture situation.

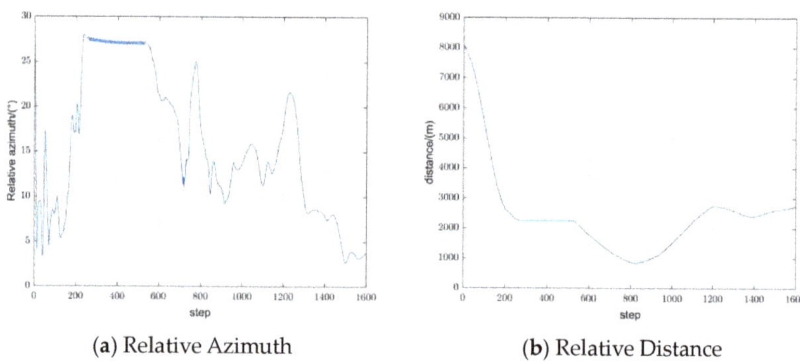

(a) Relative Azimuth (b) Relative Distance

Figure 13. Relative situation between opponent and our UAV in active capture situation.

The selection of meta-policy in active capture situation is shown in Figure 14. Under the situation, our UAV begins to choose QP. At about 200 steps, because our UAV falls into the opponent's threat zone in the process of pursuing the opponent, it chooses SC. At about 830 steps, our UAV chooses QP and finally wins.

4.3.2. Pursued Situation

The pursued situation means that at the beginning of the operation, our relative azimuth is small, and the opponent's relative azimuth is large. The overall situation is that the opponent UAV is at the tail of our UAV. A simulation will be performed in this situation. The initial state of the pursued situation is shown in Table 6.

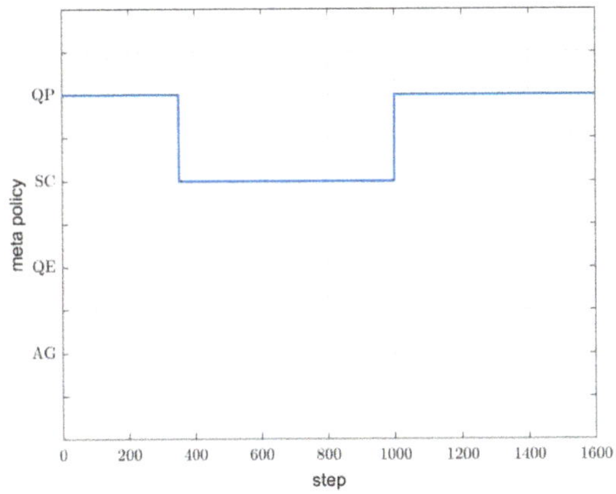

Figure 14. Meta-policy selection in active capture situation.

Table 6. Initialization information in pursued situation.

	X (km)	Y (km)	Z (km)	Pitch (°)	Yaw (°)	Speed (m/s)	Distance (km)	Azimuth (°)
Our UAV	2.1	2.9	3.0	2	50	70	8.01	165.1
Opponent UAV	−1.8	8.7	3.6	1	70	90		

The trajectory in pursued situation is shown in Figure 15. From Figure 15a, it can be seen that after the simulation begins, our UAV increases its pitch angle and reduces the distance between the opponent and us. When approaching the height of the opponent, our UAV gradually reduces its pitch angle to prevent our UAV from being higher than the height of the opponent. It can be seen from Figure 15b that after the simulation begins, our UAV quickly adjusts the heading angle to reduce the relative azimuth of the opponent. After adjusting the azimuth, our UAV maneuvers toward the opponent and shortens the distance between the two sides. To further reduce the relative azimuth of the opponent, our UAV begins to rotate the heading angle and maintain the tail-chasing situation of our UAV against the opponent.

From Figure 16a, we can see that at the beginning of the simulation, the relative azimuth is very large, and our UAV is in a state of being pursued. After that, our UAV quickly reduces the relative azimuth and changes its unfavorable situation by maneuvering. After meeting the angle requirements of the threat zone, the simulation continues, and the azimuth has been meeting this requirement ever since.

The distance curve is shown in Figure 16b. It can be seen that at the beginning of the simulation because our UAV is pursued by the opponent, our initial speed is faster. Therefore, the distance between the opponent and our UAV will increase at the beginning. After that, our UAV quickly changes the heading angle and maneuvers toward the opponent. At the same time, the distance between the opponent and our UAV can always be maintained in this range after meeting the distance requirements of the threat zone.

The selection of meta-policy in the pursued situation is shown in Figure 17. Under the pursued situation, our UAV begins to choose the SC. At about 220 steps, our UAV chooses the AG to occupy a favorable position. At about 1100 steps, our UAV chooses the QP and successfully captures the opponent.

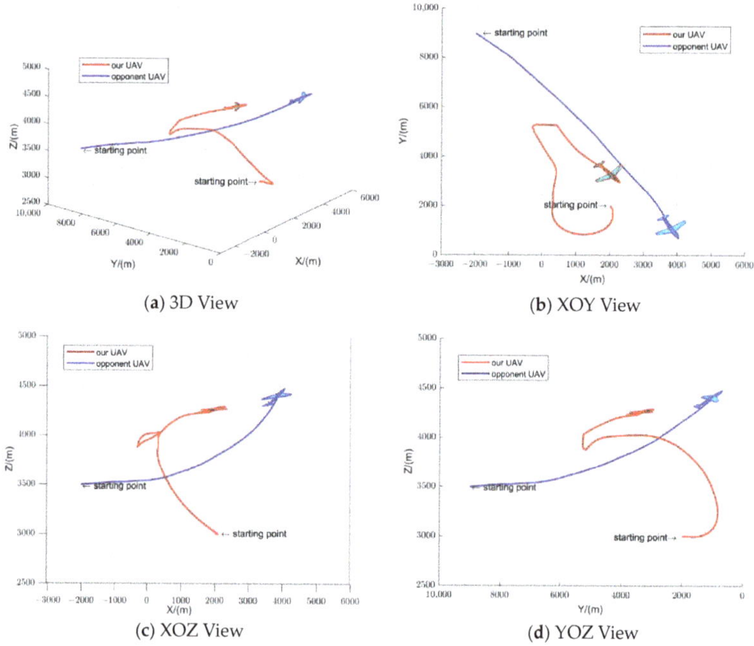

(a) 3D View (b) XOY View (c) XOZ View (d) YOZ View

Figure 15. The flight trajectory in pursued situation.

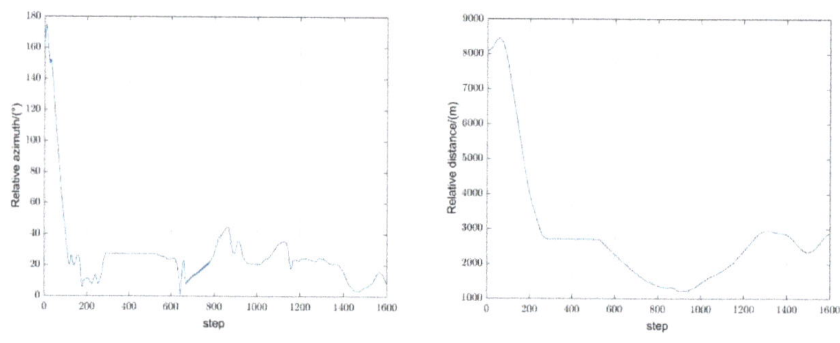

(a) Relative Azimuth (b) Relative Distance

Figure 16. Relative situation between opponent and our UAV in pursued situation.

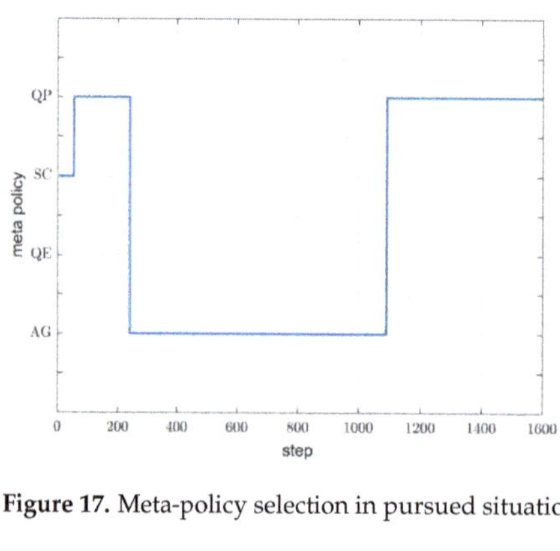

Figure 17. Meta-policy selection in pursued situation.

5. Conclusions

Based on the background of intelligent decision-making in the UAV pursuit-evasion game, this paper proposes a hierarchical maneuver decision-making method based on a PG option. The framework of maneuver decision-making is divided into two parts: the bottom is the four meta-policies of advantage game (AG), quick escape (QE), situation change (SC) and quick pursuit (QP), and the top is the policy selector trained by the PG algorithm. Delay evaluation and expert experience are introduced to effectively solve the problem of frequent meta-policy switching. The simulation and experimental results have shown that the method can choose reasonable meta-policies to control UAV maneuvering under different pursuit-evasion situations to capture the opponent. Some future work includes adding stationary or moving obstacles to simulation scenarios to make simulation more complex.

Author Contributions: Funding acquisition, B.L.; Methodology, H.Z.; Resources, B.L. and G.W.; Validation, H.Z., P.H. and K.Y.; Writing—original draft, B.L. and H.Z.; Writing—review and editing, G.W. and E.N. All authors have read and agreed to the published version of the manuscript.

Funding: This research was partially funded by Project supported by the National Nature Science Foundation of China under grant no. 62003267, the Fundamental Research Funds for the Central Universities under grant no. G2022KY0602, the Technology on Electromagnetic Space Operations and Applications Laboratory under grant no. 2022ZX0090, the Key Research and Development Program of Shaanxi Province under grant no. 2023-GHZD-33, and the key core technology research plan of Xi'an under grant no. 21RGZN0016.

Data Availability Statement: Data sharing not applicable.

Conflicts of Interest: The authors declare no conflict of interest.

References

1. Chen, B. Research on AI Application in the Field of Quadcopter UAVs. In Proceedings of the 2020 IEEE 2nd International Conference on Civil Aviation Safety and Information Technology (ICCASIT), Weihai, China, 14–16 October 2020; pp. 569–571.
2. Li, B.; Gan, Z.; Chen, D.; Sergey Aleksandrovich, D. UAV Maneuvering Target Tracking in Uncertain Environments Based on Deep Reinforcement Learning and Meta-Learning. *Remote Sens.* **2020**, *12*, 3789. [CrossRef]
3. Li, B.; Song, C.; Bai, S.; Huang, J.; Ma, R.; Wan, K.; Neretin, E. Multi-UAV Trajectory Planning during Cooperative Tracking Based on a Fusion Algorithm Integrating MPC and Standoff. *Drones* **2023**, *7*, 196. [CrossRef]
4. Liu, X.; Su, Y.; Wu, Y.; Guo, Y. Multi-Conflict-Based Optimal Algorithm for Multi-UAV Cooperative Path Planning. *Drones* **2023**, *7*, 217. [CrossRef]
5. Li, S.; Wu, Q.; Du, B.; Wang, Y.; Chen, M. Autonomous Maneuver Decision-Making of UCAV with Incomplete Information in Human-Computer Gaming. *Drones* **2023**, *7*, 157. [CrossRef]
6. Zhang, H.; He, P.; Zhang, M.; Chen, D.; Neretin, E.; Li, B. UAV Target Tracking Method Based on Deep Reinforcement Learning. In Proceedings of the 2022 International Conference on Cyber-Physical Social Intelligence (ICCSI), Nanjing, China, 18–21 November 2022; pp. 274–277.
7. Alanezi, M.A.; Haruna, Z.; Sha'aban, Y.A.; Bouchekara, H.R.E.H.; Nahas, M.; Shahriar, M.S. Obstacle Avoidance-Based Autonomous Navigation of a Quadrotor System. *Drones* **2022**, *6*, 288. [CrossRef]
8. Shahid, S.; Zhen, Z.; Javaid, U.; Wen, L. Offense-Defense Distributed Decision Making for Swarm vs. Swarm Confrontation While Attacking the Aircraft Carriers. *Drones* **2022**, *6*, 271. [CrossRef]
9. Awheda, M.D.; Schwartz, H.M. A fuzzy reinforcement learning algorithm using a predictor for pursuit-evasion games. In Proceedings of the 2016 Annual IEEE Systems Conference (SysCon), Orlando, FL, USA, 18–21 April 2016; pp. 1–8.
10. Gao, K.; Han, F.; Dong, P.; Xiong, N.; Du, R. Connected Vehicle as a Mobile Sensor for Real Time Queue Length at Signalized Intersections. *Sensors* **2019**, *19*, 2059. [CrossRef] [PubMed]
11. Alexopoulos, A.; Kirsch, B.; Badreddin, E. Realization of pursuit-evasion games with unmanned aerial vehicles. In Proceedings of the 2017 International Conference on Unmanned Aircraft Systems (ICUAS), Miami, FL, USA, 13–16 June 2017; pp. 797–805.
12. Gan, Z.; Li, B.; Neretin, E.; Sergey Aleksandrovich, D. UAV Maneuvering Target Tracking based on Deep Reinforcement Learning. *J. Phys.* **2021**, *1958*, 12015. [CrossRef]
13. Yu, F.; Zhang, X.; Li, Q. Determination of The Barrier in The Qualitatively Pursuit-evasion Differential Game. In Proceedings of the 2018 IEEE CSAA Guidance, Navigation and Control Conference (CGNCC), Xiamen, China, 10–12 August 2018; pp. 1–6.
14. Pan, Q.; Zhou, D.; Huang, J.; Lv, X.; Yang, Z.; Zhang, K.; Li, X. Maneuver decision for cooperative close-range air combat based on state predicted influence diagram. In Proceedings of the 2017 IEEE International Conference on Information and Automation (ICIA), Macao, China, 18–20 July 2017; pp. 726–731.

15. Mikhail, K.; Vyacheslav, K. Notes on the pursuit-evasion games between unmanned aerial vehicles operating in uncertain environments. In Proceedings of the 2021 International Conference Engineering and Telecommunication (En&T), Dolgoprudny, Russia, 24–25 November 2021; pp. 1–5.
16. Han, Z. The Application of Artificial Intelligence in Computer Network Technology. In Proceedings of the 2021 2nd International Seminar on Artificial Intelligence, Networking and Information Technology (AINIT), Shanghai, China, 15–17 October 2021; pp. 632–635.
17. Zhu, X.; Wang, Z.; Li, C.; Sun, X. Research on Artificial Intelligence Network Based on Deep Learning. In Proceedings of the 2021 2nd International Conference on Information Science and Education (ICISE-IE), Chongqing, China, 26–28 November 2021; pp. 613–617.
18. Lyu, L.; Shen, Y.; Zhang, S. The Advance of Reinforcement Learning and Deep Reinforcement Learning. In Proceedings of the 2022 IEEE International Conference on Electrical Engineering, Big Data and Algorithms (EEBDA), Changchun, China, 25–27 February 2022; pp. 644–648.
19. Li, W.; Wu, J.; Chen, J.; Lia, K.; Cai, X.; Wang, C.; Guo, Y.; Jia, S.; Chen, W.; Luo, F.; et al. UAV countermeasure maneuver decision based on deep reinforcement learning. In Proceedings of the 2022 37th Youth Academic Annual Conference of Chinese Association of Automation (YAC), Beijing, China, 19–20 November 2022; pp. 92–96.
20. Zhang, Y.Z.; Xu, J.L.; Yao, K.J.; Liu, J.L. Pursuit missions for UAV swarms based on DDPG algorithm. *Acta Aeronaut. Astronaut. Sin.* **2020**, *41*, 314–326.
21. Zhang, R.; Zong, Q.; Zhang, X.; Dou, L.; Tian, B. Game of Drones: Multi-UAV Pursuit-Evasion Game With Online Motion Planning by Deep Reinforcement Learning. *IEEE Trans. Ind. Neural Netw. Learn. Syst.* **2022**. [CrossRef] [PubMed]
22. Fu, X.; Zhu, J.; Wei, Z.; Wang, H.; Li, S. A uav pursuit-evasion strategy based on ddpg and imitation learning. *Int. J. Aerosp. Eng.* **2022**, *2022*, 1–14. [CrossRef]
23. Sun, Y.; Yan, C.; Lan, Z.; Lin, B.; Zhou, H.; Xiang, X. A Scalable Deep Reinforcement Learning Algorithm for Partially Observable Pursuit-Evasion Game. In Proceedings of the 2022 International Conference on Machine Learning, Cloud Computing and Intelligent Mining (MLCCIM), Xiamen, China, 5–7 August 2022; pp. 370–376.
24. Vlahov, B.; Squires, E.; Strickland, L.; Pippin, C. On Developing a UAV Pursuit-Evasion Policy Using Reinforcement Learning. In Proceedings of the 2018 17th IEEE International Conference on Machine Learning and Applications(ICMLA), Orlando, FL, USA, 17–20 December 2018; pp. 859–864.
25. Li, Z. A Hierarchical Autonomous Driving Framework Combining Reinforcement Learning and Imitation Learning. In Proceedings of the 2021 International Conference on Computer Engineering and Application (ICCEA), Kunming, China, 25–27 June 2021; pp. 395–400.
26. Cheng, Y.; Wei, C.; Sun, S.; You, B.; Zhao, Y. An LEO Constellation Early Warning System Decision-Making Method Based on Hierarchical Reinforcement Learning. *Sensors* **2023**, *23*, 2225. [CrossRef] [PubMed]
27. Qiu, Z.; Wei, W.; Liu, X. Adaptive Gait Generation for Hexapod Robots Based on Reinforcement Learning and Hierarchical Framework. *Actuators* **2023**, *12*, 75. [CrossRef]
28. Li, Q.; Jiang, W.; Liu, C.; He, J. The Constructing Method of Hierarchical Decision-Making Model in Air Combat. In Proceedings of the 2020 12th International Conference on Intelligent Human-Machine Systems and Cybernetics (IHMSC), Hangzhou, China, 22–23 August 2020; pp. 122–125.
29. Bacon, L.; Harb, J.; Precup, D. The option-critic architecture. In Proceedings of the 31st AAAI Conference on Artificial Intelligence, San Francisco, CA, USA, 4–9 February 2017; pp. 1726–1734.
30. Wu, Y.; Sun, G.; Xia, X.; Xing, M.; Bao, Z. An Improved SAC Algorithm Based on the Range-Keystone Transform for Doppler Rate Estimation. *IEEE Geosci. Remote Sens. Lett.* **2013**, *10*, 741–745.
31. Gao, M.; Chang, D. Autonomous Driving Based on Modified SAC Algorithm through Imitation Learning Pretraining. In Proceedings of the 2021 21st International Conference on Control, Automation and Systems (ICCAS), Jeju, Republic of Korea, 12–15 October 2021; pp. 1360–1364.
32. Xiao, T.; Qi, Y.; Shen, T.; Feng, Y.; Huang, L. Intelligent Task Offloading Method for Vehicular Edge Computing Based on Improved-SAC. In Proceedings of the 2022 IEEE 5th Advanced Information Management, Communicates, Electronic and Automation Control Conference (IMCEC), Chongqing, China, 16–18 December 2022; pp. 1720–1725.
33. Zhu, Q.; Su, S.; Tang, T.; Xiao, X. Energy-efficient train control method based on soft actor-critic algorithm. In Proceedings of the 2021 IEEE International Intelligent Transportation Systems Conference (ITSC), Indianapolis, IN, USA, 19–22 September 2021; pp. 2423–2428.
34. Ota, K.; Jha, D.K.; Kanezaki, A. Training larger networks for deep reinforcement learning. *arXiv* **2021**, arXiv:2102.07920v1.

drones

MDPI

Article

Research on Drone Fault Detection Based on Failure Mode Databases

Defei Hou [1], Qingran Su [2,*], Yi Song [3] and Yongfeng Yin [4]

[1] School of Cyber Science and Engineering, Southeast University, Nanjing 211189, China; 230199096@seu.edu.cn
[2] School of Computer Science and Engineering, Beihang University, Beijing 100191, China
[3] School of Reliability and System Engineering, Beihang University, Beijing 100191, China; 1641208406@buaa.edu.cn
[4] School of Software, Beihang University, Beijing 100191, China; yyf@buaa.edu.cn
* Correspondence: suqingran@buaa.edu.cn; Tel.: +86-18810723977

Abstract: Drones are widely used in a number of key fields and are having a profound impact on all walks of life. Working out how to improve drone safety through fault detection is key to ensuring the smooth execution of tasks. At present, most research focuses on fault detection at the component level as it is not possible to locate faults quickly from the global system state of a UAV. Moreover, most methods are offline detection methods, which cannot achieve real-time monitoring of UAV faults. To remedy this, this paper proposes a fault detection method based on a fault mode database and runtime verification. Firstly, a large body of historical fault information is analyzed to generate a summary of fault modes, including fault modes at the system level. The key safety properties of UAVs during operation are further studied in terms of system-level fault modes. Next, a monitor generation algorithm and code instrumentation framework are designed to monitor whether a certain safety attribute is violated during the operation of a UAV in real time. The experimental results show that the fault detection method proposed in this paper can detect abnormal situations in a timely and accurate manner.

Keywords: drone safety; software failure mode databases; runtime verification

Citation: Hou, D.; Su, Q.; Song, Y.; Yin, Y. Research on Drone Fault Detection Based on Failure Mode Databases. *Drones* **2023**, 7, 486. https://doi.org/10.3390/drones7080486

Academic Editors: Bo Li, Chunwei Tian, Daqing Chen and Ming Yan

Received: 27 June 2023
Revised: 22 July 2023
Accepted: 24 July 2023
Published: 25 July 2023

1. Introduction

Drones have become an indispensable component of modern equipment systems, and as the number of components in these systems continues to increase, so does their complexity. Consequently, the requirements for system quality—particularly reliability, maintainability, and functionality—are also becoming increasingly stringent [1]. In recent years, there have been numerous instances of serious consequences and economic losses caused by quality problems in aerospace and other fields. As a result, safety and reliability have become key areas of focus when it comes to drones [2]. Sensor data fusion [3], fault detection [4], fault-tolerant estimation [5], and fault-tolerant control [6] are all important means of ensuring UAV safety. Among these methods, researchers tend to place particular emphasis on fault detection, which can be broadly divided into two categories: model-based approaches and data-driven approaches [7].

The traditional way to detect drone failures is through model-based approaches that use mathematical models to analyze UAV safety. These models can be further subdivided into qualitative or quantitative methods (Table 1). However, because model-based approaches often focus only on specific components, they may not always detect failures in other parts of the system. This means that, if an unmanned aerial system suddenly fails due to its complexity, resulting in higher costs, multiple model pairs may need to be deployed. Therefore, in this study, we mainly study faults at the system level. Building a fault mode database using a large amount of fault data can facilitate the analysis of fault modes and fault propagation processes at the system level. The analysis process can help fault diagnosis personnel quickly locate the problem and save diagnostic time. At present, the research on failure mode databases lacks depth. Drone systems are complex systems with multiple tasks and configurable elements. There are both common and unique failure modes among different systems. The evaluation center where the authors of the current study work has a large body of failure data. However, due to the lack of organization and induction of these data, there has been no in-depth analysis of the mechanisms of common faults. How to use historical fault data to generate fault mode databases and utilize these databases to better achieve fault detection is a problem worth studying.

Table 1. Model-based approaches to Fault Detection.

Authors	Year	Fault Location	Used Methods
D'Amato et al. [8]	2021	sensor	sensor fusion algorithm based on particle filter
Fu et al. [9]	2021	sensor, actuator	adaptive observer to realize sensor and brake fault detection
Maqsood et al. [10]	2021	sensor	sensor fault detection and isolation method for quadrotor aircraft
Miao et al. [11]	2021	sensor	adaptive nonlinear proportional integral (PI) observer
Nejati et al. [12]	2021	actuator	three-level central difference Kalman filter (RThSCDKF)
Sun et al. [13]	2021	pitot tube	two identical synthetic air data systems
Cao et al. [14]	2022	actuator	interval observer and extended state observer
Gai et al. [15]	2022	elevator, event-triggered intervals	dynamic event-triggered Hi/H∞ optimization method
Lin et al. [16]	2022	sensor	crash probability density (CPD)

Another typical method used for drone fault detection is the data-based method. In recent years, the reduction in the difficulty of data collection and analysis has driven the development of the field of data analysis. With the support of various high-speed processors, the means of mining required information from data have become increasingly powerful. Similarly, data-based methods can be divided into two categories: quantitative and qualitative. Quantitative methods include those based on statistical theory and neural networks. Qualitative methods typically include expert systems, fuzzy logic, pattern recognition, and qualitative trend analysis. Table 2 lists some typical studies, including the location of fault detection and the methods used.

Table 2. Data-driven approaches to Fault Detection.

Authors	Year	Fault Location	Used Methods
Altinors et al. [17]	2021	motor, propeller	method combining statistical feature extraction and machine learning
Park et al. [18]	2021	GPS spoofing, DoS attack, rudder, elevator, aileron, engine	unsupervised learning
Souza et al. [19]	2021	motor	signal analysis technique based on chaos
Zheng et al. [20]	2022	actuator, aero engine, equipment	data assembly annotation method based on flight data and BIT records
Cabahug et al. [21]	2022	actuator	k-means clustering algorithm

Working out how to optimize the original data or improve the relevant algorithms has also become a major research trend in recent years. Zhang et al. [22] proposed a robust deformed denoising CNN (RDDCNN) to solve the offset pixels of feature maps from noisy images. This approach can effectively improve data quality and support data-based UAV fault diagnosis. Tian et al. [23] proposed a multi-stage image denoising CNN with a wavelet transform (MWDCNN) via three stages. A dynamic convolution is used in a CNN to address the limitations in depth and width of lightweight CNNs to achieve good denoising performance.

The performance of data-based methods mainly depends on the quality of the original data. The fault data of drones is often multi-level linkage and cross-influence data, which is not conducive to analyzing complex types of faults. In addition, data-based methods are often offline and difficult to analyze in real time. This method cannot support the real-time safety monitoring of drones. It is more of a means of analysis after accidents, and cannot detect and handle abnormal situations during the operation of drones. Therefore, this method cannot guarantee the safety of drones during operation.

The above two methods mainly focus on component-level fault detection, rather than system-level fault analysis and localization. In this study, our goal is to fully explore and utilize drone fault data accumulated over the past year to gain a deeper understanding of drone system-level fault mechanisms. The aims of this study are: (1) To fully utilize a large amount of drone fault data, establish a fault mode database, and identify the occurrence and propagation modes of drone faults. To further extract fault features and attributes from fault modes and condense them into safety attributes during drone operation for subsequent runtime verification processes. (2) To design and implement a monitoring generation algorithm and code detection framework for real-time monitoring of whether drones violate a certain safety attribute during operation. By using the runtime monitoring method, the operation process of unmanned aerial vehicles can be monitored in real time and faults can be quickly detected.

The rest of the paper is organized as follows: Section 2 presents the related works about failure mode databases and runtime verification. Research on security detection based on failure mode databases is included in Section 3. Section 4 describes the implementation of a monitor generation framework and linear timing logic extension. The Clang-based code instrumentation framework is explained in Section 5. Experimental design and methodological evaluation are described in Section 6. Finally, Section 7 concludes the paper.

2. Related Works

2.1. Failure Mode Databases

In recent years, failure mode databases have become a research hot spot in database system engineering, knowledge engineering, and other fields, and this approach has been applied in many fields. Failure mode databases are structured, easy-to-use, comprehensive, and organized failure clusters—a collection of interrelated patterns stored, organized, managed, and used in computer memory in a certain way in order to solve the needs of problems in some fields. These failure modes include theoretical knowledge related to the domain, factual data, and heuristic knowledge derived from expert experience. UAV failure mode databases are of great significance to the study of reliability trends. How to form a real scientific failure mode database is one of the key contents of the research.

A common solution to this problem is to extract UAV failure patterns from the records contained in a UAV accident database by cleaning, parsing, and normalizing the fault data. For example, by obtaining accident records from different countries and analyzing the causes of accidents. Suitable data sources include:

- NASA and the FAA Aviation Safety Reporting System (ASRS) [24];
- U.S. Aviation Safety Communique (SAFECOM) data [25];
- Australian Transport Safety Bureau (ATSB) aviation safety investigations and reports [26];
- UK Air Accident Investigation Branch (AAIB) data [27].

The attribute fields in the database should include the data source, accident year, month, location, flight phase, aircraft model, flight hour history, flight time of the mission prior to the accident, altitude, weather, mission being performed, root cause, and accident outcome. At present, the research on UAV accident databases mainly focuses on the collection and analysis of accident data. Wild et al. [28] collected and analyzed 152 incident and accident records, but only 40 records deal with UAV accidents, and the data date back to 2006–2015. More recently, a total of 160 accident reports from 2015 to 2020 were collected and documented in an integrated database [29].

At present, the study of failure mode databases lacks depth. UAV systems are complex systems with multiple tasks and configurable elements. There are both common and unique failure modes among different systems. In addition, the reliability level of UAVs in different regions varies greatly due to manufacturing differences. The evaluation center where the authors of this current study work has a large amount of fault data of various types of UAVs in China, but these data have not been sorted and summarized and the mechanisms of common faults have not been thoroughly analyzed. How to make use of historical fault data to generate a fault mode database and realize fault detection better is a problem worth studying. UAVs face the following major problems:

(1) There has been a lack of unified analysis and summarizing of accumulated equipment failure data and underutilization of the advantages of historical data.
(2) Lack of effective fault identification, characterization, and matching methods to provide guidance for the whole life cycle activities of UAVs. Implementing such methods could effectively avoid the introduction of software defects and avoid the occurrence of failures, thus improving the reliability of unmanned aerial systems.

In this paper, the construction of a fault module database provides a set of solutions from cause analysis to normalized fault representation. Based on this database, we can have a deeper understanding of the development trends and failure modes of UAVs and improve the reliability and safety of UAVs.

2.2. Runtime Verification

Runtime verification techniques are formal verification techniques, which use logical formulas to describe properties and transform them into formal structures. Runtime verification tends to check that there is a poor path, that is, up to the current time, whether the system's running path is within safety parameters. After being proposed, this method has received continuous attention from academia and industry. One of its characteristics is that it plays a role in the system software—it is in the real run environment used to monitor the system—so it can find potential defects that the traditional software testing method misses. It complements classical verification techniques (for example, theorem proving and model checking) and provides a more practical method for the verification of system running trajectories. At the cost of limited execution coverage, runtime validation provides accurate information about the runtime behavior of the monitored system for subsequent analysis. The object of action can be a software system, a hardware or information physical system, a sensor network, or any system where dynamic behavior can be observed. The following is a brief overview of some recent research on runtime validation.

Abbas et al. [30] introduced private runtime verification. Liu et al. [31] introduced the idea of incremental verification and proposed an incremental probabilistic model-checking method based on heuristics. Y-Rozier et al. [32] explore the connection and difference between simulation and runtime verification. Stockmann et al. [33] proposed architecture runtime verification. Teixeira et al. [34] studied the properties of a runtime verification-specified API and wrote a minimalist specification language. Bicevskis et al. [35] discuss data quality checking during the execution of a business process by using runtime verification. Lee et al. [36] developed an effective model-checking method for IoT system operation verification. Legunsen et al. [37] proposed an aware runtime verification technique. Geng et al. [38] introduced a new smart tagging method to verify the completion of tasks involving one-to-many and many-to-one dependencies. Ring et al. [39] significantly reduced the size of the state space of the verification process and reduced the complexity of the detection process. Ye et al. [40] proposed an adaptive runtime verification method based on multi-agent systems. Tsigkanos et al. [41] proposed a service-oriented software architecture and technical framework to support the runtime verification of decentralized edge-dense systems. Hu et al. [42] proposed a runtime verification method based on the Robotic Operating System (ROS). Tracy et al. [43] introduced an open-source framework to achieve efficient and high-performance runtime monitoring. Ye et al. [44] developed a new approach based on approximate computation to achieve sufficiently fast and accurate repeated execution of security verification. Kong et al. [45] proposed a method aimed at monitoring traces that reveal the runtime state of the software. Jung et al. [46] proposed an automatic runtime prioritization method based on a classification tree. Miranda et al. [47] proposed a method to automatically detect whether the errors reported by the monitor are real errors. The authors of [48] proposed the concept of UAV safe operation monitoring and the operational limits to be monitored. A prominent example of such operational limitations is geofencing. Geofencing uses virtual fencing to prevent drones from entering restricted airspace. Felipe et al. [49] proposed a solution based on stream runtime verification, which offers a high-level declarative language to describe sophisticated monitors together with guarantees on execution time and memory usage. They showed how monitors can be combined with temporal planning not only to monitor assumptions but also to support mitigation and remediation in UAV missions. Bonnah et al. [50] presented a rewriting-based algorithm for runtime monitoring of safety requirements expressed in TWTL for specifying time-bounded serial tasks.

Runtime verification is widely used in academia and industry to ensure the reliability and security of the system, whether it is before deployment, testing, verifying, debugging, or after deployment. However, the setting of monitoring conditions still lacks a solid basis. Therefore, this paper intends to analyze the failure mechanism of UAVs through the UAV failure mode database to extract the monitoring conditions. According to the above monitoring conditions, the UAV is monitored in real time to improve its reliability.

3. Research on Safety Detection Based on Failure Mode Databases

Drones play a vital role in modern equipment, showing the characteristics of increasing scale and complexity. In the face of more serious quality and reliability problems, the guarantee of equipment software presents new challenges. Therefore, in order to give full play to the value of existing fault data, this paper collects and analyzes UAV faults based on the collected fault data, generates a fault mode database, and identifies and locates the faults during the operation of the UAV by combining this database with runtime verification technology.

The overall research framework of this paper is shown in Figure 1.

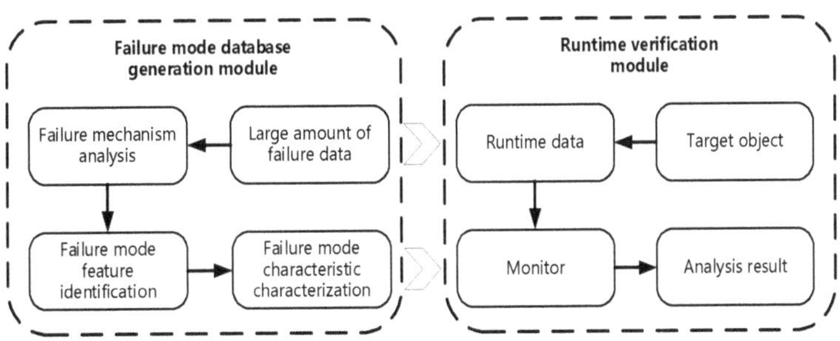

Figure 1. Overall frame diagram.

(1) First, a software-based failure mode database is constructed. The evaluation center at which the authors of the current study work has a large body of fault data; in order to better realize its value, it is necessary to summarize it into a fault mode database. In

this paper, fault mechanism analysis is carried out to summarize the failure causes and fault propagation modes of unmanned aerial vehicle systems. Then, the failure mode feature identification method is used to extract the failure characteristics. Finally, the method of characterizing the failure mode characteristics is studied. After extracting the attributes of each characteristic, the features of each attribute of the fault data are described. According to the results of the above process, the safety attributes that must be observed during the operation of the UAV are derived and the detection content is provided for use by the verification process during the subsequent operation.

(2) Second, runtime verification is implemented. To achieve this, linear time logic is used to describe safety attributes. Spot is used to complete the conversion process from LTL to the monitor, with some modifications being made to Spot according to actual needs. When combined with the safety attribute description derived in the failure mode database building step, this process allows the running state of the UAV to be monitored in real time.

3.1. Failure Mechanism Analysis

A system defect is a condition in which the system does not conform to specified requirements and behaves in ways inconsistent with the desired way of use. System defects, which are introduced by human errors, are static inherent properties of the system and are present across the whole system life cycle. During system execution, defects are activated under certain conditions. A system fault is an abnormal condition in a component, device, or subsystem. System failure occurs when a system fault cannot be handled by fault-tolerant technology, resulting in the loss of all or part of the functions of the system during operation and deviation from the expected normal state.

According to the process of occurrence, three necessary and sufficient conditions for system failure caused by system defects can be described:

- The actual input the system receives causes the system defect to be executed;
- An executed defect causes a change in the state of the data after the location of the system defect (failure is triggered);
- Faulty system data is passed as a result (the user perceives a fault).

These three necessary and sufficient conditions also reflect the whole process of the fault produced by the defect. This process is sequential, after the system executes a defect, the defect causes the data state after the defect to be infected, and finally, after several iterations, the wrong data is output by the system. Figure 2 describes the mechanism of system failure and reflects the conditions of fault identification and discovery. These three conditions correspond to the probability that the input leads to the execution of the position (execution probability), the probability that the position mutation leads to the change of the data state, (contagion probability), and the probability that the changed data state leads to a change in the program output (propagation probability). Only when these three conditions are met can the defect produce a fault and be identified.

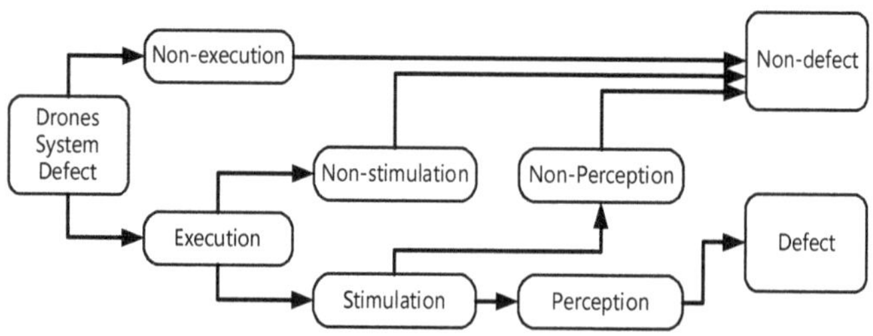

Figure 2. System failure mechanism.

3.2. Failure Mode Feature Identification

The system fault pattern feature identification process needs requires a features analysis of function, execution, interface, and related constraints. Through functional analysis, execution analysis, and interface analysis, the understanding of the identified features is strengthened, which lays a foundation for improving the accuracy of system fault pattern matching. Failure data description begins with original attribute extraction.

The complete system properties are extracted using the method given in the IEEE 1044–2009 standard. At the same time, the ideas of the GJB 437-88, GJB/Z 1391-2006, and GJB 841-90 standards are referenced. The attributes are modified from the perspective of fault control and management. The extracted attributes are normalized so that modifying one attribute does not affect other attributes. Finally, attribute fields that consider efficiency and quality that are accurate and easy to reuse later are extracted.

First, the fault ID and fault description attributes in IEEE 1044–2009 are used and the system (subsystem and subsystem), product model, product batch, and collection date are added as the basic information of fault data. Second, the fault state attribute is used to give the overall expression of the fault. The severity and impact attributes are used and then changed to fault-level and potential-impact attributes. If the type attribute is used, it is changed to a fault-type attribute. Third, the injection activity and discovery activity attributes are used. From the perspective of fault discovery and introduction, it is modified into the attributes of the fault discovery and introduction stages. Next, the fault or failure performance, modification content, and modification measure attribute are given. Finally, the root cause of a fault is considered and the attribute of the cause of the fault is added. A list of the attributes generated from fault data is given in Table 3.

Table 3. Attribute extraction of system fault data representation.

Number	Attribute	Definition
1	ID	The unique number of the fault
2	Affiliation	System to which the fault belongs
3	Description	Overall description of the fault
4	Discovery	System development phase in which faults are found
5	Introduction	The system development phase of introducing faults
6	Factor	Human factors that introduce faults
7	Type	A type defined for the characteristics of the system
8	Level	Severity of the fault
9	Manifestation	System failure caused by fault excitation
10	Potential impact	Danger or effect caused by failure

3.3. System Failure Mode Characteristic Characterization Method

According to the characteristics of different system fault patterns, measurement parameters, data acquisition methods, and deviation thresholds are designed. These provide the basis for accurate system fault pattern matching.

After extracting the characteristic attributes, the features of each attribute of the fault data must be described. The attributes are the same from data entry to data entry, but the characteristics of each attribute are different. Below, we describe the model and standardized description of system fault data.

For specific fault attributes, the attribute's characteristics must be described and the attribute's value range must be specified. The characteristics determined in this process should be consistent with the run characteristics of the system and can reflect the characteristics of system faults. How to improve reusability and meet the use needs of various developers should also be considered.

The existing feature description method of system fault data is mainly natural language, but this approach is prone to subjectivity. Due to differences in how individuals think, fault descriptions have a high degree of individuation and poor portability. In this paper, formalized and normalized expressions are developed to describe the characteristics of attributes. The different ways of describing attribute characteristics are given in Table 4.

Table 4. Attribute description of system fault data.

Number	Attribute	Definition
1	ID	DT—"Acquisition date"—"Serial number"
2	Affiliation	Which system
3	Description	Semi-formal description: "ZZ faults were introduced into YY by XX".
4	Discovery	(Requirement analysis/Design/Coding implementation/Test verification/Use and maintenance) phase
5	Introduction	(Requirement/Design/Coding) phase
6	Factor	Redundancy/Negligence/Forgetting/ Inadequate consideration
7	Type	Input class/Output class/Processing logic class/Working status class
8	Level	Key/Important/Ordinary/Other
9	Manifestation	The fault was triggered at XX, resulting in XXXX.
10	Potential impact	The failure caused by this fault may cause XXXX.

Attribute description is the key to fault description. Through the feature description method, the ambiguity caused by the use of over-subjective natural language is solved, which is conducive to the quantitative analysis of data and improvement of the value of fault data utilization. The main contribution of fault data attribute feature definition to fault data description is the provision of fault type features based on input class, output class, processing logic class, and working state class. These four classes are also the core and main features of any fault description, which is conducive to improving the efficiency of collecting and managing subsequent fault data. Finally, after fault data attribute extraction and provision of the feature description method, the system fault data description method is provided. The system fault pattern library is sorted according to the fault description method and subsequent management becomes rule-based. Table 5 provides the failure modes related to drone functions. Due to space constraints, only some of the failure modules are listed here as an illustration.

Table 5. Failure modules at the functional level.

Faulty Module	Fault Submodule
Power safety management	Fuel pump control
	Fuel monitoring and handling
	Oil delivery monitoring and handling
	Aerial stop
	Aerial start
	Engine off
Electrical safety management	Pitch control
	Radio altimeter power control
	Emergency power control
	Grid-connected control of transmission network
	Electrical fault handling
	Critical fault handling
Flight safety management	Destroyed key processing
	Link redundancy management
	Link status judgment and processing
	Switch of right of control
	Sensor Management
	Flight control computer management
	Aircraft weight management

4. Monitor Generation Framework Based on Linear Temporal Logic

In this study, we take unmanned aerial vehicle systems as the research object, focusing on safety monitoring technology based on runtime verification, with a focus on solving the problem of how to generate the required monitors and how to connect them with source code. This section focuses on the former, which is how to generate monitors and achieve target monitoring operations. Figure 3 shows the overall flowchart of the framework.

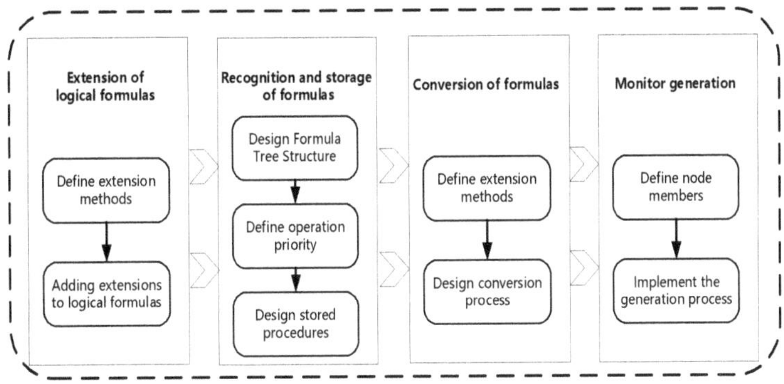

Figure 3. Monitor Generation Framework.

The entire framework can be divided into four parts. First, based on the different test objects and requirements documents, the properties that the tested system should meet during runtime are organized and described in natural language. In order for computers to handle these safety requirements, they need to be transformed into formal descriptions. In this study, we chose to use linear temporal logic to describe safety attributes, and we extended LTL to enhance its expressive ability, achieving the effect of using more concise forms to describe richer meanings and enhancing the accuracy of property descriptions. Second, in this study, we define a formula tree structure and corresponding conversion algorithms, which will be used by computers to store formulas after reading them. Once again, the conversion rules for extended LTL formulas were defined and the extended formulas were simplified according to this rule to obtain the transformed LTL formula tree. Next, the conversion algorithm is used to convert the logical formula into the corresponding security requirement monitor. The monitor will receive data from the tested object for analysis and perform state migration based on the analysis results. If the data meet the safety requirements, the monitor can execute the task smoothly. If the data violate the safety requirements, the monitor will discover this issue and issue a warning. The purpose of a warning is to draw the attention of the operator, who can promptly view system information, handle problems, and reduce the probability of accidents.

The following sections will introduce the approach in detail: Section 4.1 describes the extension of the meaning of linear temporal logic, Section 4.2 describes the formula recognition and storage process, Section 4.3 describes the formula conversion process, and Section 4.4 describes the monitor generation process.

4.1. Extension of Linear Temporal Logic

There are four commonly used logical connectives in linear temporal logic, namely, G, F, U, and X. G is a unary operator, indicating that the formula must be constant; F is a unary operator, indicating that the formula will be established in the future; X is a unary operator, indicating that the next cycle formula needs to be established; and U is a binary operator, which means that the formula on the left should always be true until the formula on the right is true. In order to enhance the expressive power of LTL, in this study, we extend LTL to express richer meanings with more refined formulas and increase the flexibility and accuracy of the formulas. In this study, we add a time point or interval after the temporal operators G, F, U, and X. The meaning of [i] is that the operator should hold in the i-th cycle, while [i: j] means that the operator should hold at the beginning of a cycle and continue for j cycles. For ease of understanding, the following specific examples are given:

(1) After expansion, the expression ability of the formula has been improved and more concise formulas can be used to express richer meanings; the flexibility has also been improved.
    (1) Assuming a represents an event, then a must hold true in all states, which is somewhat like an intersection. Only when a holds true in all states can it be considered true and as long as a does not hold true in one state, it does not hold true. Operators have their usage scenarios, but they may not be as flexible. If you want a to hold true in the next three states, it doesn't matter if the other states hold true, as the G operator cannot.
    (2) However, after adding a time interval, G[1:3]a can be used to indicate that a holds in the next three states. If traditional LTL is used, this must be represented with Xa&XXa&XXXa. If there are more than three states, the length difference of the formula will be even greater.
(2) After expansion, the expression accuracy of the formula has also been improved.
    (1) Fa indicates that a will always hold in the future, which is somewhat like a union. As long as a holds in one state, then Fa holds. The problem with the F operator is that it has a high degree of uncertainty. We request that a be true in the future, but we don't know when it will be true—it may be the 100th or 1000th state, etc.
    (2) However, after adding a time interval, it is possible to define a time range and require that the F formula be valid within that time range. F[1:4]a means that a should hold at least once in the following four states for the F formula to hold. With time constraints, the meaning of the formula becomes clearer and effectively improves the uncertainty of the F operator.

It can be seen that the extended LTL formula can express richer meanings with shorter formulas, increase the accuracy of the formula, and be more intuitive to use. Because the writing of logical formulas requires manual completion, this extension can reduce the burden on operators, reduce the difficulty of formula writing, and make the process of formula writing easier to complete.

4.2. Recognition and Storage of Formulas

The property specification in this study is represented by LTL, which has been extended in the previous section to simplify the writing difficulty of LTL formulas. How the LTL formula is processed is described below. To accomplish this, in this study, we define an LTL formula tree and corresponding conversion algorithm.

When obtaining an LTL formula, it is necessary to first store it for subsequent operations. Choosing a suitable data structure to store the formula can make subsequent processing more convenient. The operators in the LTL formula include logical operators and temporal operators and there is an order of priority to their execution. In addition, parentheses can also be used in the LTL formula to express high priority. The operation results of high-priority subformulas may serve as input variables for other subformulas. Based on the above points, this paper selects a tree structure to store formulas and defines a formula tree structure. All leaf nodes of the tree are propositional variables and all non-leaf nodes are operators. Because trees naturally have classification properties, they are suitable for splitting formulas, and the root node of one subtree is the leaf node of another subtree, which is similar to the "operation results of high-priority subformulas will serve as input variables for other subformulas" mentioned earlier.

Before executing the storage algorithm, you need to define the priority of the logical connective and temporal operator in the LTL formula to avoid priority conflicts during algorithm execution. Operator priority follows the following three rules:

(1) The priority of a unary operator is higher than that of a binary operator.
(2) The unary logical connective has the same priority as the unary temporal operator and the binary logical connective has the same priority as the binary temporal operator.
(3) Operators within parentheses have higher priority than those outside parentheses. The more layers of parentheses, the higher the priority.

Based on the above rules, in this study, we provide a ranking of priorities, as shown in Table 6. The priority of the three tables decreases from top to bottom, with operators in the same table having the same priority. Whatever operator has the highest relative position, has the highest priority.

(1) If the currently read object is a propositional variable, add it to the propositional variable stack.
(2) If the currently read object is an operator, compare the symbol priority of that operator with the top of the operator stack.
    (1) If the top of the stack is empty, the current operator is directly pushed onto the stack.
    (2) If the priority of the operator is higher than or equal to the top of the stack operator, then the operator is added to the operator stack.
    (3) If the priority of the operator is lower than the top of the stack operator, the top of the stack operator will exit the stack.
        (1) If the outbound operator is a unary operator, take an object from the propositional variable stack and construct a subtree. Using a unary operator as the parent node and an object as the right child node. Add the subtree to the propositional variable stack and repeat step (2) until the operator is added to the stack.
        (2) If the outbound operator is a binary operator, take two objects from the propositional variable stack and construct a subtree. The binary operator is used as the parent node, the object that first exits the stack is used as the right child node and the object that exits the stack later is used as the left child node. Add the subtree to the propositional variable stack and repeat step (2) until the operator is added to the stack.
(3) If the current operator is a left parenthesis, it is directly pushed onto the stack.
(4) If the current operator is a right parenthesis, the top of the stack operator will exit the stack.
    (1) If the stack operator is a left parenthesis, the current round of processing ends and the next object is read in
    (2) If the stack out operator is a unary operator, take an object from the propositional variable stack and construct a subtree. Using a unary operator as the parent node and an object as the right child node. Add the subtree to the propositional variable stack and repeat step (4) until (4.1) is established.
    (3) If the outbound operator is a binary operator, take an object from the propositional variable stack and construct a subtree. The binary operator is used as the parent node, the object that first exits the stack is used as the right child node and the object that exits the stack later is used as the left child node. Add the subtree to the propositional variable stack and repeat step (4) until (4.1) is established.

After the above process, the tree representation structure corresponding to the LTL formula can be generated.

Table 6. Operator precedence.

	Prioritization
High ↑ Low	¬ (Negation) G (Global) F (Future) X (Next) U (Until) R (Release) ∧ (And) ∨ (Or) → (Contain)

4.3. Conversion of Formulas

In the previous section, an LTL formula tree was used to store extended LTL formulas. The description of how the formula is transformed to reduce the complexity of the temporal operators is given below.

Table 7 gives the conversion rules applied in this study. In the table, f_1 is the category of formulas to be converted and f_2 is the specific conversion operation. The conversion process is to operate on the LTL formula tree generated in the previous section. The rules in this study are defined recursively and the tree structure naturally has recursive properties, so the two are easy to combine.

Table 7. Conversion rules.

Conversion Rules
$f_1 = G[i]a$
$f_1 = F[i]a$
$f_2 = \begin{cases} X(F[i-1]a) & \text{when } i > 0 \\ a & \text{when } i = 0 \end{cases}$
$f_1 = X[i]a$
$f_2 = \begin{cases} X(X[i-1]a) & \text{when } i > 0 \\ a & \text{when } i = 0 \end{cases}$
$f_1 = a\ U[i]\ b$
$f_2 = \begin{cases} X(U[i-1]a) & \text{when } i > 0 \\ a \vee b & \text{when } i = 0 \end{cases}$
$f_1 = G[i:j]a$
$f_2 = \begin{cases} X(G[i-1:j]a) & \text{when } i > 0 \\ a \wedge X(G[0:j-1]a) & \text{when } i = 0 \wedge j > 0 \end{cases}$
$f_1 = F[i:j]a$
$f_2 = \begin{cases} X(F[i-1:j]a) & \text{when } i > 0 \\ a \vee X(F[0:j-1]a) & \text{when } i = 0 \wedge j > 0 \end{cases}$
$f_1 = X[i:j]a$
$f_2 = \begin{cases} X(X[i-1:j]a) & \text{when } i > 0 \\ X(X[0:j-1]a) & \text{when } i = 0 \wedge j > 0 \end{cases}$
$f_1 = a\ U[i:j]\ b$
$f_2 = \neg G[i:j]\neg b \wedge G[0:i]a \wedge G[i:i](((a\ U\ b) \vee Ga) \wedge \neg G\neg b)$

To explain the G[i] conversion process, the following example is provided:

For formula G[i], when i is greater than 0, the formula needs to wait for i cycles to be established. In this study, we deal with i through recursion, adding an x operator before the formula to be processed each time and then recursing through the $i-1$ layer until the recursion is completed when i = 0. The corresponding operation in the tree is to generate a node representing the X operator to replace G[i] with G[i − 1] as its right child node and then recursively process the right child node.

4.4. Monitor Generation

In the previous process, the conversion of the formula was completed. Next, it is necessary to convert the LTL formula into a monitor. In this study, formulas are processed from the outside to the inside, where 'inside' and 'outside' refers to priority, with external operators having lower priority and internal operators having higher priority. Correspondingly, in the formula tree, the lower the level of an operator node, the closer it is to the outside, while the higher the level, the closer it is to the inside. For a subformula, the outermost operator is the root node of the corresponding subtree, and accessing the root node is simpler than accessing the left and right subtrees, indicating that the LTL formula tree plays a positive role.

In this study, we use $\Leftarrow$ to assign values to each member of a node. For example, New $\Leftarrow \{\varphi\}$ means adding formula φ to the New list of the current node. The program processes the Nodes list structure accordingly, where the nodes in Nodes and a special node init together form all the states of the monitor. Here, init is the initial state of the monitor, and Nodes is initialized as empty. In the following algorithm, function new_ID() creates a unique node ID value every time it is called. Function Neg() is taken as the Inverse function, $\text{Neg}(\text{A}) = \neg\text{A}, \text{Neg}(\neg\text{A}) = \text{A}, \text{Neg}(\text{True}) = \text{False}, \text{Neg}(\text{False}) = \text{True}$, where A is a proposition. Figure 4 shows the process of generating a monitor.

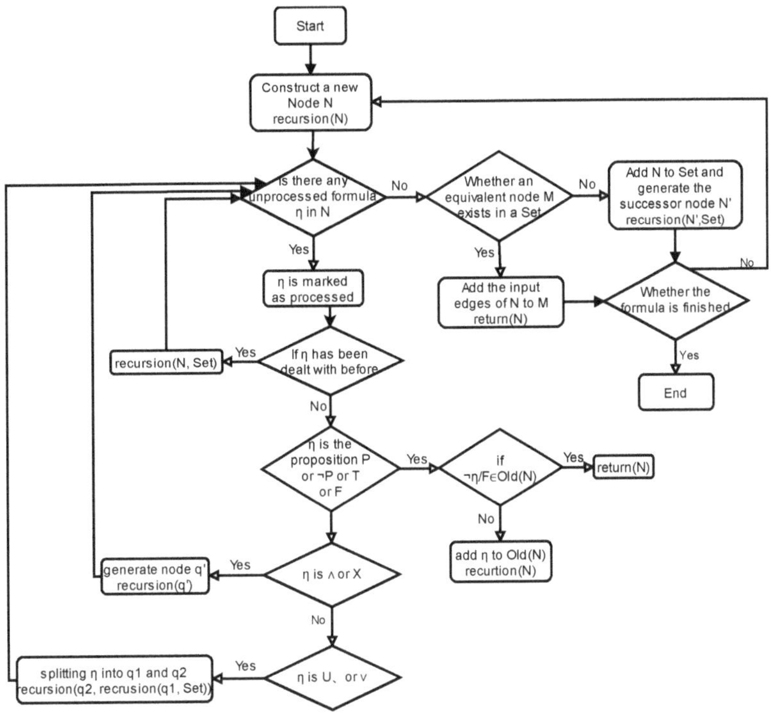

Figure 4. Monitor Generation Process.

5. Clang-Based Code Instrumentation Framework

This section introduces the relevant content of the code instrumentation framework, and the overall process is shown in Figure 5. First, a definition of instrumentation requirements is required. This process involves writing instrumentation requirements files that describe the characteristics of the functions to be instrumented and the content to be instrumented. The second step is to match the insertion position, which requires writing a suitable matcher based on the insertion requirements. The matcher's function is to find the insertion point that meets the requirements given in the target program. Finally, the execution of the instrumentation action will add interception functions at the matched instrumentation points and modify function calls in the target program. The process is detailed below. Section 5.1 introduces the definition of pile insertion requirements and Section 5.2 introduces the matching of pile insertion positions.

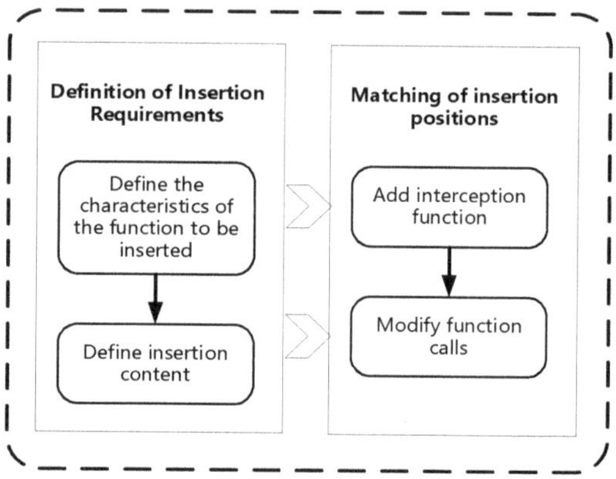

Figure 5. Code instrumentation framework.

5.1. Definition of Stake Insertion Requirements

The instrumentation requirement file needs to include several elements. First, the characteristic of the function to be instrumented (return value type, name, parameter) needs to be included. This is used to locate the relevant position in the source code. The more detail included about the feature, the more accurate the matching position will be. Second, the content and the corresponding location of the pile to be inserted need to be included. The content refers to the specific code to be inserted and the pile location refers to the code that can be inserted before and after the pile insertion point.

In this study, we investigated two methods of stake insertion: (1) to directly stake the definition of the objective function, and (2) to plug in a brand new function, which will call the target function and modify the call statement of the target function to the call statement of the new function to achieve interception of the target function. The first method involves directly modifying the function code—if you need to modify the logic of the function itself, you can consider using this method. The second method does not directly modify the original function code, but instead wraps the function in a layer and adds some statements before and after the function call to achieve certain functions, which are not strongly related to the target function itself. If there is such a requirement, then the second method is more suitable. Due to the purpose of instrumentation in this study being to pass relevant information about function calls to the monitor, the function of sending messages does not belong to the internal logic of the function. Therefore, this study chose to use the second method for instrumentation in the source code.

The specific format is shown in Table 8. The first line describes the characteristics of the objective function, that is, the desired return value type is return_ Type, while function_name is function_ Intercept functions with name and parameters as parameters. The function body is divided into three parts. The first part is parameter information, which is used to record the additional parameters required for the interception function. The second part is the code that needs to be instrumented before executing the target function. The third part is the code that needs to be instrumented after executing the target function. The call to the target function occurs between the second and third parts. If the target function has a return value, it will be converted into a temporary variable to save the return value, which will be returned after the interception function is executed.

Table 8. Format of Stake Insertion Requirements File.

Format
Return_type function_name (parameters) {
// new params
Define the parameters required for instrumentation
// before
Specifies the items to be inserted before function name is executed
// after
Specifies the items to be inserted after function name is executed
}

Table 9 gives a simple example of defining the requirements for instrumentation. In the table, lines 1–4 are the objective functions, 6–15 are the content of the instrumentation requirements file, and 16–22 are the interception functions. For example, there is a function func1 that prints "The func1 is called." (lines 1–4) when called. Now, it needs to be intercepted. The first line of the instrumentation requirement provides that the return value type of the current instrumentation's objective function is int, named func1, and has no formal parameters (line 5). The empty statement in the new params section indicates that the interceptor function does not require additional parameters (line 8). The before section is a print statement that will be executed before func1 is called (line 12). The after section is a print statement that will be executed after func1 is called (line 14). After obtaining this information, the instrumentation module will generate an interception function. The name of the interception function is the target function name preceded by "new_", indicating that this is an interception function for func1 (line 17). The return value type of the interceptor function is consistent with that of the target function. The parameters include at least the formal parameters of the target function and the Function pointer of the target function. The return value type of the interception function is consistent with the target function. If the target function has a return value, a temporary variable will be used to store its return value when calling the target function. After the interception function is executed, the value will be returned. In addition, the instrumentation function will replace the call statement of the target function in the source code with the call statement of the intercepting function to achieve interception of the target function.

Table 9. Example of Definition of Stake Insertion Requirements.

Example
1. int func1(){
2. printf("The funcl is called.");
3. return 0;
4. }
5. int funcl(){
6. // before
7. printf("Before funcl is called.");
8. // after
9. printf("After funcl is called.");
10. }
11. int new funcl(int (*func1)()){
12. printf("Before funcl is called."); // before
13. int result = funcl();
14. printf("After funcl is called."); // after
15. return result;
16. }

After obtaining the requirement file, it is necessary to read it for parsing, including checking for any errors and extracting the instrumentation information from it. Due to the fact that the format of the instrumentation requirement file in this study references the writing method of the C programming language, it is possible to use the Clang compiler to perform grammar checks on the requirement file. If an error is found, the information will not be extracted and the error needs to be corrected first. This is because, if an erroneous requirement is parsed, it may introduce errors into subsequent instrumentation and monitoring processes, affecting the correct execution of subsequent processes. If no errors are detected, the parsing module will extract the instrumentation information for storage and provide it to the instrumentation execution module.

5.2. Matching of Pile Insertion Positions

5.2.1. Selection of Matching Methods

Code written using AST Matcher is more streamlined and has higher readability; the matching logic is also more natural. Therefore, in this study, we chose the AST Matcher method for node matching. The class diagram of node matching and code instrumentation using the AST Matcher method used in this study is shown in Figure 6. There are three core classes. The MyFrontendAction class is obtained by inheriting ASTFrontendAction, the MyASTConsumer object is created by the CreateASTConsumer method of the parent class, and the Member variable of the Rewriter class is declared. MyASTConsumer inherits from the ASTConsumer class and uses the HandleTranslationUnit method to obtain translation units. At the same time, it is declared that the Member variable of the MatchFinder class is used to match the insertion position and the Member variable of the MyHandler class is used for code insertion. The MyHandler class inherits from the MatchCallback class and processes the position matched by the MatchFinder object by calling the parent class run method.

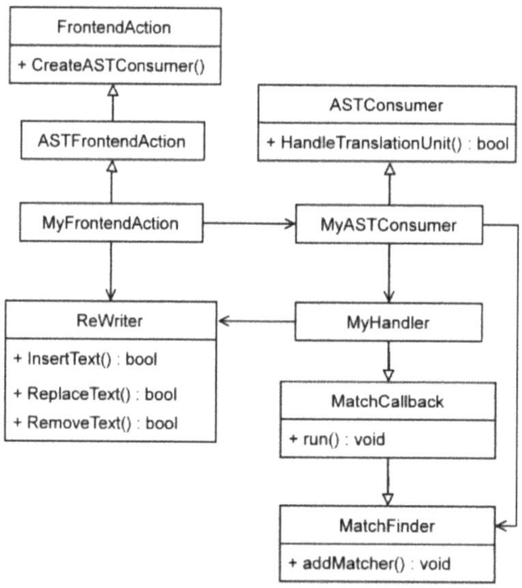

Figure 6. Code instrumentation class diagram.

5.2.2. Implementation of Pile Insertion Action

In the previous section, the matching of the target nodes has been achieved and their positions in the AST have been found. Now, they must be staked. Clang's Rewriter class provides the ability to add, delete, and replace source code, making it possible for code instrumentation and refactoring. When using the AST Matcher method, matching and processing nodes often occur in pairs. A matcher is added through addMatcher in the AST Consumer to find the desired node. At the same time, developers need to derive their own

ASTNodeHandler from the MatchFinder:: MatchCallback class, override the run method of the parent class, and implement node processing.

Taking the Func. c file as an example, the definition and call section of the Func1 function have been successfully matched in the previous section. Next, we will intercept it through instrumentation. The main process is to generate a new interceptor function, with "new_" added before the original function name to indicate that it is a newly generated interceptor function. The parameters of the interceptor function need to include all the parameters of the original function and the Function pointer of the original function so that the original function can be called inside the interceptor function. The function body of the interception function is divided into three parts: The first part is the before part of the instrumentation requirement file. The second part is the call to the original function. If the original function has a return value, a temporary variable will be used to temporarily store its return value and it will be returned at the end of the intercepting function. The third part is the after section of the instrumentation requirements document. At the same time, all calls to the original function in the source code need to be replaced by calls to the interceptor function and the Function pointer of the original function needs to be passed as a parameter to the interceptor function.

6. Experimental Analysis

6.1. Experiment Settings

The object under test in this paper is a certain type of digital drone, which includes drone code and console code. The basic attributes of the drone can be set to simulate different flight tasks, such as airport altitude, drone weight, and flight path. The drone system fault mode database contains many types of faults. We select four fault categories of fault to test in this experimental study, with each category representing a safety attribute. The categories are common attribute faults, an abnormal state, external interference, and combined faults. The common attribute experiment tests whether the drones can follow the flight plan to fly, that is, whether it can enter the specified flight stage within the specified time range and maintain its flight state. The abnormal state experiment tests whether some of the drone sensors are abnormal, which may be due to component failure or vulnerability attack. The external interference experiment tests whether the signals received by some sensors of the drones are affected by external interference, which would lead to abnormal sensor data. The combined faults experiment is a comprehensive experiment that considers multiple safety-requirement-triggering conditions. These four safety attributes are translated into LTL formulas and then implemented as monitors to analyze the characteristics of the drone during each flight cycle. Upon completion of the flight, sensor data and monitor status throughout the flight are displayed in graphical form, with time on the horizontal axis and sensor data or monitor status on the vertical axis.

6.2. Extraction of Drone Safety Requirements

Before monitoring a drone flight, it is necessary to first extract the safety requirements related to the drone, namely, the conditions and states that the drone should satisfy during operation. This study is based on using the fault mode database to extract safety requirements related to drone functionality. In this study, we summarize the power, electrical, power system, engine, communication link, and other aspects closely related to the safety of unmanned aerial vehicles during the ground start, takeoff, and flight stages. A total of 26 categories and about 80 safety requirements are proposed and corresponding safety requirement formulas are written. Due to space limitations, Table 10 only lists some typical safety requirements that affect drone safety.

Table 10. Drone safety requirements.

Operation Phase	Security Requirements Description
Ground waiting	The ground waiting phase enters the takeoff ground segment when the following conditions are met simultaneously: The route and airport altitude have been bound; Received takeoff instructions. G((phase = 0 & air_line = 1 & pressure_altitude = 1 & takeoff) → phase = 1)
Takeoff ground segment	In the takeoff ground segment, when the airspeed is greater than 30 m/s and the altitude is greater than 15 m, turn to the takeoff departure zone. G((phase = 1 & airspeed > 30 & height > 15) → phase = 2)
Takeoff departure zone	In the takeoff departure zone, when the flight altitude is greater than the safe altitude of 80 m, enter the climb phase 1. G((phase = 2 & height > 80) → phase = 3)

Some common attributes of the drone need to be changed to ensure that the drone can fly safely and complete its mission successfully. Therefore, in this study, we define some protocols to detect whether the drone is flying normally according to the expected plan. Taking the altitude data as an example, the safety attribute is defined. After taking off, the drones should reach the flight height of 200 m within 900 cycles and keep to this altitude within the range of (195,205) m for the rest of the flight until the flight state becomes "descending". In addition, another highly related safety requirement is tested simultaneously. When the altitude of the drone is less than 300 m, the radio altimeter must be activated and remain active (altitude status value is 2). The formal formula is as follows.

$$G(F[0,900](\text{height} \geq 200)\&(\text{phase} = \text{fixed\_height} \rightarrow ((195 \leq \text{height} \leq 205)\ U\ (\text{phase} = \text{descent}))));G(\text{height} < 300 \rightarrow \text{height\_state} = \text{radio})$$

Figure 7 shows the altitude data, altimeter status, and monitor results during a common attributes test flight. As can be seen, the drone did not reach a height of 200 m after 900 cycles and the monitor detected an anomaly (setting the result to 1). Then, during the level flight phase, the altitude of the drone remained around 200 m, so the monitor did not detect any abnormalities. The drone remained at an altitude of less than 300 m during flight, so the radio altimeter remained active and met safety requirements. The monitor did not detect any abnormalities.

6.3. Abnormal State Experiment

During the operation of drones, they may encounter some abnormal situations, including internal component failures or machine vulnerabilities that are exploited by a hostile actor and maliciously attacked, resulting in unexpected results in the range or trend of drone sensor data, leading to abnormal flight status and ultimately to the drone crashing.

The drone obtains altitude data through altitude sensors. If the sensor malfunctions or is subjected to external attacks (attackers constantly send incorrect altitude data to the drone, causing it to make incorrect judgments about the current altitude), the drone will continuously adjust its pitch angle to achieve the desired altitude. Therefore, this experiment detects whether the drone is experiencing abnormal conditions by determining the fluctuation of the drone's pitch angle. The safety attributes are defined as, when the drone is in level flight, the variance of the pitch angle data for the first 300 cycles ending with the current cycle must be calculated and its value should not continue to exceed 5. The formal formula is as follows.

$$G((\text{state} = \text{fixed\_height}) \rightarrow (\text{VAR}_{300}(\text{pitch}) < 5))$$

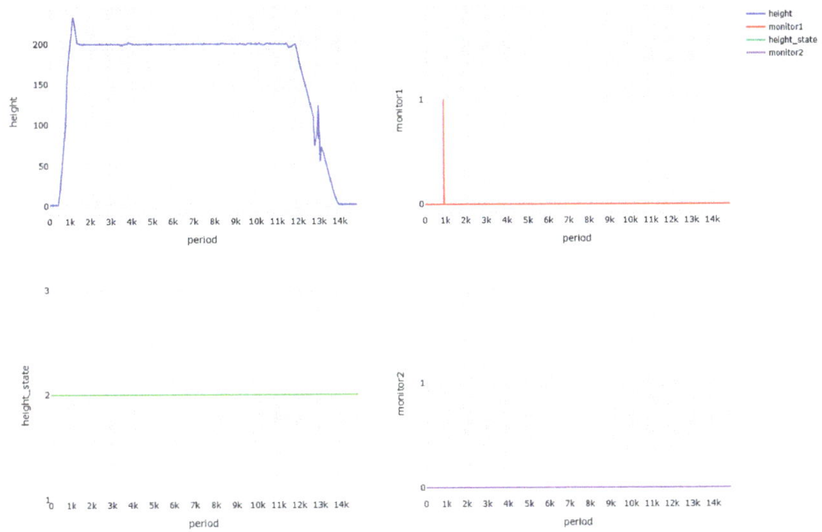

Figure 7. Height and monitor results of a common attributes test flight.

Figure 8 shows the pitch angle data, variance, and monitor detection results during the flight. It can be seen that the pitch angle oscillates at about 5000 cycles, its variance exceeds 5 at about 5100 cycles, and the monitor value detects an abnormal situation after the pitch angle variance is greater than 5 for 300 consecutive cycles.

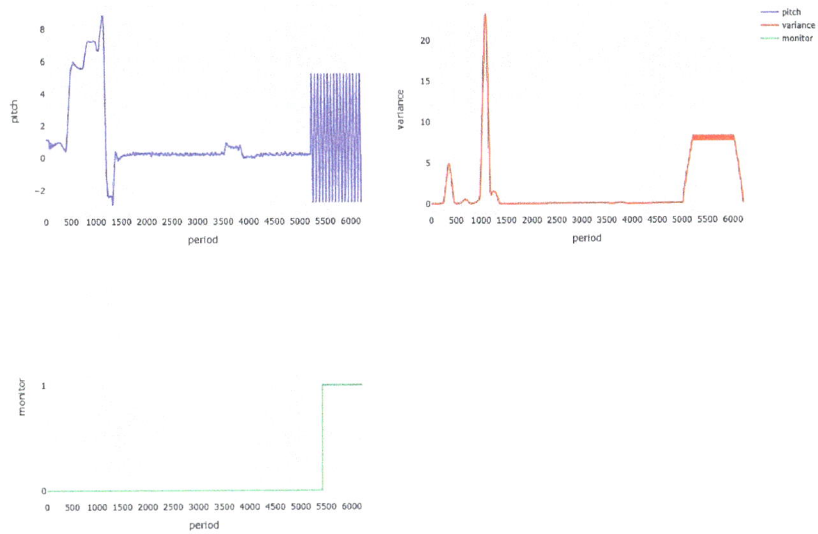

Figure 8. Pitch angle and monitor results for an abnormal state test flight.

6.4. External Interference Experiment

Currently, under normal circumstances, many drones will use GPS navigation to locate their position to realize flight path planning and control. Therefore, the flight process of drones is closely related to GPS data. If the GPS signal is interfered with, the normal flight of drones will be affected. The use of GPS decoys is a type of attack that interferes with the normal flight of drones by preventing them from receiving real GPS signals and making them receive false GPS signals. A jammer intercepts the real GPS signal and emits many false GPS signals. Since a drone always selects GPS signals from the strongest source, the real signal can be drowned in noise if the fake signal from the jammer is strong enough. So the drone can only receive false signals, which indirectly control the drone's navigation system and trick the drone into flying to the wrong place. If the drone deviates from its normal course, its roll angle changes, so roll angle data can be analyzed to detect GPS jamming attacks. In this experiment, several positioning points are set for the drone. After the drone reaches the positioning point in a straight flight, it will turn and after the direction is adjusted, it will fly in a straight line again to the next positioning point.

A drone's GPS receiver is vulnerable to external attack, but other sensors such as gyroscopes, levels, and barometric altimeters are generally difficult to attack. Changes in gyroscope roll can determine when a drone veers off course after a GPS decoy attack. Since the roll angle of the drone will not change greatly when it flies along a straight line, it will change only when it turns to the navigation point after some time. Therefore, the following safety attributes are defined. When the drone is flying, it will be given 1000 cycles to adjust the direction after reaching a navigation point. Then the absolute value of the roll angle of the drones shall be less than 5 until it reaches the next anchor point. The formal formula is shown as follows:

$$\mathrm{G}(\mathrm{arrive} \rightarrow (\mathrm{X}[1000]\mathrm{G}(-5 < \mathrm{rool} < 5)\ \mathrm{U}\ \mathrm{arrive}))$$

Figure 9 shows the roll angle data and monitor detection results during the flight process, with four positioning points reaching approximately 1000, 3000, 9000, and 10,000 cycles. It can be seen that, in the interval between the drone reaching the positioning point for 1000 cycles and reaching the next positioning point, when the absolute value of the roll angle is greater than 5 (approximately over 12,000 cycles), the monitor detects an abnormal situation. When the absolute value of the roll angle is less than 5, the monitor returns to normal.

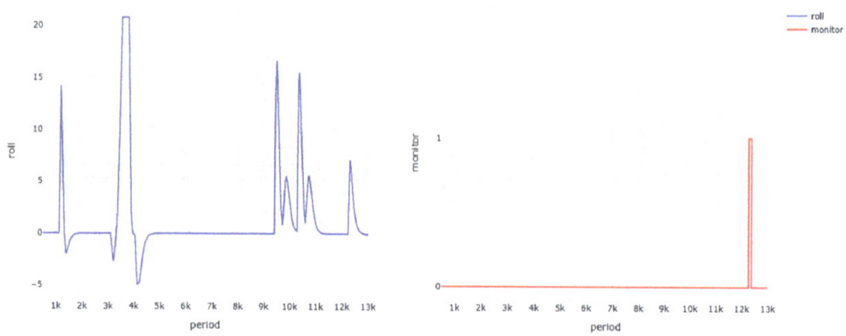

Figure 9. Roll angle and monitor results for external interference test flight.

6.5. Combined Faults Experiment

During the flight of unmanned aerial vehicles, some changes in safety status cannot be judged by a single condition, but require the monitoring of a combination of multiple conditions and the corresponding safety requirements are also composed of multiple events. For example, the following safety requirements exist to determine whether the drones can normally enter the air stop state when the conditions are met:

(1) The speed is less than 1000 RPM;
(2) The lubricating oil pressure is less than 1.5 bar;
(3) The landing gear is in a compressed state;
(4) Flight altitude greater than 60 m;
(5) The flight phase is not a ground hold.

When all the above conditions are met simultaneously, then

(1) Judge the engine's air stop;
(2) When the flight phase is before glide down 1, place a falling flag.

The corresponding security requirement formula is:

$$G((rotate\_speed < 1000 \; \& \; oil\_press < 1.5 \; \& \; undercarriage = 0 \; \& \; height > 60 \\ \& \; phase \neq ground\_wait) \rightarrow (engine\_status \\ = (flameout\_air \mid pre\_start \mid start \mid start\_estimate) \; \& \; phase = drift))$$

Figure 10 shows the values of various attributes and monitor results in the air parking monitoring experiment. From the graph, it can be seen that for around 3000 cycles, the data of the preconditions met the requirements for air parking. At this time, the value of the engine status changed to 1 (air parking sign), but the value during the flight phase did not change to 13 (falling sign). Therefore, the monitor discovered an abnormal situation. Subsequently, the engine was restarted and after the successful restart, the engine status returned to normal. The preconditions were no longer met and the monitor returned to normal.

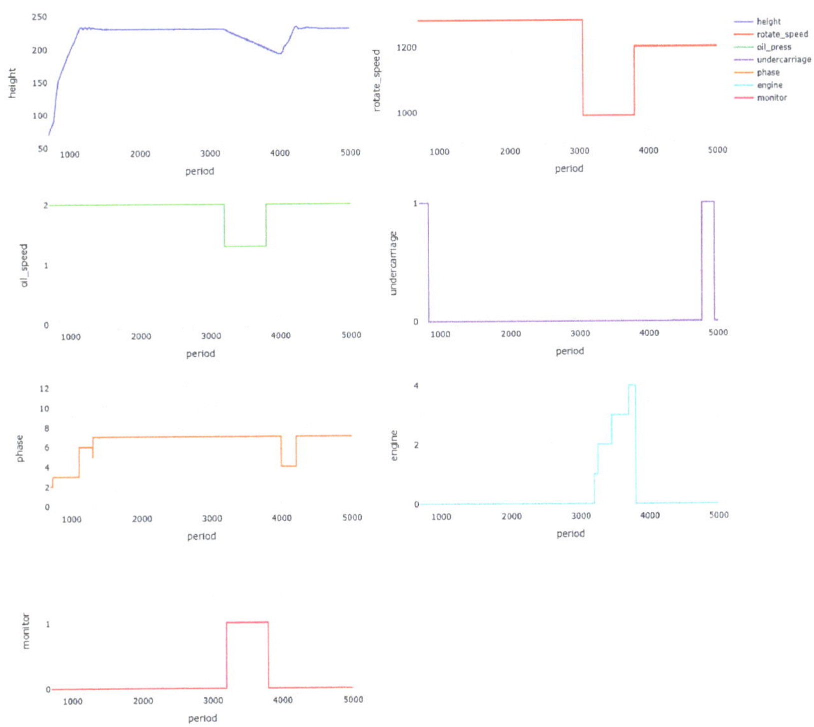

Figure 10. Airborne parking monitoring results.

7. Conclusions

UAVs are widely used in various fields and their safety and reliability are of great concern. In order to improve the safety of UAVs, in this paper, we propose a fault detection mechanism based on a combination of a fault mode database and runtime verification. According to the fault type of the UAV identified from the fault mode database, the key safety requirements and attributes of the UAV during operation can be extracted. Combining this information with runtime verification, the key safety attributes of the UAV during operation are monitored to achieve real-time monitoring and monitoring.

Experimental results show that the proposed method is valid and effective. It can realize efficient and automated real-time monitoring of a drone in flight, timely detect abnormal conditions of the drone, and provide early warning. The method proposed in this paper can effectively improve the efficiency of safety monitoring in the process of UAV operation to significantly improve the safety level of UAVs. In future research, the extraction process of security attributes will be further enriched. Working out how to extract a safety attribute that is closer to the specific fault type experienced by the drone is a shortcoming of this paper. At present, only system-level security attributes can be monitored and it is still very difficult to analyze the causes of anomalies. In the future, fault modes at the subsystem level will be analyzed to extract more detailed security attributes. Safety attributes will also be combined with more accurate fault location, as it is of great importance to realize accurate fault location alongside rapid anomaly detection.

Author Contributions: Conceptualization, D.H. and Q.S.; methodology, Q.S.; software, Y.S.; validation, D.H., Q.S. and Y.S.; investigation, Y.S.; resources, Y.Y.; data curation, D.H.; writing—original draft preparation, Q.S.; writing—review and editing, Q.S.; visualization, Q.S.; supervision, Y.Y.; project administration, Y.Y.; funding acquisition, Y.Y. All authors have read and agreed to the published version of the manuscript.

Funding: This research received no external funding.

Data Availability Statement: The data presented in this study are available on request from the corresponding author. The data are not publicly available due to privacy.

Conflicts of Interest: The authors declare no conflict of interest.

References

1. Tian, C.; Zhang, X.; Lin, J.C.W.; Zuo, W.; Zhang, Y.; Lin, C.W. Generative adversarial networks for image super-resolution: A survey. *arXiv* **2022**, arXiv:2204.13620.
2. Jiang, Y.; Gao, Y.; Song, W.; Li, Y.; Quan, Q.U.A.N. Bibliometric analysis of drones swarms. *J. Syst. Eng. Electron.* **2022**, *32*, 55–67.
3. Restás, A. Drone Applications Fighting COVID-19 Pandemic—Towards Good Practices. *Drones* **2022**, *6*, 15. [CrossRef]
4. Sibanda, M.; Mutanga, O.; Chimonyo, V.G.P.; Clulow, A.D.; Shoko, C.; Mazvimavi, D.; Dube, T.; Mabhaudhi, T. Application of Drone Technologies in Surface Water Resources Monitoring and Assessment: A Systematic Review of Progress, Challenges and Opportunities in the Global South. *Drones* **2021**, *5*, 84. [CrossRef]
5. Shakhatreh, H.; Sawalmeh, A.H.; Al-Fuqaha, A.; Dou, Z.; Almaita, E.; Khalil, I.; Othman, N.S.; Khreishah, A.; Guizani, M. Unmanned Aerial Vehicles (UAVs): A Survey on Civil Applications and Key Research Challenges. *IEEE Access* **2019**, *7*, 48572–48634. [CrossRef]
6. Doherty, P.; Rudol, P. A UAVs search and rescue scenario with human body detection and geolocalization. In *AI 2007: Advances in Artificial Intelligence*; Orgun, M.A., Thornton, J., Eds.; Springer: Berlin/Heidelberg, Germany, 2007; pp. 1–13.
7. Puchalski, R.; Giernacki, W. UAVs fault detection methods, state-of-the-art. *Drones* **2022**, *6*, 330. [CrossRef]
8. D'Amato, E.; Nardi, V.A.; Notaro, I.; Scordamaglia, V. A Particle Filtering Approach for Fault Detection and Isolation of UAVs IMU Sensors: Design, Implementation and Sensitivity Analysis. *Sensors* **2021**, *21*, 3066. [CrossRef]
9. Fu, X.; Geng, X. Fault Estimation and Robust Fault-tolerant Control for Singular Markov Switching Systems with Mixed Time-Delays and UAVs Applications. *J. Control Eng. Appl. Inform.* **2021**, *23*, 53–66.
10. Maqsood, H.; Taimoor, M.; Ullah, Z.; Ali, N.; Sohail, M. Novel Sensor Fault Detection and Isolation for an Unmanned Aerial Vehicle. In Proceedings of the 2021 International Bhurban Conference on Applied Sciences and Technologies (IBCAST), Islamabad, Pakistan, 12–16 January 2021; pp. 486–493.
11. Miao, Q.; Wei, J.; Wang, J.; Chen, Y. Fault Diagnosis Algorithm Based on Adjustable Nonlinear PI State Observer and Its Application in UAVs Fault Diagnosis. *Algorithms* **2021**, *14*, 119. [CrossRef]
12. Nejati, Z.; Faraji, A.; Abedi, M. Robust Three Stage Central Difference Kalman Filter for Helicopter Unmanned Aerial Vehicle Actuators Fault Estimation. *Int. J. Eng.* **2021**, *34*, 1290–1296.
13. Sun, K.; Gebre-Egziabher, D. Air data fault detection and isolation for small UAS using integrity monitoring framework. *Navigation* **2021**, *68*, 577–600. [CrossRef]
14. Cao, L.; Yang, X.; Wang, G.; Liu, Y.; Hu, Y. Fault detection based on extended state observer and interval observer for UAVs. *Aircr. Eng. Aerosp. Technol.* **2022**, *71*, 1759–1771. [CrossRef]
15. Gai, W.; Li, S.; Zhang, J.; Zheng, Y.; Zhong, M. Dynamic Event-Triggered Hi/H∞ Optimization Approach to Fault Detection for Unmanned Aerial Vehicles. *IEEE Trans. Instrum. Meas.* **2022**, *71*, 1–11. [CrossRef]
16. Lin, C.E.; Shao, P.C. Failure analysis for an unmanned aerial vehicle using safe path planning. *J. Aerosp. Inf. Syst.* **2020**, *17*, 358–369. [CrossRef]

17. Altinors, A.; Yol, F.; Yaman, O. A sound based method for fault detection with statistical feature extraction in UAVs motors. *Appl. Acoust.* **2021**, *183*, 108325. [CrossRef]
18. Park, K.H.; Park, E.; Kim, H.K. Unsupervised Fault Detection on Unmanned Aerial Vehicles: Encoding and Thresholding Approach. *Sensors* **2021**, *21*, 2208. [CrossRef]
19. Souza, J.S.; Bezerril, M.C.; Silva, M.A.; Veras, F.C.; Lima-Filho, A.; Ramos, J.G.; Brito, A.V. Motor speed estimation and failure detection of a small UAVs using density of maxima. *Front. Inf. Technol. Electron. Eng.* **2021**, *22*, 1002–1009. [CrossRef]
20. Zheng, K.; Jia, G.; Yang, L.; Wang, J. A Compound Fault Labeling and Diagnosis Method Based on Flight Data and BIT Record of UAVs. *Appl. Sci.* **2021**, *11*, 5410. [CrossRef]
21. Cabahug, J.; Eslamiat, H. Failure Detection in Quadcopter UAVs Using K-Means Clustering. *Sensors* **2022**, *22*, 6037. [CrossRef]
22. Zhang, Q.; Xiao, J.; Tian, C.; Chun-Wei Lin, J.; Zhang, S. A robust deformed convolutional neural network (CNN) for image denoising. *CAAI Trans. Intell. Technol.* **2022**, *8*, 331–342. [CrossRef]
23. Tian, C.; Zheng, M.; Zuo, W.; Zhang, B.; Zhang, Y.; Zhang, D. Multi-stage image denoising with the wavelet transform. *Pattern Recognit.* **2023**, *134*, 109050. [CrossRef]
24. Aviation Safety Reporting System. NASA, FAA. Available online: https://asrs.arc.nasa.gov/ (accessed on 20 February 2022).
25. Aviation Safety Reporting System. The Department of the Interior (DOI) and the U.S. Forest Service (USFS). Available online: https://www.safecom.gov/about (accessed on 20 February 2022).
26. Australian Transport Safety Bureau (ATSB). Available online: https://www.atsb.gov.au/ (accessed on 20 July 2023).
27. Air Accidents Investigation Branch (AAIB). Available online: https://www.gov.uk/government/organisations/air-accidents-investigation-branch (accessed on 20 July 2023).
28. Wild, G.; Murray, J.; Baxter, G. Exploring civil drone accidents and incidents to help prevent potential air disasters. *Aerospace* **2016**, *3*, 22. [CrossRef]
29. Real-World Faults and Their Injection into Autonomous Unmanned Aerial Vehicles. Available online: https://haotianchen.net/project/adfi (accessed on 20 July 2023).
30. Abbas, H. Work-in-Progress: Private Runtime Verification. In Proceedings of the 2019 International Conference on Embedded Software (EMSOFT), New York, NY, USA, 13–18 October 2019; pp. 1–2.
31. Liu, Y.; He, C. A Heuristics-Based Incremental Probabilistic Model Checking at Runtime. In Proceedings of the 2020 IEEE 11th International Conference on Software Engineering and Service Science (ICSESS), Beijing, China, 16–18 October 2020; pp. 355–358.
32. Rozier, K.Y. From Simulation to Runtime Verification and Back: Connecting Single-Run Verification Techniques. In Proceedings of the 2019 Spring Simulation Conference (SpringSim), Tucson, AZ, USA, 29 April–2 May 2019; pp. 1–10.
33. Stockmann, L.; Laux, S.; Bodden, E. Architectural Runtime Verification. In Proceedings of the 2019 IEEE International Conference on Software Architecture Companion (ICSA-C), Hamburg, Germany, 25–26 March 2019; pp. 77–84.
34. Teixeira, L.; Miranda, B.; Rebêlo, H.; d'Amorim, M. Demystifying the Challenges of Formally Specifying API Properties for Runtime Verification. In Proceedings of the 2021 14th IEEE Conference on Software Testing, Verification and Validation (ICST), Porto de Galinhas, Brazil, 12–16 April 2021; pp. 82–93.
35. Bicevskis, J.; Bicevska, Z.; Nikiforova, A.; Oditis, I. Towards Data Quality Runtime Verification. In Proceedings of the 2019 Federated Conference on Computer Science and Information Systems (FedCSIS), Leipzig, Germany, 1–4 September 2019; pp. 639–643.
36. Lee, E.; Seo, Y.-D.; Kim, Y.-G. A Cache-Based Model Abstraction and Runtime Verification for the Internet-of-Things Applications. *IEEE Internet Things J.* **2020**, *7*, 8886–8901. [CrossRef]
37. Legunsen, O.; Zhang, Y.; Hadzi-Tanovic, M.; Rosu, G.; Marinov, D. Techniques for Evolution-Aware Runtime Verification. In Proceedings of the 2019 12th IEEE Conference on Software Testing, Validation and Verification (ICST), Xi'an, China, 22–27 April 2019; pp. 300–311.
38. Geng, T.; Njilla, L.; Huang, C.-T. Smart Markers in Smart Contracts: Enabling Multiway Branching and Merging in Blockchain for Decentralized Runtime Verification. In Proceedings of the 2021 IEEE Conference on Dependable and Secure Computing (DSC), Aizuwakamatsu, Japan, 30 January–2 February 2021; pp. 1–8.
39. Ring, M.; Bornebusch, F.; Luth, C.; Wille, R.; Drechsler, R. Verification Runtime Analysis: Get the Most Out of Partial Verification. In Proceedings of the 2020 Design, Automation & Test in Europe Conference & Exhibition (DATE), Grenoble, France, 9–13 March 2020; pp. 873–878.
40. Ye, X.; Liu, W.; Wang, N. Runtime Verification of Multi-Agent Self-Adaptive System. In Proceedings of the 2021 IEEE 24th International Conference on Computer Supported Cooperative Work in Design (CSCWD), Dalian, China, 5–7 May 2021; pp. 12–17.
41. Tsigkanos, C.; Bersani, M.M.; Frangoudis, P.A.; Dustdar, S. Edge-Based Runtime Verification for the Internet of Things. *IEEE Trans. Serv. Comput.* **2021**, *15*, 16.
42. Hu, C.; Dong, W.; Yang, Y.; Shi, H.; Zhou, G. Runtime Verification on Hierarchical Properties of ROS-Based Robot Swarms. *IEEE Trans. Reliab.* **2020**, *69*, 674–689. [CrossRef]
43. Tracy, T.; Tabajara, L.M.; Vardi, M.; Skadron, K. *Runtime Verification on FPGAs with LTLf Specifications*; TU Wien Academic Press: Vienna, Austria, 2020; pp. 36–46.
44. Ye, M.; Feng, X.; Wei, S. Runtime Hardware Security Verification Using Approximate Computing: A Case Study on Video Motion Detection. In Proceedings of the 2019 Asian Hardware Oriented Security and Trust Symposium (AsianHOST), Xi'an, China, 16–17 December 2019; pp. 1–6.

45. Kong, S.; Lu, M.; Li, L.; Gao, L. Runtime Monitoring of system Execution Trace: Method and Tools. *IEEE Access* **2020**, *8*, 114020–114036. [CrossRef]
46. Jung, B.; Kruse, P.M. Runtime Prioritization with the Classification Tree Method for Test Automation. In Proceedings of the 2020 IEEE International Conference on Software Testing, Verification and Validation Workshops (ICSTW), Porto, Portugal, 24–28 October 2020; pp. 376–379.
47. Miranda, B.; Lima, I.; Legunsen, O.; d'Amorim, M. Prioritizing Runtime Verification Violations. In Proceedings of the 2020 IEEE 13th International Conference on Software Testing, Validation and Verification (ICST), Porto, Portugal, 24–28 October 2020; pp. 297–308.
48. Schirmer, S.; Torens, C. Safe operation monitoring for specific category unmanned aircraft. In *Automated Low-Altitude Air Delivery: Towards Autonomous Cargo Transportation with Drones*; Springer: Berlin/Heidelberg, Germany, 2021; pp. 393–419.
49. Gorostiaga, F.; Zudaire, S.; Sánchez, C.; Schneider, G.; Uchitel, S. Assumption monitoring of temporal task planning using stream runtime verification. In *International Symposium on Leveraging Applications of Formal Methods*; Springer: Cham, Switzerland, 2022; pp. 397–414.
50. Bonnah, E.; Hoque, K.A. Runtime monitoring of time window temporal logic. *IEEE Robot. Autom. Lett.* **2022**, *7*, 5888–5895. [CrossRef]

drones

MDPI

Article

Cooperative Standoff Target Tracking Using Multiple Fixed-Wing UAVs with Input Constraints in Unknown Wind

Zhong Liu [1,*], Lingshuang Xiang [2] and Zemin Zhu [1]

[1] School of Computer Science, Huanggang Normal University, Huanggang 438000, China; jsjzzm@hgnu.edu.cn
[2] ChangJiang Industry Investment Group Co., Ltd, Wuhan 430062, China; xianglingshuang@cjchanye.com
* Correspondence: liuzhong@hgnu.edu.cn; Tel.: +86-158-2973-2829

Abstract: This paper investigates the problem of cooperative standoff tracking using multiple fixed-wing unmanned aerial vehicles (UAVs) with control input constraints. In order to achieve accurate moving target tracking in the presence of unknown background wind, a coordinated standoff target tracking algorithm is proposed. The objective of the research is to steer multiple UAVs to fly a circular orbit around a moving target with prescribed intervehicle angular spacing. To achieve this goal, two control laws are proposed, including relative range regulation and space phase separation. On one hand, a heading rate control law based on a Lyapunov guidance vector field is proposed. The convergence analysis shows that the UAVs can asymptotically converge to a desired circular orbit around the target, regardless of their initial position and heading. Through a rigorous theoretical proof, it is concluded that the command signal of the proposed heading rate controller will not violate the boundary constraint on the heading rate. On the other hand, a temporal phase is introduced to represent the phase separation and avoid discontinuity of the wrapped space phase angle. On this basis, a speed controller is developed to achieve equal phase separation. The proposed airspeed controller meets the requirements of the airspeed constraint. Furthermore, to improve the robustness of the aircraft during target tracking, an estimator is developed to estimate the composition velocity of the unknown wind and target motion. The proposed estimator uses the offset vector between the UAV's actual flight path and the desired orbit, which is defined by the Lyapunov guidance vector field, to estimate the composition velocity. The stability of the estimator is proved. Simulations are conducted under different scenarios to demonstrate the effectiveness of the proposed cooperative standoff target tracking algorithm. The simulation results indicate that the temporal-phase-based speed controller can achieve a fast convergence speed and small phase separation error. Additionally, the composition velocity estimator exhibits a fast response speed and high estimation accuracy.

Keywords: multiple UAVs; cooperative control; target tracking; Lyapunov guidance vector field; unknown background wind

Citation: Liu, Z.; Xiang, L.; Zhu, Z. Cooperative Standoff Target Tracking Using Multiple Fixed-Wing UAVs with Input Constraints in Unknown Wind. *Drones* **2023**, *7*, 593. https://doi.org/10.3390/drones7090593

Academic Editor: Pablo Rodríguez-Gonzálvez

Received: 22 August 2023
Revised: 11 September 2023
Accepted: 12 September 2023
Published: 20 September 2023

1. Introduction

Tracking a moving ground target is one of the important capabilities of UAVs [1]. Making the tracking process automatic and free of human intervention is essential for relieving the burden on UAV operators and improving the efficiency and safety of UAV missions. The goal of this paper is to develop a control scheme that allows multiple fixed-wing UAVs to cooperatively track a moving ground target in unknown windy conditions.

Target tracking using multiple fixed-wing UAVs remains a challenge. On one hand, the motion of fixed-wing UAVs is subject to various input constraints. To prevent stalling, a fixed-wing aircraft cannot hover and must maintain a positive forward airspeed. Accordingly, the UAVs must fly in a circle around the ground-based moving target [2]. On the other hand, to avoid collisions and to ensure that the sensors can cover the target, the UAVs need to be evenly distributed around the target and maintain a certain phase interval [3]. Standoff tracking is a possible solution for target tracking using a team of fixed-wing UAVs.

In this pattern, the UAVs keep a certain distance (called the standoff radius) from the target and move in a circular motion (termed the standoff circle) at a proper altitude relative to the target.

In cooperative target tracking missions with unknown wind, three main technical issues should be considered: (1) Relative distance regulation, which focuses on how to enable the UAVs to converge to a circular orbit with a prescribed standoff radius around the target by controlling their headings [4]. (2) Intervehicle phase separation, which focuses on distributing the hovering UAVs uniformly over a standoff circle with a certain angular phase difference by controlling their airspeeds [5]. (3) Background wind resistance, which focuses on how to achieve robust stability for UAVs performing standoff tracking in the presence of wind and the target's motion [6].

Relative distance regulation, which aims to steer the UAV to a circular orbit around the target, is the key to achieving standoff tracking of a moving target with a single UAV. Typically, a guidance law is proposed to regulate the position of the UAV on a predefined circular path. The path usually guides the UAV to circle around a ground-based moving target at a constant distance. Consequently, the UAV trajectory can be expressed as a circle with a predefined standoff radius in the target's frame. There are various types of the guidance law, including reference point guidance (RPG) [7–9], Lyapunov guidance vector field (LGVF) [10–14], and so on. Based on a predefined target tracking path, the standoff tracking problem of a ground-based moving target can be converted into a path following problem [7]. For example, in [8,9], a nonlinear guidance law is proposed to achieve path following for a curved path. In this approach, each point on the curved path is designated as a reference point, and a lateral acceleration command is generated to drive the UAV to the reference point. However, because the ground-based target's moving speed is usually much slower than that of the UAV, the RPG method cannot be directly applied to the standoff target tracking problem. As a new form of potential field, the LGVF is introduced to guide the UAVs to achieve standoff target tracking. In [10], an LGVF is proposed for hovering maneuvers around a stationary target. This approach also enables moving target tracking, but it may lead to slow convergence due to constant curvature in the LGVF. To shorten the convergence time, the authors in [11] combine the tangent with the LGVF, while in [12,13], the authors add the circulation parameter c into the original LGVF. The shape of the LGVF can be adaptively adjusted by changing the circulation parameter. In this way, a faster convergence to the standoff circle can be achieved due to a higher contraction component. Based on the works of [12,13], an offline optimal parameter searching method was presented for selecting the optimal guidance function to shorten the convergence time in [14]. However, this approach has such a heavy computational load that it cannot be extended to real-time application scenarios. Although these above methods have been verified to be feasible for the single UAV standoff tracking problem, they only focus on the optimization of the LGVF without considering the input constraints of fixed-wing UAVs, such as heading rate limitations. If the curvature of the LGVF is too large, the actual trajectory of a UAV cannot converge to a standoff circle due to the saturated rudders. Therefore, it is still necessary to design a control law for regulating the relative distance while satisfying the turning rate limitation.

The performance of target localization algorithms is significantly impacted by the relative sensor-target geometry. Observation configurations, or different sensor-target geometries, produce varying uncertainty ellipses of the target location algorithm. It is worth considering which observation configuration can yield the best target localization results. By minimizing the Cramer–Rao lower bound (CRLB), which provides a lower bound on the estimator performance, the uncertainty in the estimation process can be reduced. Therefore, in [15] they utilize the determinant of the CRLB to determine the observation configuration that results in a minimal measure of the uncertainty ellipse. According to the conclusions in [15], if only two UAVs perform a standoff tracing mission, the intersection angle subtended at the target (called the phase separation angle) by two UAVs will be $(\frac{\pi}{2})$; if the number of UAVs is $N \geq 3$, the phase separation angle between adjoining UAVs will be $\left(\frac{2\pi}{N}\right)$. An optimal configuration should position the UAVs at equal angular intervals around the perimeter of the standoff circle. This requires phase separation in the coordinated standoff tracking problem. Various phase separation methods have been proposed, including model predictive control (MPC) [16,17], sliding mode control (SMC) [18,19], conical pendulum motion [20], consensus algorithm [21–24], and so on. In [16,17], a nonlinear MPC framework for coordinated standoff tracking by two UAVs is proposed The optimal control outputs for speed and turning rate are generated by minimizing the sum of weighted cost functions that include the standoff-distance regulation error and phase separation error in a receding horizon. However, as the number of UAVs increases, the computational complexity and iterative time required for searching optimal results also increase. In [18], a hovering algorithm based on sliding mode control is presented to control the virtual leader's position on a standoff circle centered at the ground-based moving target. However, the SMC method suffers from chattering due to the discontinuity of the signum function in the control law. To eliminate chattering, the signum function is replaced by a saturation function in Ref. [19]. However, both Refs. [18,19] ignore the vehicle airspeed constraint. The airspeed of the fixed-wing UAV is restricted within lower and upper bounds. In order to satisfy the airspeed boundary constraint, Ref. [20] proposes that the UAV reduces its speed by decreasing the standoff radius when flying on the right-hand side of the target and its airspeed reaches the upper bound. Conversely, the UAV increases the standoff radius to increase its airspeed when its airspeed reaches the lower bound. However, this approach causes the distance between the UAV and the target to oscillate, leading to failure in achieving the control objectives of the cooperative standoff target tracking problem. Different from the integrated controller in [20], the heading control channel and velocity control channel are decoupled in [21–24]. The space phase angle is chosen as the coordination variable of the consensus algorithm, allowing for the design of cooperative airspeed controllers for the UAVs. The phase separation angles among the UAVs can asymptotically converge, reflecting equal space separation. However, the airspeed constraint is not taken into account in the space phase separation method. Furthermore, the phase separation angles between $[-\pi, \pi)$ are discontinuous, and will lead to oscillating airspeed control input. This is not beneficial for reducing the space phase separation error. Therefore, it is critical to design a phase separation controller that provides a smooth signal output and meets the airspeed constraint of UAVs in the coordinated standoff tracking problem.

Most of the research mentioned above assumes an ideal non-wind environment or a known constant background wind. However, in practice, background wind is usually present and it affects the performance of the UAVs, especially when the background wind is unknown. In [25], a robust term is added to the standoff tracking control law to obtain disturbance rejection for wind gusts. However, the response time of the robust controller is too long and unsuitable for target tracking requiring high maneuverability. To quantitatively describe wind dynamics, a simple conservative model (e.g., sine function as in [26] or linear model as in [27]) or a more sophisticated one (e.g., stochastic as in [28]) can be used. In [28], the Dryden model [29] and Davenport model [30] are used to describe the dynamics of wind at high altitudes or near the ground, respectively. The unscented Kalman filter (UKF) is introduced to estimate the velocity of background wind. However, the accuracy of the wind speed estimation largely depends on the accuracy of the constructed wind dynamic model. In reality, wind is stochastic and time-varying, which makes it difficult to model. In [31], an adaptive estimator is utilized to estimate the wind velocity. In the case of a stationary target without wind, the vehicle trajectory converges to a circular orbit (called a perfect trajectory or orbit) by implementing a designed standoff tracking control law. However, for a moving target with wind, the actual trajectory cannot converge to the perfect one due to the disturbance caused by the wind and the target's motion. The wind velocity estimation can be obtained by reducing the offset between the actual vehicle trajectory and the perfect trajectory. However, the convergence rate of the estimator is slow because it only uses the radial distance of the offset. Therefore, designing a wind velocity estimator that has a high accuracy and fast convergence rate remains challenging.

This paper addressed the challenges of cooperative standoff target tracking using multiple fixed-wing UAVs with input constraints and the considerations of unknown background wind and target motion. Controllers satisfying the input constraints, such as heading rate and airspeed limitation, are designed to guarantee a team of fixed-wing UAVs can perform efficiently during coordinated standoff tracking missions. The major contributions of the paper are as follows: (a) A heading rate control law based on LGVF is proposed. To satisfy the limitation of the heading rate, the minimum allowable standoff radius is formulated. It is proved that this proposed heading rate controller can guarantee that the UAV can asymptotically converge to a standoff circle hovering over the target under arbitrary initial conditions of the position and heading. (b) To avoid discontinuities in the space phase angle, a new term called the temporal phase is proposed to represent the phase separation. An airspeed control law is introduced to steer a team of UAVs maintaining an optimal observation configuration, which requires distribution around the standoff circle with equal phase separation. The proofs for satisfying the airspeed limitation and global convergence using the proposed speed controller are provided. (c) The target's motion and background wind are regarded as external disturbances. The offset vector caused by external disturbances between the actual trajectory and the perfect/desired orbit is utilized to estimate the composition velocity of wind and the target's motion. It is proved that the estimated result asymptotically converges to the true value of the composition velocity.

The remainder of this paper is structured as follows. Section 2 presents the problem formulation, including assumptions made in this study, a UAV kinematic model with control input constraints, and control objectives. Section 3 discusses the problem of standoff tracking using a single UAV. Based on the LGVF, a lateral controller with a heading rate input constraint is proposed to regulate the position of a UAV on a circular orbit around the target under the condition of an arbitrary initial position and heading. In Section 4, we introduce a term called temporal phase to represent the spatial distribution of the UAVs on the standoff circle and propose cooperative controllers with an airspeed input constraint to achieve the desired temporal phase separation. In Section 5, an online estimator is developed to adapt the proposed heading rate and airspeed controllers to the case of a moving target in the presence of wind. Section 6 presents a more detailed control and coordination architecture of standoff target tracking. The computational complexity and in-vehicle communication are analyzed in more detail. Simulation and experimental results are demonstrated in Section 7, followed by a summary and conclusions in Section 8.

2. Problem Formulations

Without loss of generality, the following assumptions are made to render the problem simpler and well posed.

Assumption 1: *The UAVs fly at a constant altitude, and the target moves in a two-dimensional plane, ignoring its height.*

Assumption 2: *The position of the moving target is assumed to be known.*

Assumption 3: *The communication between the UAVs is ideal, without any restrictions such as limited communication range, packet loss, and delay.*

Assumption 4: *The UAV is equipped with a low-level autopilot that holds constant altitude, and follows the command inputs of the speed and heading rate.*

Remark 1: *It is assumed that the true value of the target's position is known, and this is used to estimate the composition velocity of the target's motion and background wind. However, the velocities of the target and wind are unknown in this paper. We keep the target's location out of the scope since we aim at providing a solid formulation concerning the problem of coordinated standoff tracking of a ground target. The next endeavor of this work should explore the possibility of using an onboard observation sensor such as a camera to facilitate tracking.*

2.1. UAV Model

Under the above assumptions, the inertial frame in a two-dimensional plane is constructed. The x- and y-axes point east and north, respectively. As shown in Figure 1, the kinematic of the UAV is expressed as follows:

$$\dot{x} = v_s cos\psi + w_x; \dot{y} = v_s sin\psi + w_y; \dot{\psi} = u \tag{1}$$

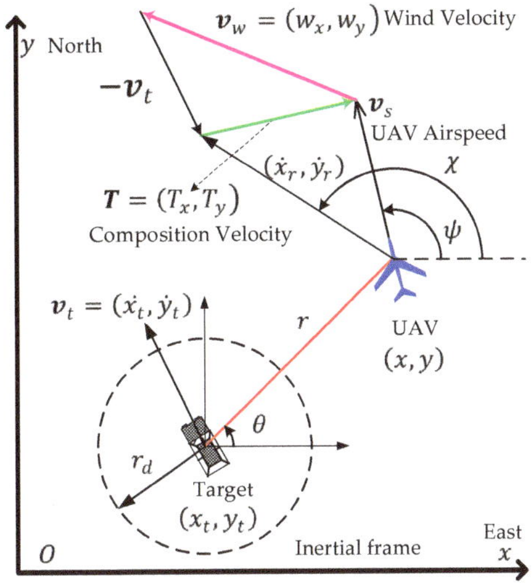

Figure 1. Geometry of ground target tracking in wind.

In Equation (1), $v_w = (w_x, w_y)$ is the background wind velocity. $(x, y) \in R^2$ is the two-dimensional position of the UAV in the inertial frame. $\psi \in [-\pi, \pi)$ is the UAV heading. v_s is the true air speed (TAS) of the UAV. u is the heading rate of the UAV. $U = (v_s, u)^T$ is the control input signal followed by the low-level autopilot of the UAV. The airspeed and the heading rate should be enforced with the following input constrains.

$$0 < v_{min} \leq v_s \leq v_{max} \tag{2}$$

$$|u| \leq \omega_{max} \tag{3}$$

In Equations (2) and (3), v_{min} and v_{max} are the minimum and maximum airspeed, respectively. ω_{max} is the upper bound of the heading rate.

2.2. The Ground-Based Moving Target Model

The ground-based moving target (GMT) is regarded as a mass point whose position, velocity, and acceleration with respect to the inertial frame are denoted by (x_t, y_t), $(\dot{x}_t, \dot{y}_t)$, and $(\ddot{x}_t, \ddot{y}_t)$, respectively. Concerning the target, we define only its state variables without providing any further information regarding its kinematic model. Thus, our approach covers the general case of a ground-based target's motion.

2.3. The Relative Motion Model

The relative motion with background wind is expressed as follows:

$$\dot{x}_r = v_s cos\psi + w_x - \dot{x}_t; \dot{y}_r = v_s sin\psi + w_y - \dot{y}_t; \dot{\psi} = u \tag{4}$$

The target's motion and wind are regarded as external disturbances. These two velocities can be combined into a single velocity term, which is called the composition velocity and denoted as $T = (T_x, T_y) = (\dot{x}_t - w_x, \dot{y}_t - w_y)$. Equation (4) can be rewritten as follows:

$$\dot{x}_r = v_r cos\chi; \dot{y}_r = v_r sin\chi; \dot{\chi} = \lambda_u(\psi)u \tag{5}$$

In Equation (5), $\chi \in [-\pi, \pi)$ is the relative course angle. v_r is the relative speed. $\lambda_u(\psi)$ is the parameter of relative motion in the model. They are calculated as follows:

$$v_r = \sqrt{v_s^2 + T_x^2 + T_y^2 - 2v_s(T_x cos\psi + T_y sin\psi)} \tag{6}$$

$$\chi = arctan\left(\frac{v_s sin\psi - T_y}{v_s cos\psi - T_x}\right) \tag{7}$$

$$\lambda_u(\psi) = \frac{v_s^2 - v_s(T_x cos\psi + T_y sin\psi)}{v_r^2} \geq \frac{1}{2} \tag{8}$$

The relative motion, as shown in Equation (5), can be expressed in polar coordinates as

$$\begin{bmatrix} \dot{r} \\ r\dot{\theta} \end{bmatrix}_{UAV} = \begin{bmatrix} \dot{x}_r cos\theta + \dot{y}_r sin\theta \\ -\dot{x}_r sin\theta + \dot{y}_r cos\theta \end{bmatrix} = \begin{bmatrix} v_r cos(\chi - \theta) \\ v_r sin(\chi - \theta) \end{bmatrix} \tag{9}$$

The distance vector between the UAV and the target is $r = (x_r, y_r)$, the relative distance is $r = \|r\|$, and the observation phase is $\theta \in [-\pi, \pi)$.

2.4. The Objectives of the Control Problem

As shown in Figure 2, there is a team of UAVs (A_i, $i = 1, 2, \ldots, N$) performing a cooperative standoff tracking mission for a ground-based moving target in unknown background wind. In the process of tracking, on one hand, each UAV needs to circle around the target to maintain a constant relative distance. On the other hand, the hovering UAVs are distributed around the target with equal phase separation to avoid collisions and maximize the coverage of sensors.

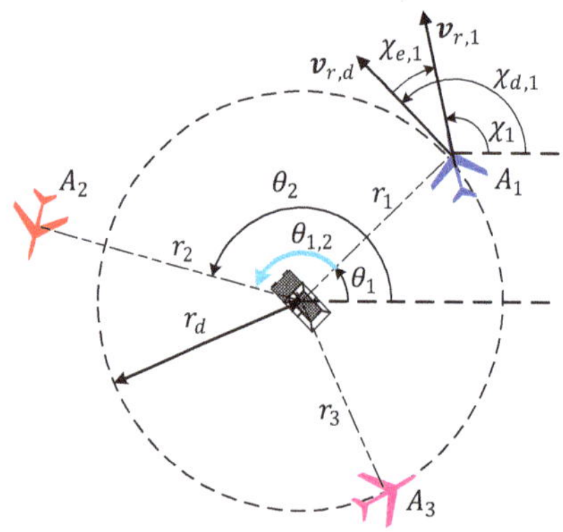

Figure 2. Illustration of cooperative standoff target tracking.

The following three control objectives need to be achieved for the multi-UAV cooperative standoff tracking control problem.

(i) Control objective 1: relative distance regulation. The relative distance r_i between the UAV A_i and the target should be converged to the desired standoff radius r_d.

$$\Delta r_i = r_i - r_d \longrightarrow 0 \tag{10}$$

(ii) Control objective 2: relative course convergence. The relative course χ_i of the UAV A_i should be converged to the desired relative course $\chi_{d,i}$ to maintain circular motion around the target.

$$\chi_{e,i} = \chi_i - \chi_{d,i} \longrightarrow 0 \tag{11}$$

(iii) Control objective 3: intervehicle phase separation. The phase separation $\theta_{i,j}$ between A_i and its neighbor A_j should be converged to the optimal configuration θ_d.

$$\Delta\theta_i = \theta_{i,j} - \theta_d = \theta_j - \theta_i - \theta_d \longrightarrow 0 \tag{12}$$

According to Ref. [15], the optimal observation configuration θ_d is defined as follows:

$$\theta_d = \begin{cases} \pi/2, N = 2 \\ 2\pi/N, \ N > 2 \end{cases} \tag{13}$$

Therefore, when studying the multi-UAV cooperative standoff tracking control problem, the following three issues are mainly considered in this paper.

(a) How to design the lateral control law subject to the heading rate constraint (Equation (3)) to regulate the relative course of A_i so that A_i can fly along the standoff circle with radius r_d around the target. This would mean that objectives 1 and 2 would both be achieved.
(b) How to design the longitudinal control law subject to the airspeed limitation (Equation (2)) to guarantee that the hovering UAVs are distributed over the standoff circle with equal phase separation. This would mean that control objective 3 would be achieved.
(c) How to estimate the composition velocity of wind and target in order to improve the performance of the UAVs during the tracking in the presence of wind and the target's motion.

3. Standoff Tracking Using a Single UAV

3.1. Guidance Law Based on LGVF

In [10], the desired relative course is generated from an LGVF that guides the UAV to circle around the target with a predefined standoff radius r_d. The LGVF for a standoff target tracking can be described as follows:

$$\begin{bmatrix} \dot{x}_r \\ \dot{y}_r \end{bmatrix}_{GF} = -v_r \begin{bmatrix} \left(\frac{x_r}{r}\right)\left(\frac{r^2-r_d^2}{r^2+r_d^2}\right) + \left(\frac{y_r}{r}\right)\left(\frac{2rr_d}{r^2+r_d^2}\right) \\ -\left(\frac{x_r}{r}\right)\left(\frac{2rr_d}{r^2+r_d^2}\right) + \left(\frac{y_r}{r}\right)\left(\frac{r^2-r_d^2}{r^2+r_d^2}\right) \end{bmatrix} \tag{14}$$

Define the following vector field angle $\phi_{GF} \in [0, \pi)$:

$$cos\phi_{GF} = \frac{r_d^2 - r^2}{r^2 + r_d^2}; sin\phi_{GF} = \frac{2rr_d}{r^2 + r_d^2} \tag{15}$$

Equation (15) can be expressed in polar coordinates as follows:

$$\begin{bmatrix} \dot{r} \\ r\dot{\theta} \end{bmatrix}_{GF} = \begin{bmatrix} cos\theta & sin\theta \\ -sin\theta & cos\theta \end{bmatrix} \begin{bmatrix} \dot{x}_r \\ \dot{y}_r \end{bmatrix}_{GF} = v_r \begin{bmatrix} cos\phi_{GF} \\ sin\phi_{GF} \end{bmatrix} \tag{16}$$

The following guidance law is introduced to guide the UAV to fly along the LGVF in Equation (16):

$$\chi_d = \theta + \phi_{GF} \tag{17}$$

where the observation phase θ is defined by

$$\theta = \begin{cases} arctan\left(\frac{y_r}{x_r}\right), & if\, r > 0 \\ \chi, & if\, r = 0 \end{cases} \tag{18}$$

The feasibility of the guidance law Equation (17) is given by Theorem 1.

Theorem 1: *If the guidance law Equation (17) is applied to the relative motion model of the UAV as shown in Equation (9), the relative distance between the UAV and the target asymptotically converges to the predefined standoff radius, i.e.,* $r \longrightarrow r_d$ *as* $t \longrightarrow +\infty$.

Remark 2: *The proof of Theorem 1 is seen in [10]. It can be observed from Theorem 1 that in order to achieve standoff tracking, it is required that the relative course* χ *should be always aligned with the desired relative course* χ_d *generated by the guidance law Equation (17).*

3.2. Heading Rate Controller Design Subject to the Input Constraints

According to the conclusion of Theorem 1, if χ is always equal to χ_d, the UAV eventually converges to the standoff circle with the desired radius r_d, and performs the standoff target tracking mission successfully. However, in general, there exists an angle error between χ and χ_d initially and the relative course χ cannot be directly controlled. Thus, we need to design a heading rate controller to guarantee that the relative course χ eventually converges to χ_d. In other words, the proposed heading rate controller indirectly controls χ by regulating the UAV heading directly.

The relative course error $\chi_e \in [-\pi, \pi)$ can be defined by

$$\chi_e = \chi - \chi_d \tag{19}$$

Differentiating Equation (17) with respect to t leads to the following equation:

$$\dot{\chi}_d = \dot{\theta} + \dot{\phi}_{GF} \tag{20}$$

According to Equation (9), the dynamic of the azimuth θ is obtained as

$$\dot{\theta} = \frac{v_r}{r} sin(\chi - \theta) \tag{21}$$

Differentiating Equation (15) with respect to t:

$$\dot{\phi}_{GF} = \frac{2r_d}{r^2 + r_d^2} v_r cos(\chi - \theta) \tag{22}$$

The desired relative course rate can be obtained by substituting Equations (21) and (22) into Equation (20).

$$\dot{\chi}_d = \frac{v_r}{r}[sin(\chi_e + \phi_{GF}) + sin\phi_{GF} cos(\chi_e + \phi_{GF})] \tag{23}$$

It can be observed from Equation (23) that when $r \to 0$, $\dot{\chi}_d \to +\infty$. Thus, in order to facilitate the engineering application, let $\dot{\chi}_d = \frac{4v_r}{r_d}$ when $r = 0$. Based on the relative course error χ_e, Equation (19), and the desired relative course rate $\dot{\chi}_d$ Equation (23), the heading rate controller is designed as follows:

$$u = \begin{cases} \omega, & if|\omega| \leq \omega_{max} \\ sgn(\omega)\omega_{max}, & if|\omega| > \omega_{max} \end{cases}; \omega = -k\chi_e + \frac{\dot{\chi}_d}{\lambda_u(\psi)}; k > 0 \tag{24}$$

It is observed from Equation (24) that the heading rate controller consists of two main parts: a feedback term $\left(\frac{\dot{\chi}_d}{\lambda_u(\psi)}\right)$ and a feedforward term $(-k\chi_e)$. $k > 0$ represents the feedback gain. By implementing the heading rate controller as shown in Equation (24), the relative course χ of the UAV is indirectly controlled to follow the desired one χ_d generated by the LGVF. Therefore, in order to ensure that χ_d can be followed by the proposed controller, the allowable lower bound of ω_{max} must be determined. Before discussing the allowable lower bound of ω_{max}, Lemma 1 and Lemma 2 are presented.

Lemma 1: *If the airspeed of the UAV is faster than the composition velocity, i.e.,* $v_s^2 \geq T_x^2 + T_y^2$, *the following inequality is true.*

$$\frac{v_r}{\lambda_u(\psi)} \leq \frac{\left(v_s + \sqrt{T_x^2 + T_y^2}\right)^2}{v_s} \tag{25}$$

Remark 3: *The proof of Lemma 1 is seen in Appendix A. Lemma 1 can be used to determine the bound of* $\dot{\chi}_d$ *if the heading rate input constraint* ω_{max} *is provided. The relevant conclusions are shown in Lemma 2.*

Lemma 2: *If the heading rate control law given by Equation (24) is applied to the relative motion model of the UAV as shown in Equation (9), there exists a constant* $k_1 : 0 < k_1 \leq \frac{k}{2}$, *such that:* ① $\dot{\chi}_d \geq -\lambda_u(\psi)\omega_{max} + k_1 sin\chi_e, \chi_e \in [0, \pi)$; ② $\dot{\chi}_d \leq \lambda_u(\psi)\omega_{max} + k_1 sin\chi_e, \chi_e \in [-\pi, 0)$.

Remark 4: *The proof of Lemma 2 is seen in Appendix B. Lemma 2 provides the bound of the desired heading rate, i.e.,* $|\dot{\chi}_d - k_1 sin\chi_e| \leq \lambda_u(\psi)\omega_{max}, \chi_e \in [-\pi, \pi)$.

Based on Lemmas 1 and 2, the allowable bound of ω_{max} is determined in Theorem 2.

Theorem 2: *In order to ensure that the UAV successfully performs standoff target tracking by implementing the heading rate controller as shown in Equation (24), and the input signalunever violates the heading rate constraint, i.e.,* $|u| \leq \omega_{max}$, *the allowable lower bound of* ω_{max} *must satisfy the following condition.*

$$\omega_{max} \geq \omega_{inf} \triangleq \frac{4(v_s + \sqrt{T_x^2 + T_y^2})^2}{v_s r_d} \tag{26}$$

Proof: The proof is discussed in two cases in terms of the relative distance.

(a) $r = 0$. According to the proposed heading rate control law as shown in Equation (24), when $r = 0$, $\chi_e = 0$. Now, the desired relative course rate is $\dot{\chi}_d = \frac{4v_r}{r_d}$. Then, according to Lemma 1, the following can be derived:

$$u = \left(\frac{4}{r_d}\right)\left(\frac{v_r}{\lambda_u(\psi)}\right) \leq \left(\frac{4}{r_d}\right)\left(\frac{\left(v_s + \sqrt{T_x^2 + T_y^2}\right)^2}{v_s}\right) \triangleq \omega_{inf} \tag{27}$$

It can be observed from Equation (27) that when $r = 0$, if $\omega_{max} \geq \omega_{inf}$, then $|u| \leq \omega_{max}$ is always true. This means that when $r = 0$, the proposed heading rate control satisfies the input constraint as shown in Equation (3).

(b) $r > 0$. According to Lemma 2, the desired heading rate $\dot{\chi}_d$ is bounded, and the following inequalities can be derived:

$$
\begin{aligned}
&|\dot{\chi}_d - k_1 sin\chi_e| \leq \lambda_u(\psi)\omega_{max} \\
&\Rightarrow \left|\frac{\dot{\chi}_d}{\lambda_u(\psi)} - 2k_1 sin\chi_e\right| \leq \omega_{max} \\
&\Rightarrow \left|\frac{\dot{\chi}_d}{\lambda_u(\psi)} - ksin\chi_e\right| \leq \omega_{max} \\
&\Rightarrow \left|\frac{\dot{\chi}_d}{\lambda_u(\psi)} - k\cdot\chi_e\right| \leq \omega_{max} \\
&\Rightarrow |u| \leq \omega_{max}
\end{aligned}
\tag{28}
$$

It can be observed from Equation (28) that when $r > 0$, $|u| \leq \omega_{max}$ is always true. Thus, the proof for Theorem 2 is completed. □

Remark 5: *Theorem 2 provides the formulated inequality between the standoff radius r_d and the maximum heading rate ω_{max} as follows:*

$$
r_d \geq r_{d,min} \triangleq \frac{4\left(v_s + \sqrt{T_x^2 + T_y^2}\right)^2}{v_s\omega_{max}} \tag{29}
$$

In other words, due to the input constraint $|u| \leq \omega_{max}$, when the UAV performs the standoff target tracking mission with a constant airspeed v_s, the predefined standoff radius r_d has an allowable lower bound. The minimum allowable standoff radius can be calculated according to Equation (29).

3.3. Stability Analysis of Saturated Heading Rate Control Law

Although Theorem 2 proves that the proposed heading rate control law satisfies the input constraint $|u| \leq \omega_{max}$, whether χ converges to χ_d or not will be investigated in the following. In addition, if $\chi_e \to 0$ as $t \longrightarrow +\infty$, the convergence speed of χ_e also needs further discussion. Thus, Lemma 3 is presented.

Lemma 3: *If the UAV described by the relative motion model as shown in Equation (9) implements the heading rate control law given by Equation (24), there exists a positive constant $0 < k_2 \leq \frac{k}{2} \leq k_1$ such that ① $\dot{\chi}_e \geq -k_2 sin\chi_e > 0$, $\chi_e \in [-\pi, 0)$; ② $\dot{\chi}_e \leq -k_2 sin\chi_e < 0$, $\chi_e \in [0, \pi)$.*

Remark 6: *The proof of Lemma 3 is seen in Appendix C. According to Lemma 3, it can be derived that*

$$
\chi_e\dot{\chi}_e \leq -k_2\chi_e sin\chi_e \leq 0 \tag{30}
$$

Equation (30) shows that for every initial relative course error $\chi_e(t_0) \in [-\pi, \pi)$, the relative course error $\chi_e \to 0$ as $t \to +\infty$. Thus, the convergence of the proposed heading rate control law is proved in Lemma 3. In addition, Lemma 3 also gives a lower bound on the convergence rate $\dot{\chi}_e$, which is defined as follows:

$$
|\dot{\chi}_e| \geq |\dot{\chi}_e|_{inf} = k_2|sin\chi_e|,\ \chi_e \in [-\pi, \pi) \tag{31}
$$

By solving the differential inequality shown in (31), the relative course error χ_e varies according to the following inequality:

$$
\left|tan\left(\frac{\chi_e}{2}\right)\right| \leq \left|tan\left(\frac{\chi_e(t_0)}{2}\right)\right|e^{-k_2 t} \tag{32}
$$

Equation (32) can be used to analyze the bound of the relative distance r between the UAV and the target. The relevant conclusions are shown in Lemma 4.

Lemma 4: *Given an arbitrary initial relative course error $\chi_e(t_0) \in [-\pi, \pi)$, if the UAV described by the relative motion model as shown in Equation (9) implements the heading rate control law given by Equation (24), the relative distancerhas an upper boundr_{sup}as follows:*

$$r_{sup} = r_0 + \frac{v_s + T}{k_2} ln \left| \frac{tan\left(\frac{\chi_{e0}}{2}\right)}{tan\left(\frac{\alpha_0}{2} - \frac{\pi}{4}\right)} \right| \tag{33}$$

where $T = \sqrt{T_x^2 + T_y^2}$ and $\alpha_0 = arctan\left(\frac{r_0}{r_d}\right)$. The proof of Lemma 4 is seen in Appendix D.

Based on Lemma 3 and Lemma 4, Theorem 3 presents a complete proof for the stability of the proposed heading rate control law with control input constrains.

Theorem 3: *Given an arbitrary initial relative course error $\chi_{e0} \in [-\pi, \pi)$, if the UAV described by the relative motion model as shown in Equation (9) implements the heading rate control law given by Equation (24), the trajectory of the UAV asymptotically converges to the standoff circle, i.e., $r \longrightarrow r_d$ and $\chi \to \chi_d$ as $t \longrightarrow +\infty$.*

Proof: Let $\mu > 0$, and a Lyapunov function be introduced as follows:

$$V = \frac{1}{2}(r - r_d)^2 + \frac{1}{2}\mu\chi_e^2 \tag{34}$$

Differentiating Equation (34) with respect to time t, the outcome is

$$\dot{V} = (r - r_d)\dot{r} + \mu\chi_e\dot{\chi}_e \tag{35}$$

(a) If $|\chi_e| \in \left[\frac{\pi}{2}, \pi\right]$, according to Lemma 4, we obtain $(r - r_d)\dot{r} = (r - r_d)v_r cos(\chi - \theta)$ $\leq (r_{sup} + r_d)(v_s + T)$. And according to Lemma 3, we have $\mu\chi_e\dot{\chi}_e \leq \mu\chi_e(-k_2 sin\chi_e)$ $\leq \mu\left(\frac{\pi}{2}\right)(-k_2 sin\chi_{e0})$. Together, they yield that

$$\dot{V} \leq (r_{sup} + r_d)(v_s + T) - \mu k_2\left(\frac{\pi}{2}\right)sin\chi_{e0} \tag{36}$$

It can be observed that from Equation (36), if $\mu \geq \mu_1 \triangleq \frac{2(r_{sup}+r_d)(v_s+T)}{k_2\pi sin\chi_{e0}}$, then $\dot{V} \leq 0$.

(b) If $|\chi_e| \in \left[0, \frac{\pi}{2}\right]$, according to Lemma 3, one obtains

$$\mu\chi_e\dot{\chi}_e \leq -\mu k_2\chi_e sin\chi_e \tag{37}$$

According to the relative motion model of the UAV as shown in Equation (9), we have

$$\dot{r} = -\left(\frac{r^2 - r_d^2}{r^2 + r_d^2}\right)v_r cos\chi_e - v_r sin\phi_{GF} sin\chi_e \tag{38}$$

Substituting (37) and (38) into (35) yields

$$\dot{V} \leq -\frac{(r - r_d)^2(r + r_d)v_r}{r^2 + r_d^2}cos\chi_e - v_r(r - r_d)\, sin\phi_{GF} sin\chi_e - \mu k_2\chi_e sin\chi_e \tag{39}$$

For the second term in Equation (39), it holds that

$$-v_r(r - r_d)\, sin\phi_{GF} sin\chi_e \leq 2v_r|r - r_d|\left|sin\frac{\chi_e}{2}\right| \tag{40}$$

For the third term in Equation (39), it holds that

$$\chi_e sin\chi_e \geq 2sin^2\left(\frac{\chi_e}{2}\right) \tag{41}$$

Substituting Equations (40) and (41) into Equation (39) yields

$$\dot{V} \leq -\frac{(r-r_d)^2(r+r_d)v_r}{r^2+r_d^2} + 2v_r|r-r_d|\left|sin\frac{\chi_e}{2}\right| - 2\mu k_2 sin^2\left(\frac{\chi_e}{2}\right) \quad (42)$$

Solving Equation (42) by completing the square, it is easy to know that if $\mu \geq \mu_2 \triangleq \frac{(r_{sup}^2+r_d^2)}{k_2(r_{sup}+r_d)}$, then

$$\dot{V} \leq -2\mu k_2 sin^2\left(\frac{\chi_e}{2}\right) - \left\{\left[\frac{(r+r_d)}{r^2+r_d^2}\right]^{\frac{1}{2}}|r-r_d| - \left[\frac{(r^2+r_d^2)}{r+r_d}\right]^{\frac{1}{2}}\left|sin\left(\frac{\chi_e}{2}\right)\right|\right\}^2 \leq 0 \quad (43)$$

Therefore, when $\mu \geq max\{\mu_1, \mu_2\}$, the time derivative of the Lyapunov function is always non-positive, i.e., $\dot{V} \leq 0$. It can be observed from Equation (43) that $\dot{V} = 0$ implies that $r = r_d$ and $\chi_e = 0$. According to LaSalle's invariance principle, it can be concluded that $r \longrightarrow r_d$ and $\chi \rightarrow \chi_d$ as $t \longrightarrow +\infty$. This means that the trajectory of the UAV asymptotically converges to the standoff circle. Thus, the proof of Theorem 3 is completed. □

Remark 7: *According to Theorem 3, it is concluded that by implementing the proposed heading rate control law, the relative distance and the relative course of the UAV converge to the desired values. This means that the control objectives 1 and 2 proposed in Section 2.2 are achieved, and the UAV tracks the moving target while maintaining the desired standoff distance successfully.*

4. Cooperative Standoff Tracking Using Multiple UAVs

When a team of UAVs is used to track a ground-based moving target, coordination between aircraft is necessary to avoid collisions and to maximize sensor coverage of the target. A possible solution to this coordination problem is the so-called "phase separation" approach whereby the UAVs fly along the standoff circle with an equal intervehicle phase separation angle with respect to their neighbors.

Ref. [25] proposes a space phase separation algorithm (SPSA) to achieve the desired angular spacing, as illustrated in Figure 3. The UAVs are distributed counterclockwise in the standoff circle according to the ascending sequence of their unique identity numbers $A_i(i = 1, 2, \ldots N)$. Each UAV A_i independently calculates its airspeed control input $v_{s,i}$ based on its own phase angle θ_i, the phase angle $\theta_{left} = \theta_{i-1}$ of its left neighbor A_{i-1}, and the phase angle $\theta_{right} = \theta_{i+1}$ of its right neighbor A_{i+1}. For A_1, its left neighbor is A_N and its right neighbor is A_2. For A_N, its left neighbor is A_{N-1} and its right neighbor is A_1.

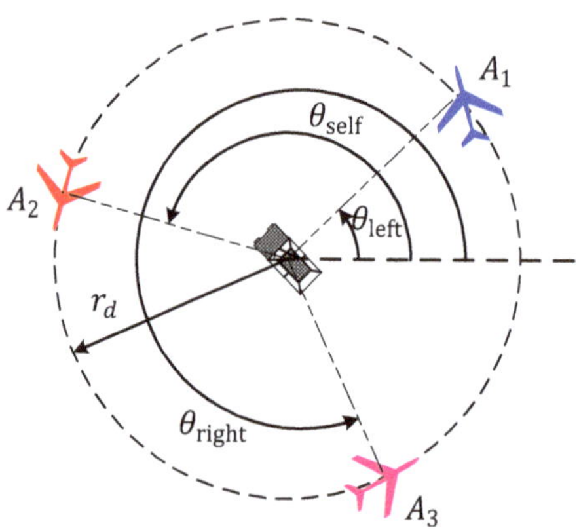

Figure 3. Illustration of phase separation control for UAVs.

In [25], the space phase separation angle of A_i, $\Delta\theta_i \in [-\pi, \pi)$, is defined as $\Delta\theta_i = \theta_i - \theta_{left}$. Similarly, the phase separation angle of the right neighbor A_{i+1} is defined as $\Delta\theta_{right} = \theta_{right} - \theta_i$. The space phase separation error of A_i, $e_{\theta,i} \in [-\pi, \pi)$ is defined as $e_{\theta,i} = \Delta\theta_i - \theta_d$. Similarly, the space phase separation error of the right neighbor A_{i+1} is defined as $e_{\theta,right} = \Delta\theta_{right} - \theta_d$. Thus, the airspeed control input of A_i can be designed as $v_{s,i} = k_\theta(e_{\theta,right} - e_{\theta,i})r_i + v_{sd}$. Where $k_\theta > 0$, r_i represents the relative distance between A_i and the target. v_{sd} represents the desired airspeed when the team of UAVs hover around the target along a standoff circle synchronously. In this paper, v_{sd} is called the standoff speed for short.

However, the SPSA method proposed in [25] has two disadvantages. (i) The airspeed limitation, as shown in Equation (2), is not considered in SPSA; (ii) the space phase separation $\Delta\theta_i$ and corresponding error $e_{\theta,i}$ are between $[-\pi, \pi)$, and thus they are discontinuous. This discontinuity will lead to oscillations in the airspeed control input, resulting in poor tracking performance of the UAVs. This shortcoming has been confirmed by simulation results shown in Figure 9 in [25]. In order to avoid the discontinuity of the space phase angle, the space phase is replaced by a new notion called the temporal phase, which can also be used to represent the distribution of the UAVs on the standoff circle. Based on the temporal phase, a new temporal phase separation algorithm (TPSA) is proposed to achieve the desired temporal phase separation in a cooperative standoff target tracking mission.

4.1. Airspeed Controller Design for Temporal Phase Separation

In the SPSA method, the angle θ_i is utilized to represent the space phase of the UAV A_i. The space phase describes the distribution of the UAVs which remain in a circle around the target. Similarly, we can also introduce the temporal phase to equivalently describe the distribution of the UAVs. The temporal phase of A_i is defined as follows:

$$\tau_i(\theta_i) = \begin{cases} \frac{2\pi}{T_\tau}\left[\int_0^{\theta_i+2\pi}\left(\frac{r_d}{v_{rd}}\right)d\theta\right] - \pi, & if\ \theta_i \in [-\pi, 0) \\ \frac{2\pi}{T_\tau}\left[\int_0^{\theta_i}\left(\frac{r_d}{v_{rd}}\right)d\theta\right] - \pi & if\ \theta_i \in [0, \pi) \end{cases} \tag{44}$$

where $\tau_i(\theta_i) \in [-\pi, \pi)$ represents the temporal phase of A_i. v_{rd} is the desired relative speed of A_i with respect to the target and can be expressed as follows.

$$v_{rd} = \left[v_{sd}^2 + T_x^2 + T_y^2 - 2v_{sd}(T_x cos\psi_d + T_y sin\psi_d)\right]^{\frac{1}{2}} \tag{45}$$

where ψ_d represents the desired heading when A_i flies along the standoff circle.

$$\psi_d = arcsin\left(\frac{T_y cos\chi_d - T_x sin\chi_d}{v_{sd}}\right) + \chi_d \tag{46}$$

In Equation (44), T_τ represents the time required for the UAV to complete a circle of flight along the standoff circle with the desired airspeed v_{sd}. T_τ is used to normalize the flight time of the UAV and is defined as $T_\tau = \int_0^{2\pi}\left(\frac{r_d}{v_{rd}}\right)d\theta$. It can be observed from Equation (44) that the temporal phase $\tau_i(\theta_i)$ represents the normalized time required for the UAV A_i to fly from space phase angle 0 to the current space phase angle θ_i along the standoff circle.

Therefore, the temporal phase separation between A_i and its neighbor is defined by the difference in their temporal phases.

$$\tau_{i-1,i} = \tau_{i-1} - \tau_i \tag{47}$$

Suppose there are $N(N \geq 2)$ UAVs, a leader–follower formation strategy is used in this paper. Without loss of generality, it assumes that A_1 is the leader and its airspeed is the desired one, e.g., $v_{s,1} = v_{sd}$, and is held constant. Then, A_2 follows A_1, and adjusts its airspeed to achieve the desired temporal separation with its leader A_1. Similarly, A_i follows its leader A_{i-1} and achieves the desired temporal separation by varying its airspeed. The temporal separation error is defined as follows:

$$\Delta\tau_i = \tau_{i-1} - \tau_i - \tau_d \tag{48}$$

where τ_d represents the desired separation. The airspeed input is designed as follows:

$$\begin{cases} v_{s,1} = v_{sd} \\ v_{s,i} = v_{sd} + \left(\frac{\Delta v \cdot \Delta\tau_i}{\pi}\right)\left(\frac{r_{i-1}^2 + r_d^2}{r_{i-1}^2 + r_i^2}\right), i > 1 \end{cases} \tag{49}$$

where $\Delta v > 0$ represents the airspeed increment of the UAV in one time step, and it reflects the performance of the onboard autopilot.

Before verifying that the proposed controller Equation (49) always satisfies the minimum and maximum airspeed constraints in Equation (2), Assumption 5 is introduced as follows:

Assumption 5: *There exist appropriate values of v_{sd} and Δv, so that the following constraints are satisfied, $v_{sd} - \Delta v > \sqrt{T_x^2 + T_y^2}$ and $v_{min} + \Delta v \leq v_{sd} \leq v_{max} - \Delta v$.*

Remark 8: *Generally speaking, the airspeed of the UAV is always faster than the target speed and the wind speed, thus the constraint $v_{sd} - \Delta v > \sqrt{T_x^2 + T_y^2}$ could be satisfied. In addition, Δv is dependent on the performance of the autopilot, which definitely satisfies the requirement of $v_{min} + \Delta v \leq v_{sd} \leq v_{max} - \Delta v$. Therefore, Assumption 5 is reasonable. Based on Assumption 5, the following Theorem 4 is proposed.*

Theorem 4: *If the standoff speed v_{sd} and the speed increment Δv satisfy the inequality conditions presented in Assumption 5, then the proposed controller given by Equation (49) always satisfies the minimum and maximum airspeed constraints given by Equation (2).*

Proof: According to Equation (49), let $\gamma = v_{s,i} - v_{sd} = \Delta v \cdot \alpha \cdot \beta$, where $\alpha = \frac{\Delta\tau_i}{\pi}$ and $\beta = \frac{r_{i-1}^2 + r_d^2}{r_{i-1}^2 + r_i^2}$. Due to $\Delta\tau_i \in [-\pi, \pi)$, then $|\Delta\tau_i| \leq \pi$ and $|\alpha| \leq 1$. If $r_i \geq r_d$, then $0 < \beta \leq 1$, i.e., $|\beta| \leq 1$.

$$|\gamma| = \Delta v \cdot |\alpha| \cdot |\beta| \leq \Delta v \tag{50}$$

Considering $v_{s,i} = v_{sd} + \gamma$, yields

$$v_{sd} - |\gamma| \leq (|v_{s,i}| = |v_{sd} + \gamma|) \leq v_{sd} + |\gamma| \tag{51}$$

Substituting (50) into (51) yields

$$v_{sd} - \Delta v \leq |v_{s,i}| \leq v_{sd} + \Delta v \tag{52}$$

According to Assumption 5, it is obtained that $v_{min} \leq v_{sd} - \Delta v \leq |v_{s,i}| \leq v_{sd} + \Delta v \leq v_{max}$, which implies that $0 < v_{min} \leq v_{s,i} \leq v_{max}$. Therefore, it can be concluded that the proposed airspeed control law given by Equation (49) always satisfies the minimum and maximum airspeed constraints given by Equation (2). Thus, the proof for Theorem 3 is completed. □

4.2. Stability Analysis of Airspeed Control Law

A stability analysis of the airspeed control law is provided in the following Theorem 5.

Theorem 5: *For the relative motion model of a UAV described in Equation (9), if the heading rate control law given by Equation (24) and the cooperative airspeed control laws given by Equation (49) are applied to a team of UAVs, then each aircraft A_i can achieve equal temporal separation and fly along the standoff circle, i.e., as $t \longrightarrow +\infty, \Delta\tau_i \longrightarrow 0, i = 2, 3, \ldots, N$.*

Proof: Differentiating Equation (48) with respect to time t yields

$$\Delta\dot{\tau}_i = \dot{\tau}_{i-1} - \dot{\tau}_i = \left(\frac{2r_d\pi}{T_\tau}\right)\left[\left(\frac{\dot{\theta}_{i-1}}{v_{rd,(i-1)}}\right) - \left(\frac{\dot{\theta}_i}{v_{rd,i}}\right)\right] \tag{53}$$

According to the relative motion model given by Equation (9), one obtains

$$\dot{\theta}_i = \frac{v_{r,i}}{r_i} sin\phi_{GF,i} = \frac{v_{r,i}}{r_i} \cdot \frac{2r_i r_d}{r_i^2 + r_d^2} \tag{54}$$

Suppose that aircraft A_i $(i = 2, 3, \ldots, N)$. implementing the proposed heading rate control law given by Equation (24), has been converged on the standoff circle, i.e., $r_i = r_d$. Therefore,

$$\Delta\dot{\tau}_i = \left(-\frac{2\pi}{T_\tau}\right)\left[\frac{v_{r,i} - v_{rd,i}}{v_{rd,i}} - \frac{v_{r,(i-1)} - v_{rd,(i-1)}}{v_{rd,(i-1)}}\right] \tag{55}$$

Let $\sigma = T_x cos\psi_{rd,i} + T_y sin\psi_{rd,i}$, we obtain

$$v_{r,i}^2 - v_{rd,i}^2 = (v_{s,i} - v_{sd})(v_{s,i} + v_{sd} - 2\sigma) \tag{56}$$

According to the cooperative airspeed control laws given by Equation (49), one has

$$v_{s,i} - v_{sd} = \left(\frac{\Delta v \cdot \Delta\tau_i}{\pi}\right) \tag{57}$$

Substituting Equation (57) into Equation (56) yields

$$v_{r,i} - v_{rd,i} = K_i \Delta\tau_i \tag{58}$$

where $K_i \triangleq \frac{\Delta v(v_{s,i} + v_{sd} - 2\sigma)}{\pi(v_{r,i} + v_{rd,i})}$. Substituting Equation (58) into Equation (55) and setting $v_{r,1} = v_{rd,1}$ yields

$$\begin{cases} \Delta\dot{\tau}_2 = -\frac{2\pi}{T_\tau}\left(\frac{K_2\Delta\tau_2}{v_{rd,2}} - 0\right) \\ \Delta\dot{\tau}_3 = -\frac{2\pi}{T_\tau}\left(\frac{K_3\Delta\tau_3}{v_{rd,3}} - \frac{K_2\Delta\tau_2}{v_{rd,2}}\right) \\ \vdots \\ \Delta\dot{\tau}_i = -\frac{2\pi}{T_\tau}\left(\frac{K_i\Delta\tau_i}{v_{rd,i}} - \frac{K_{(i-1)}\Delta\tau_{(i-1)}}{v_{rd,(i-1)}}\right) \end{cases} \tag{59}$$

Let $\Delta\tau = (\Delta\tau_2, \Delta\tau_3, \ldots, \Delta\tau_N)^T$ and $\varepsilon_i = \frac{2\pi K_i}{T_\tau v_{rd,i}}$, one derives

$$\Delta\dot{\tau} = -\begin{bmatrix} \varepsilon_2 & 0 & \cdots & \cdots \\ -\varepsilon_2 & \varepsilon_3 & 0 & \cdots \\ \vdots & -\varepsilon_3 & \ddots & 0 \\ \vdots & \vdots & -\varepsilon_{N-1} & \varepsilon_N \end{bmatrix} \Delta\tau \tag{60}$$

It is easy to prove that the cascade connected system shown in Equation (60) is asymptotically stable, i.e., $\Delta\tau \longrightarrow 0$ as $t \longrightarrow +\infty$. Thus, the proof for Theorem 5 is completed. □

5. Estimator for the Composition Velocity of Background Wind and Target's Motion

In the previous section, the proofs and analyses are based on the assumption that the composition velocity is accurately known a priori. However, in practical applications, this assumption is not always true. Therefore, in this section, an estimator is proposed to estimate the composition velocity in real time.

5.1. Principle of the Composition Velocity Estimator

Suppose that the composition velocity $T = (T_x, T_y)$ is bounded, and its upper bound is known a priori, denoted as $T^* = sup_{t \geq 0}\{\|T\|_2\}$. According to [31], the target motion and wind are regarded as an external disturbance. And the composition velocity of the external disturbance can be expressed in the form of a hyperbolic tangent function as follows:

$$T_x = T^* tanh\varphi_x; T_y = T^* tanh\varphi_y \tag{61}$$

The parameters of the external disturbance, φ_x and φ_y, are regarded as estimated variables to obtain the estimated results of the composition velocity. $\hat{T} = (\hat{T}_x, \hat{T}_y)$ is introduced to denote the estimation of (T_x, T_y), then it also can be expressed in the form of a hyperbolic tangent function as follows:

$$\hat{T}_x = T^* tanh\hat{\varphi}_x; \hat{T}_y = T^* tanh\hat{\varphi}_y \tag{62}$$

where $(\hat{\varphi}_x, \hat{\varphi}_y)$ is an estimated result of the external disturbance parameters. Thus, our goal is to design an estimator which can guarantee that the estimated disturbance parameters can asymptotically coverage to the true values, i.e., $(\hat{\varphi}_x, \hat{\varphi}_y) \to (\varphi_x, \varphi_y)$ as $t \longrightarrow +\infty$.

Therefore, the dynamics of the estimated relative position are formulated as follows:

$$\dot{\hat{x}}_r = v_s cos\psi - \hat{T}_x + k_3\tilde{x}_r; \dot{\hat{y}}_r = v_s sin\psi - \hat{T}_y + k_3\tilde{y}_r \tag{63}$$

where $k_3 > 0$ is a positive constant. $(\tilde{x}_r, \tilde{y}_r)$ is the estimated error in the relative position.

$$\tilde{x}_r = x_r - \hat{x}_r; \tilde{y}_r = y_r - \hat{y}_r \tag{64}$$

where $(\hat{x}_r, \hat{y}_r)$ represents the estimated relative position. Then, differentiating Equation (64) with respect to time t, and according to Equation (63), the dynamics of the relative position estimate error can be expressed as follows:

$$\dot{\tilde{x}}_r = -\tilde{T}_x - k_3\tilde{x}_r; \dot{\tilde{y}}_r = -\tilde{T}_y - k_3\tilde{y}_r \tag{65}$$

where $(\tilde{T}_x, \tilde{T}_y)$ represents the estimated error in the composition velocity:

$$\tilde{T}_x = T_x - \hat{T}_x; \tilde{T}_y = T_y - \hat{T}_y \tag{66}$$

It is worth noting that the errors $(\tilde{T}_x, \tilde{T}_y)$ could be reduced by $(\hat{\varphi}_x, \hat{\varphi}_y)$. Therefore, let $k_4 > 0$ be a positive constant, $(\hat{\varphi}_x, \hat{\varphi}_y)$ is updated as follows:

$$\dot{\hat{\varphi}}_x = -k_4\tilde{x}_r; \dot{\hat{\varphi}}_y = -k_4\tilde{y}_r \tag{67}$$

The pseudocode of the composition velocity estimation algorithm (CVEA) is illustrated in Algorithm 1.

It is worth explaining that the differences between the CVEA proposed in this paper and in [31] are shown in Figure 4. Point A represents the current position of the UAV; the corresponding azimuth angle is denoted by θ. Suppose P is the point with the corresponding azimuth angle θ on the desired obit of the LGVF, and the position of P can be calculated according to Equation (16). Generally speaking, it is hoped that the two points A and P should coincide. However, due to the effect of external disturbances, there exists an offset between the actual vehicle trajectory and the desired LGVF orbit, denoted as $\tilde{r}(\theta) = \overrightarrow{AP} = (\tilde{x}_r, \tilde{y}_r)$. The offset can be used to estimate the composition velocity. In this paper, the CVEA approach uses the whole offset vector $[\tilde{x}_r, \tilde{y}_r]^T$ to obtain the estimation of the composition velocity. However, Ref. [31] only uses the radial distance of the offset, i.e., $\tilde{r}(\theta) = \sqrt{\tilde{x}_r^2 + \tilde{y}_r^2}$, in the estimator. Theoretically speaking, because the feedback item contains more information compared to what was used in [31], the composition velocity can be estimated faster and more accurately in the proposed CVEA.

Algorithm 1 Pseudocode of the composition velocity estimation algorithm (CVEA)
Input: The estimate results $\left[\hat{T}_x^{\ominus}, \hat{T}_y^{\ominus}, \hat{\varphi}_x^{\ominus}, \hat{\varphi}_y^{\ominus}, \hat{x}_r^{\ominus}, \hat{y}_r^{\ominus}\right]^T$ at time t_{k-1}, and the measured relative position $[x_r, y_r]^T$ at the present t_k.
Output: The estimate results $\left[\hat{T}_x^{\oplus}, \hat{T}_y^{\oplus}, \hat{\varphi}_x^{\oplus}, \hat{\varphi}_y^{\oplus}, \hat{x}_r^{\oplus}, \hat{y}_r^{\oplus}\right]^T$ at time t_k.
1. Update the estimate error in the relative position as follows
$\tilde{x}_r = x_r - \hat{x}_r^{\ominus}; \tilde{y}_r = y_r - \hat{y}_r^{\ominus}$
where (x_r, y_r) can be regarded as the measurement update at time t_k; $(\hat{x}_r^{\ominus}, \hat{y}_r^{\ominus})$ represent the prior estimation of the relative position at time t_{k-1}.
2. Calculate a posterior estimate of the parameters (φ_x, φ_y) as follows
$\hat{\varphi}_x^{\oplus} = \hat{\varphi}_x^{\ominus} - k_4\tilde{x}_r \cdot \Delta t; \hat{\varphi}_y^{\oplus} = \hat{\varphi}_y^{\ominus} - k_4\tilde{y}_r \cdot \Delta t$
where $\Delta t = t_k - t_{k-1}$ represents the time step.
3. A posteriori estimation of the composition velocity (T_x, T_y) is calculated as follows
$\hat{T}_x^{\oplus} = T^* tanh\hat{\varphi}_x^{\oplus}; \hat{T}_y^{\oplus} = T^* tanh\hat{\varphi}_y^{\oplus}$
4. Update the posterior estimate of the relative position as follows
$\hat{x}_r^{\oplus} = \hat{x}_r^{\ominus} + (v_{sd}cos\psi - \hat{T}_x^{\oplus} + k_3\tilde{x}_r)\Delta t; \hat{y}_r^{\oplus} = \hat{y}_r^{\ominus} + (v_{sd}sin\psi - \hat{T}_y^{\oplus} + k_3\tilde{y}_r)\Delta t$
5. Update time $t_{k-1} \leftarrow t_k$, go to Step 1 to start the next round of estimation.

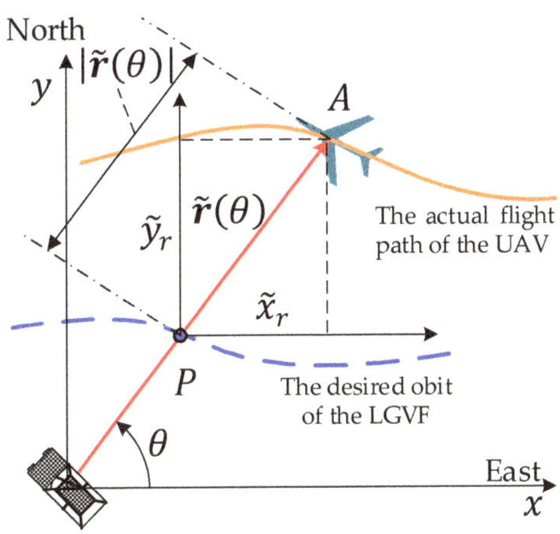

Figure 4. Principle of CVEA.

5.2. Convergence Analysis of the Composition Velocity Estimator

A convergence analysis of the composition velocity estimator is provided in the following Theorem 6.

Theorem 6: *The composition velocity estimator given by Equations (63)–(67) is asymptotically stable, i.e.,* $\hat{T}_x \to T_x$ *and* $\hat{T}_y \to T_y$ *as* $t \longrightarrow +\infty$.

Proof: Consider the following Lyapunov function:

$$\begin{aligned} V = \tfrac{1}{2}\tilde{x}_r^2 + \tfrac{1}{2}\tilde{y}_r^2 \quad &+ \tfrac{T^*}{k_4}\left[ln(cosh(\hat{\varphi}_x)) - \hat{\varphi}_x \cdot tanh(\varphi_x)\right] \\ &+ \tfrac{T^*}{k_4}\left[ln(cosh(\hat{\varphi}_y)) - \hat{\varphi}_y \cdot tanh(\varphi_y)\right] \end{aligned} \tag{68}$$

(a) Firstly, we prove that the Lyapunov function V is negative semi-definite, i.e., $\dot{V} \leq 0$.

Differentiating Equation (68) with respect to time t, and according to Equations (63)–(67) yields

$$\dot{V} = \tilde{x}_r\dot{\tilde{x}}_r + \tilde{y}_r\dot{\tilde{y}}_r - \frac{\dot{\hat{\varphi}}_x}{k_4}\tilde{T}_x - \frac{\dot{\hat{\varphi}}_y}{k_4}\tilde{T}_y = -k_3\left(\tilde{x}_r^2 + \tilde{y}_r^2\right) \leq 0 \quad (69)$$

(b) Secondly, it is shown that the Lyapunov function V is lower bounded, i.e., $V \geq V_{inf}$.

An assistant function is defined as $H(x) = ln(cosh(x)) - x\cdot tanh(x_0)$. Differentiating $H(x)$ with respect to x, we obtain $\frac{\partial H}{\partial x} = tanh(x) - tanh(x_0)$. This means that when $x = x_0$, $H(x)$ reaches the minimum, i.e., $H(x) \geq H(x_0) \geq -ln2$. Thus, when $\hat{\varphi}_x = \varphi_x$ and $\hat{\varphi}_y = \varphi_y$, the Lyapunov function V reaches the following lower bound $V \geq V_{inf} \triangleq -\frac{2T^*}{k_4}ln2$. Therefore, the function V is lower bounded.

(c) Finally, we prove that the function $\ddot{V}$ is bounded.

Differentiating Equation (69) with respect to time t, one obtains

$$\ddot{V} = -2k_3\left(\tilde{x}_r\cdot\dot{\tilde{x}}_r + \tilde{y}_r\cdot\dot{\tilde{y}}_r\right) = 2k_3\left[\tilde{x}_r\left(\tilde{T}_x + k_3\tilde{x}_r\right) + \tilde{y}_r\left(\tilde{T}_y + k_3\tilde{y}_r\right)\right] \quad (70)$$

Because $\tilde{x}_r$, $\tilde{y}_r$, $\tilde{T}_x$, $\tilde{T}_y$ are all bounded in the proposed estimator, thus the function $\ddot{V}$ is also bounded. Therefore, according to Barbalat's lemma, it is concluded that $\dot{V} \to 0$ as $t \longrightarrow +\infty$. This implies that $\tilde{x}_r \to 0$ and $\tilde{y}_r \to 0$ as $t \longrightarrow +\infty$. In addition, it can be proved that $\ddot{\tilde{x}}_r$ is also bounded, and thus we can conclude that $\dot{\tilde{x}}_r = -\tilde{T}_x - k_3\tilde{x}_r \to 0$ according to Barbalat's lemma. In other words, $\tilde{T}_x \to 0$. Similarly, it can also be proved that $\tilde{T}_y \to 0$. We can summarize that $\hat{T}_x \to T_x$ and $\hat{T}_y \to T_y$ as $t \longrightarrow +\infty$. Thus, the proof for Theorem 6 is completed. □

5.3. Limitation of the Composition Velocity Estimator

A signal flow diagram of the CVEA is shown in Figure 5. The true value of the target's position should be known when the CVEA is implemented. However, the accurate target position cannot be obtained. Thus, in order to estimate the composition velocity of the unknown wind and target's motion, Assumption 2 in Section 2 is required. The application of the proposed CVEA is limited by Assumption 2. The target's states (e.g., position, velocity, and acceleration) can be estimated using various Kalman filter algorithms, and the target's motion and wind velocity can be obtained separately. In this paper, we keep the target's location out of the scope since we aim at providing a solid formulation concerning the problem of coordinated standoff tracking of a ground target. In future work, a target state estimate algorithm will be designed and integrated into the framework of the CVEA.

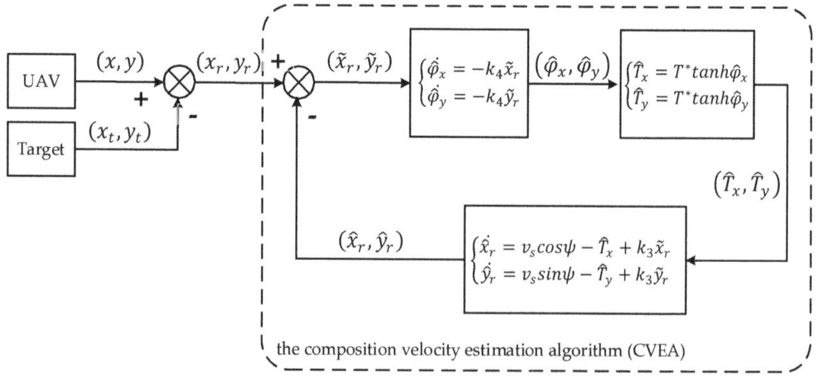

Figure 5. Signal flow diagram of CVEA.

6. Decentralized Control and Coordination Architecture of Standoff Target Tracking

A more detailed control and coordination architecture of the proposed coordinated standoff target tracking algorithm (CSTTA) is shown in Figure 6. The proposed CSTTA is scalable and does not require significant computation and communication power.

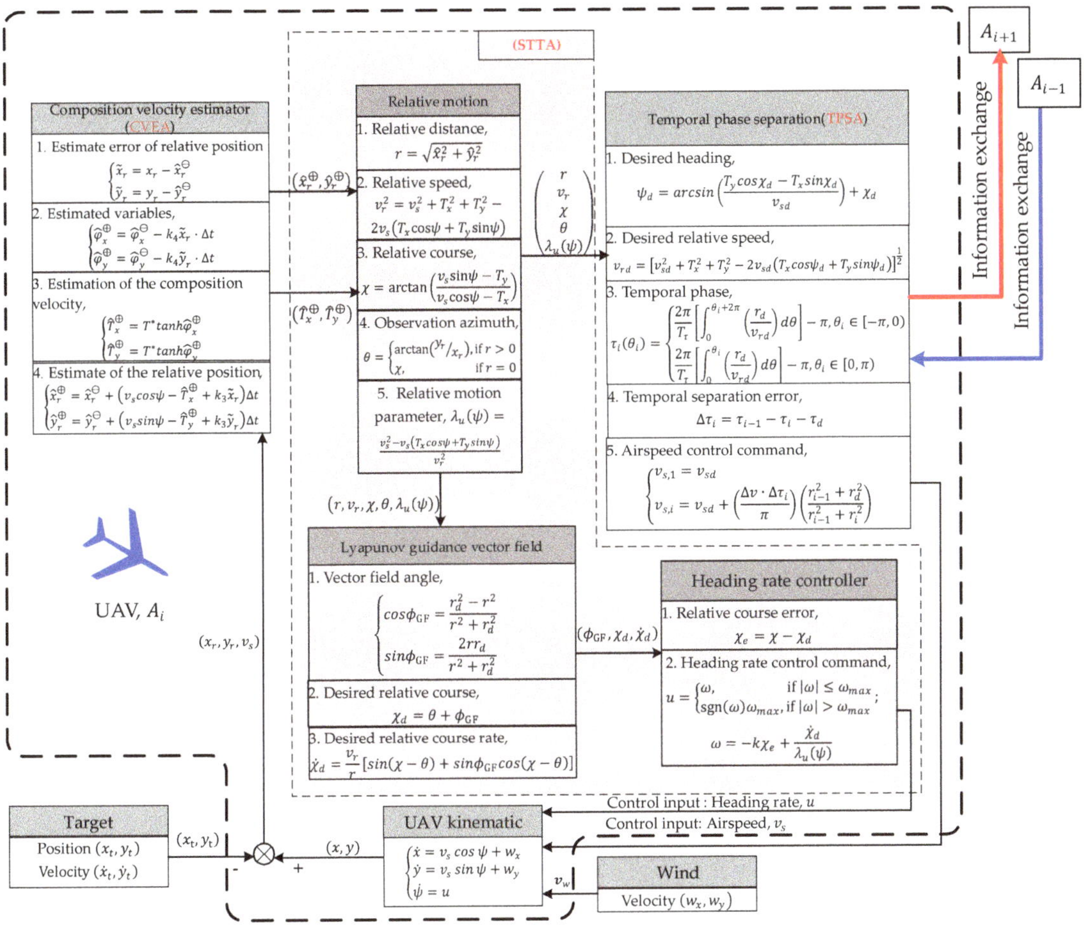

Figure 6. Decentralized control and coordination architecture of the proposed coordinated standoff target tracking algorithm (CSTTA).

On one hand, it can be observed from Figure 6 that all variables or signals are given their analytic solutions in the CSTTA. This is different from the MPC method proposed in Ref. [16]; the CSTTA proposed in this paper does not require an iterative optimization process, so it has a lower computational complexity. Therefore, the proposed CSTTA can be used in multi-UAV collaborative applications with high real-time requirements. It is worth noting that the temporal phase can be obtained by implementing Romberg integration, which has low computational complexity and will not increase the computational burden of the whole system.

On the other hand, it can be observed from Figure 6 that the control and coordination architecture is decentralized. In a distributed fashion, each UAV only transmits messages to its neighbors that are in its communication range. In the proposed CSTTA, A_i computes its own temporal separation error $\Delta\tau_i$ by exchanging information with its leader A_{i-1}. And then A_i generates its airspeed command $v_{s,i}$ and exchanges this command with its follower A_{i+1}. Therefore, the proposed CSTTA is a feasible approach to reduce the communication between UAVs.

It is worth noting that in our paper it is assumed that the communication between UAVs is perfect, without any restrictions. However, in the real-world, the wireless communication network between UAVs is vulnerable to errors and time delays, which may lead to performance degradation or even instability. In future work, we will analyze the effects of the potential communication constraints, which is a critical issue for the successful operation of multiple UAVs.

As shown in Figure 6, the proposed CSTTA method includes three parts, as follows: (i) the saturated heading rate controller, which is the standoff target tracking algorithm (STTA) for one single UAV; (ii) the airspeed controller based on the temporal phase separation algorithm (TPSA), which is used to achieve the desired temporal phase separation in cooperative standoff target tracking with multiple UAVs; (iii) the composition velocity estimation algorithm (CVEA), which is used to ensure the stability of the circular trajectory in the presence of a moving target and time-varying background wind.

7. Simulation Results

Firstly, in Section 7.1, scenario 1, that contains two UAVs and a single target, is used to verify the feasibility of the CSTTA in the presence of a moving target and background wind.

Secondly, to verify the performance of the CVEA, the simulation results of the CVEA approach are compared with those using the method used in Ref. [31]. In Section 7.2, comparative experiments are carried out in scenario 2, containing a single UAV and a target.

Finally, Section 7.3 presents scenario 3, that contains three UAVs and a single target. The performance of the proposed TPSA and the SPSA method presented in Ref. [25] are compared to verify that the TPSA has smaller separation errors and a faster convergence rate.

7.1. Scenario 1: Tracking a Moving Target Using Two UAVs

In scenario 1, there are two UAVs to perform the cooperative standoff tracking mission for a ground-based moving target in unknown background wind. The simulation conditions for scenario 1 are shown in Table 1. The kinematic constraints of the UAV are shown in Table 2. Table 3 list the initial settings of the UAVs.

Table 1. Simulation conditions (scenario 1).

Parameter	Description	Value
t_0	Initial time of simulation	0 s
Δt	Sampling time	1.0 s
t_f	Final time of simulation	400 s
T^*	Upper bound of the composition velocity	25 m/s
v_{sd}	Desired standoff airspeed	100 m/s
r_d	Desired standoff radius	1500 m

Table 2. Kinematic constraints of the UAV (scenario 1).

Parameter	Description	Value
v_{min}	Allowable minimum airspeed	60 m/s
v_{max}	Allowable maximum airspeed	160 m/s
Δv	Unit increment of airspeed	30 m/s
ω_{max}	Maximum heading rate	30°/s

Table 3. The initial settings of UAVs (scenario 1).

UAV A_i	Position (x_i,y_i)/m	Velocity v_i/(m/s)	Heading ψ_i/(°)
A_1	(700, 400)	100	135
A_2	(−1200, 600)	100	20

To verify the performance of the CSTTA method in different simulation conditions with respect to the target's motion and background wind, two groups of experiments are carried out.

- Group A: Constant velocities of target and wind;
- Group B: Time-varying velocities of target and wind.

7.1.1. Group A: Constant Velocities of Target and Wind

In group A, a constant velocity (CV) model is used to describe the motion of the ground-based target with constant velocity. The discretized equation for the CV model is expressed as

$$\begin{gathered} x_t(k+1) = F(k)x_t(k) + G(k)\omega(k) \\ F(k) = \begin{bmatrix} 1 & T_s & 0 & 0 \\ 0 & 1 & 0 & 0 \\ 0 & 0 & 1 & T_s \\ 0 & 0 & 0 & 1 \end{bmatrix}; G(k) = \begin{bmatrix} \frac{T_s^2}{2} & T_s & 0 & 0 \\ 0 & 0 & \frac{T_s^2}{2} & T_s \end{bmatrix}^T \end{gathered} \tag{71}$$

where $x_t(k) = (x_t, \dot{x}_t, y_t, \dot{y}_t)^T$. $F(k)$ is the state transition matrix and $G(k)$ is the process noise input matrix. The covariance matrix of the process noise $\omega(k)$ is $Q(k) = diag(\delta_x^2(k), \delta_y^2(k))$. $\delta_x(k)$ and $\delta_y(k)$ are the standard deviations related to the target's velocity towards the x- and y-axes. The initial settings of the target are listed in Table 4. The wind velocity is $(w_x, w_y) = (-5, -2)$ m/s.

Table 4. The initial settings of target (scenario 1: group A).

Parameter	Description	Value
(x_t, y_t)	Initial position of target	(0, 0) m
$(\dot{x}_t, \dot{y}_t)$	Initial velocity of target	(2, 3) m/s
(δ_x, δ_y)	The standard deviations of target velocity	(0.1, 0.1) m/s

If the wind velocity is $(w_x, w_y) = (-5, -2)$ m/s, and the target velocity is $(\dot{x}_t, \dot{y}_t) = (2, 3)$ m/s, then the composition velocity is $(T_x, T_y) = (7, 5)$ m/s. On one hand, according to Theorem 2 in Section 3.2, if $v_{sd} = 100$ m/s, $(T_x, T_y) = (7, 5)$ m/s, and $\omega_{max} = 0.524$ rad/s, the minimum allowable standoff radius is $r_{d,min} \approx 901\ m$. The desired standoff distance $r_d = 1500$ m $> r_{d,min}$. On the other hand, when $v_{sd} = 100$ m/s, $v_{min} = 60$ m/s, $v_{max} = 160$ m/s, and $\Delta v = 30$ m/s, it is verified that the equations $v_{sd} - \Delta v > \sqrt{T_x^2 + T_y^2}$ and $v_{min} + \Delta v \leq v_{sd} \leq v_{max} - \Delta v$ are satisfied. According to Theorem 4 in Section 4.1, the simulation conditions in group A satisfy the input constraints of the UAV.

The trajectories of the UAVs and the target in the inertial frame are shown in Figure 7. The symbol "T" represents the target, the black solid line represents the target's trajectory, and the arrow on the trajectory represents the target's motion direction. The red dotted line represents the trajectory of A_1, and the blue dash-dotted line represents the trajectory of A_2. The triangular arrow "△" indicates the heading ψ of the UAV. $A_1(t_0)$ and $A_1(t_f)$ denote the initial and final positions of A_1, and $A_2(t_0)$ and $A_2(t_f)$ denote the initial and final positions of A_2, respectively. The black arrow "→" indicates the direction of the wind velocity.

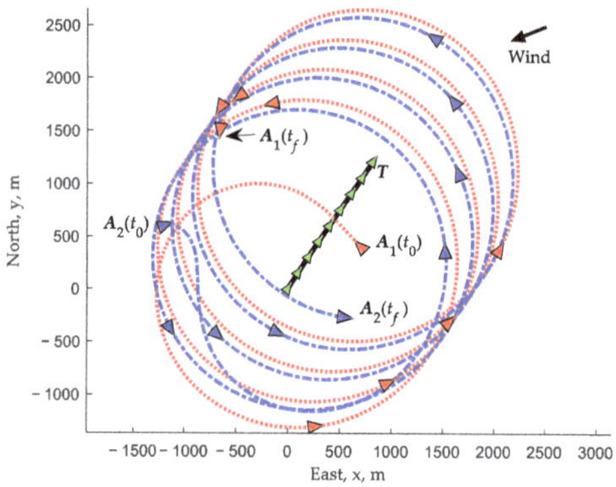

Figure 7. Trajectories of UAVs and target in inertial coordinates (scenario 1: group A).

The trajectories of the UAVs in the local target frame coordinates are shown in Figure 8, and the triangular arrow "△" represents the relative course χ of the UAVs. The UAVs initially have an arbitrary position and heading, and are eventually steered to hover around the target with a predefined standoff distance r_d = 1500 m. Meanwhile, the UAVs are converged to the desired relative course along the LGVF. In addition, the UAVs achieve the desired phase separation when all vehicles arrive at the standoff circle.

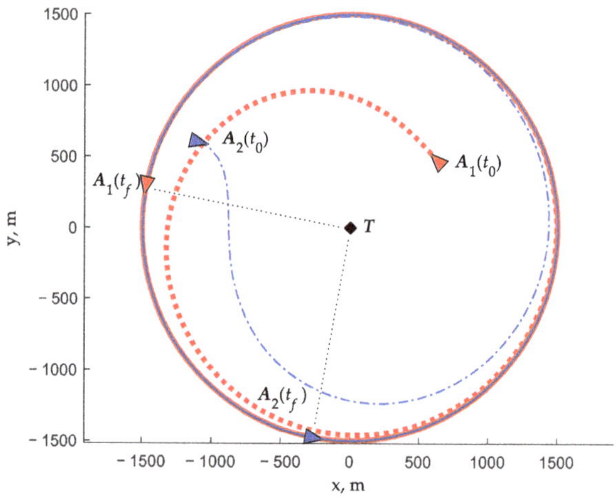

Figure 8. Trajectories of UAVs in local target frame coordinates (scenario 1: group A).

The heading rate control input u and airspeed control input v_s are shown in Figures 9 and 10, respectively. It can be seen that $|u_i| \leq \omega_{max}$ and $v_{min} \leq v_{s,i} \leq v_{max}$ $(i = 1, 2)$. This implies that the proposed heading rate controller and airspeed controller both satisfy the input constraints of the UAV. The simulation results confirm the conclusions of Theorem 2 and Theorem 4.

The estimated results of the composition velocity, $(\hat{T}_x, \hat{T}_y)$, are shown in Figure 11. And the corresponding estimated errors $(\tilde{T}_x, \tilde{T}_y)$ are presented in Figure 12. In order to show more clearly the curves of $(\hat{T}_x, \hat{T}_y)$ and $(\tilde{T}_x, \tilde{T}_y)$, Figures 11 and 12 only plot the histories of $(\hat{T}_x, \hat{T}_y)$ and $(\hat{T}_x, \hat{T}_y)$ in the time period 0~16 s. It can be seen that the estimated errors converge to 0 at about 8 s, which implies that the proposed estimator can obtain stable and accurate estimates of the composition velocity.

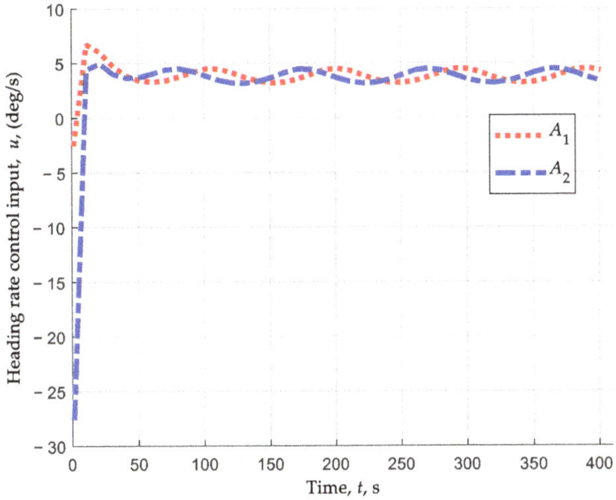

Figure 9. Heading rate control inputs of UAVs, u (scenario 1: group A).

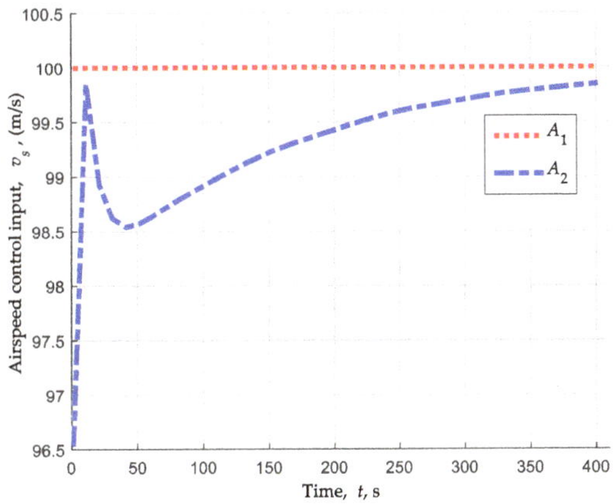

Figure 10. Airspeed control inputs of UAVs, v_s (scenario 1: group A).

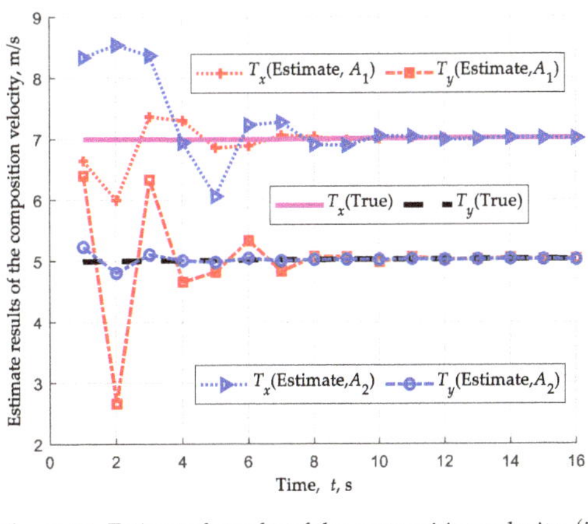

Figure 11. Estimated results of the composition velocity, $(\hat{T}_x, \hat{T}_y)$ (scenario 1: group A).

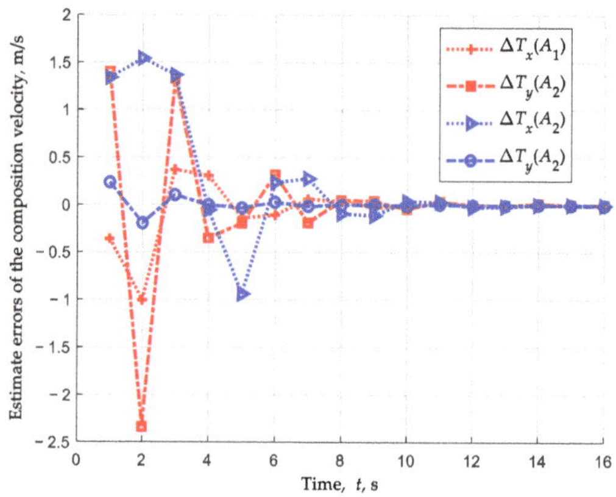

Figure 12. Estimated errors of the composition velocity, $(\tilde{T}_x, \tilde{T}_y)$ (scenario 1: group A).

The relative distances $r_i (i = 1, 2)$ between A_1 and the target and A_2 and the target are shown in Figure 13. It can be seen that the relative distances of A_1 and A_2 converge to the desired standoff radius r_d = 1500 m. The phase separation angle $\Delta\theta = \theta_2 - \theta_1$ is shown in Figure 14. It can be seen that the phase separation angle $\Delta\theta$ eventually converges to 90°. This implies that the optimal observation configuration of the UAVs is generated and maintained.

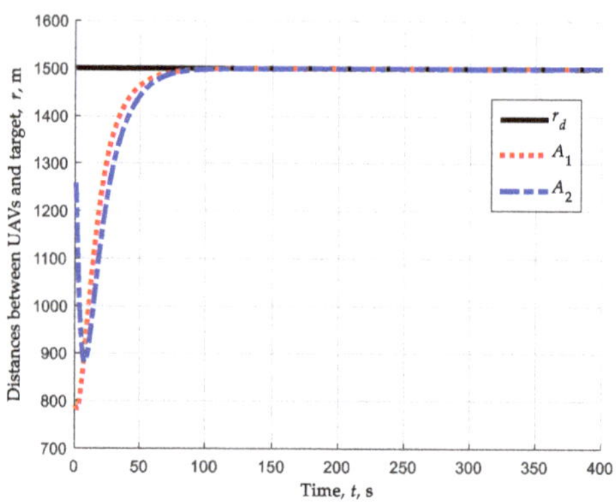

Figure 13. Distances between UAVs and target, r (scenario 1: group A).

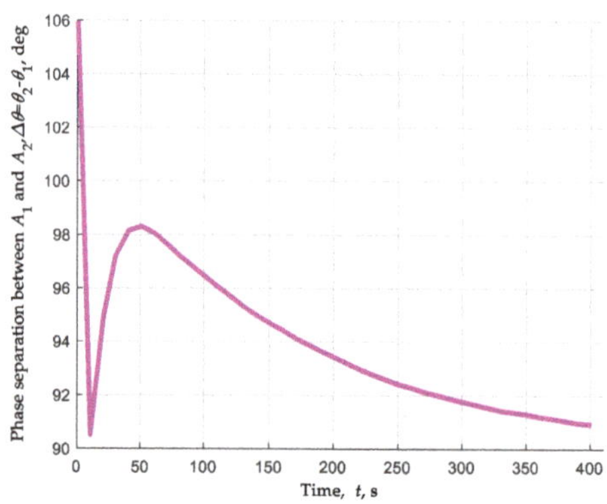

Figure 14. Phase separation between A_1 and A_2, $\Delta\theta = \theta_2 - \theta_1$ (scenario 1: group A).

7.1.2. Group B: Time-Varying Velocities of Target and Wind

A ground-based moving target has a low velocity, and they often and irregularly perform stop-and-go maneuvers with a much smaller turn radius. More agile target maneuvers are likely to have more significant higher-order derivatives, such as the third derivative of the target position, i.e., the acceleration rate or jerk of the target. A lower-order tracking model, such as the CV model, cannot adequately handle higher-order derivatives. Hence, a good model to apply to tracking a ground-based target is the jerk model. The jerk model defines the target acceleration as a correlated process exponentially decreasing in time. A discretized system equation for the jerk model for a ground target is thus expressed in the following form [16]:

$$x_t(k+1) = F(k)x_t(k) + \eta(k) \tag{72}$$

where $x_t(k) = \left(x_t, \dot{x}_t, \ddot{x}_t, y_t, \dot{y}_t, \ddot{y}_t\right)^T$. The state transition matrix $F(k)$ can be represented as follows:

$$F(k) = \begin{bmatrix} 1 & T_s & \left(e^{-\alpha T_s} + \alpha T_s - 1\right)/\alpha^2 & 0 & 0 & 0 \\ 0 & 1 & \left(1 - e^{-\alpha T_s}\right)/\alpha & 0 & 0 & 0 \\ 0 & 0 & e^{-\alpha T_s} & 0 & 0 & 0 \\ 0 & 0 & 0 & 1 & T_s & \left(e^{-\alpha T_s} + \alpha T_s - 1\right)/\alpha^2 \\ 0 & 0 & 0 & 0 & 1 & \left(1 - e^{-\alpha T_s}\right)/\alpha \\ 0 & 0 & 0 & 0 & 0 & e^{-\alpha T_s} \end{bmatrix} \tag{73}$$

where α is a correlation parameter that allows for the modeling of different classes of targets: small α for targets with relatively slow maneuvers, and large α for targets with fast and evasive maneuvers. The covariance matrix of the process noise $\eta(k)$ can be modeled as follows:

$$Q(k) = \frac{\sigma_a^2}{\alpha^4} \begin{bmatrix} q_{11} & q_{12} & q_{13} & 0 & 0 & 0 \\ q_{12} & q_{22} & q_{23} & 0 & 0 & 0 \\ q_{13} & q_{23} & q_{33} & 0 & 0 & 0 \\ 0 & 0 & 0 & q_{11} & q_{12} & q_{13} \\ 0 & 0 & 0 & q_{12} & q_{22} & q_{23} \\ 0 & 0 & 0 & q_{13} & q_{23} & q_{33} \end{bmatrix} \tag{74}$$

where σ_a is the standard deviation related to the target's acceleration, and the definitions of q_{ij} are

$$\begin{aligned} q_{11} &= \left(1 - l + 2m + \tfrac{2}{3}m^3 - 2m^2 - 4m\sqrt{l}\right) \\ q_{12} &= \alpha\left(l + 1 - \tfrac{1}{l} - 2m + m^2\right) \\ q_{13} &= \alpha^2\left(1 - l - 2m\sqrt{l}\right) \\ q_{22} &= \alpha^2\left(4\sqrt{l} - 3 - l + 2m\right) \\ q_{23} &= \alpha^3\left(l + 1 - 2\sqrt{l}\right) \\ q_{33} &= \alpha^4(1 - l) \end{aligned} \tag{75}$$

where $l = e^{-2\alpha T_s}$ and $m = \alpha T_s$. The initial settings of the target are listed in Table 5.

Table 5. The initial settings of target (scenario 1: group B).

Parameter	Description	Value
(x_t, y_t)	Initial position of target	$(0, 0)$ m
$(\dot{x}_t, \dot{y}_t)$	Initial velocity of target	$(2, 3)$ m/s
α	Correlation parameter	0.6
σ_a	The standard deviations of target acceleration	0.66 m/s^2

The characteristics of the trajectory and moving behavior of the target which are descripted by the jerk model are illustrated here through a numerical simulation. Figure 15 shows the trajectory of the moving target based on the jerk model. The time-varying speed and direction of the moving target based on the jerk model are shown in Figures 16 and 17, respectively. It is observed that the jerk is not negligible, and thus the target is maneuvering with time.

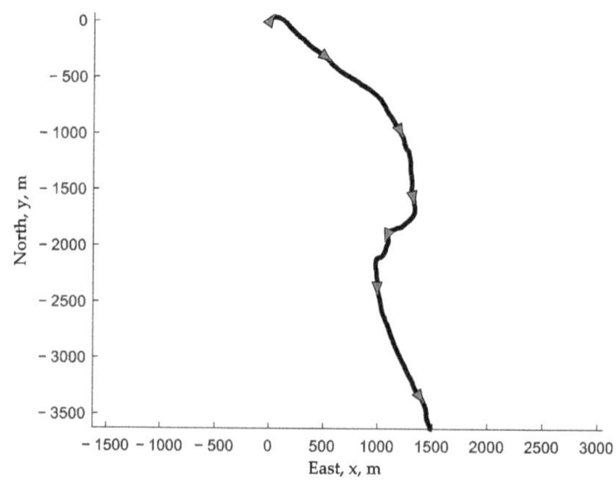

Figure 15. Trajectory of moving target with jerk model (group B).

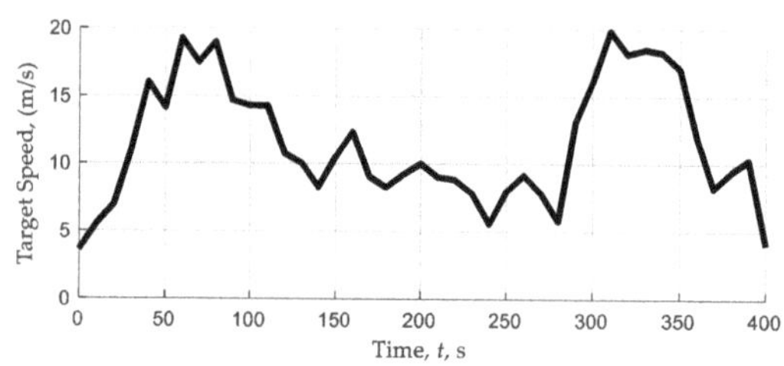

Figure 16. Speed of moving target with jerk model (scenario 1: group B).

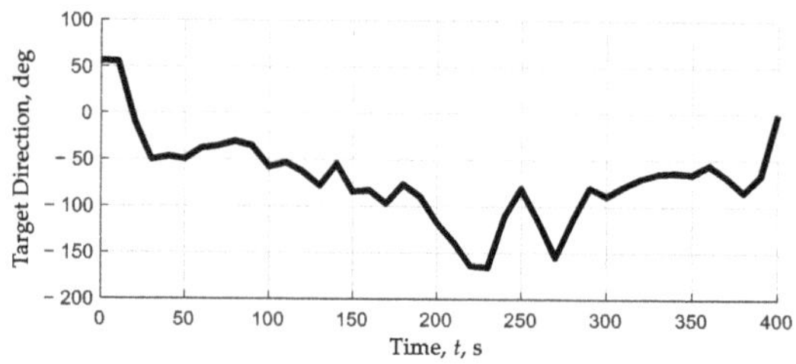

Figure 17. Direction of moving target with jerk model (scenario 1: group B).

It is worth noting that according to Theorem 2 and Theorem 4, the upper bound of the target speed must be restricted. In other words, in order to meet the input constraints of the UAV, the maximum speed of the target is not more than 20 m/s.

In addition, if the wind is time-varying, let us consider an example of a variable wind case, where the wind model is similar to the model used in [26], which is given below:

$$\begin{cases} W_x = v_w cos[\omega(t - t_0) + \phi_0] \\ W_y = v_w sin[\omega(t - t_0) + \phi_0] \end{cases} \quad (76)$$

where the constants are v_w = 5 m/s, $\omega = (\pi/180)$ rad/s, t_0 = 0 s, and $\phi_0 = 30°$.

From the settings of the target and the wind, it can be seen that, since the target speed is not more than 20 m/s and the wind speed is not more than 5 m/s, the upper bound of the composition velocity is T^* = 25 m/s in Table 1. According to Theorem 2, if v_{sd} = 100 m/s, T^* = 25 m/s, and ω_{max} = 0.524 rad/s, the minimum allowable standoff radius is $r_{d,min} \approx 1193.8$ m. The desired standoff distance, r_d = 1500 m > $r_{d,min}$, satisfies the heading rate constraint. In addition, it is verified that the inequalities $v_{sd} - \Delta v > \sqrt{T_x^2 + T_y^2}$ and $v_{min} + \Delta v \leq v_{sd} \leq v_{max} - \Delta v$ are satisfied. According to Theorem 4, the simulation conditions in group B satisfy the input constraints of the UAV.

The trajectories of the UAVs and the target in the inertial frame are shown in Figure 18. The trajectories of the UAVs in the local target frame coordinates are shown in Figure 19. The UAVs are eventually steered to hover around the target at the predefined standoff distance r_d = 1500 m, and achieve the desired phase separation on the standoff circle.

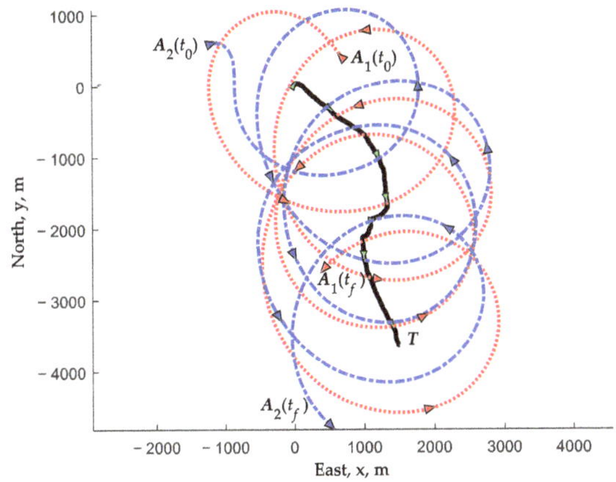

Figure 18. Trajectories of UAVs and target in inertial coordinates (scenario 1: group B).

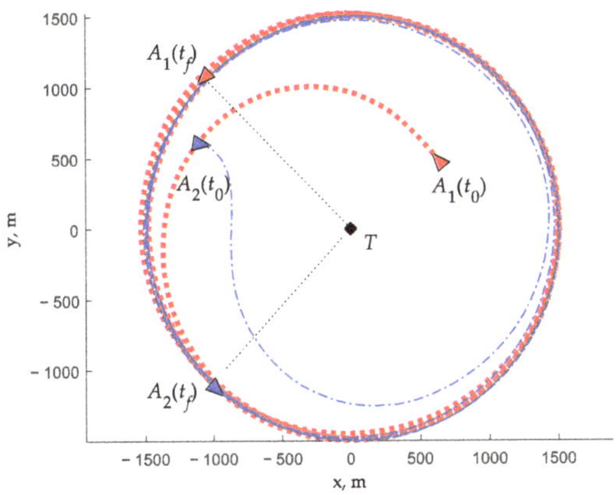

Figure 19. Trajectories of UAVs in local target frame coordinates (scenario 1: group B).

The heading rate control input u and airspeed control input v_s are shown in Figures 20 and 21, respectively. The control inputs satisfy the kinematic constraints of the UAVs.

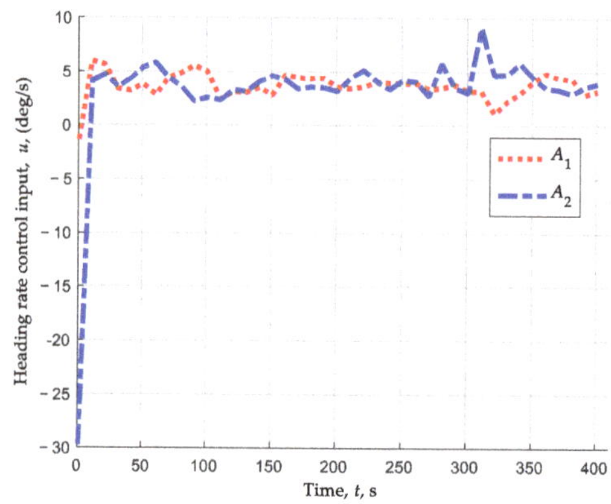

Figure 20. Heading rate control inputs of UAVs, u (scenario 1: group B).

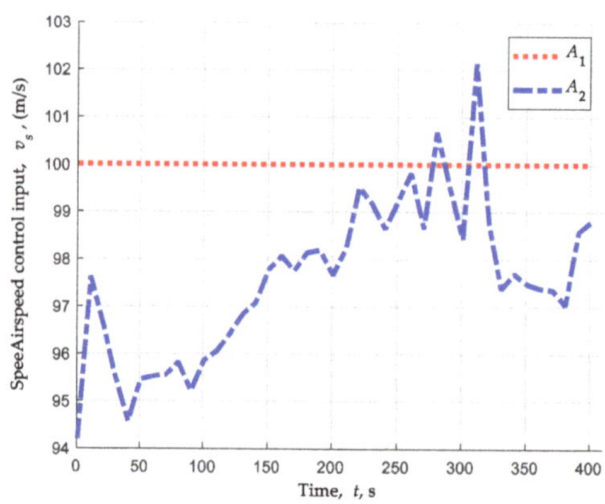

Figure 21. Airspeed control inputs of UAVs, v_s (scenario 1: group B).

The estimated results of the composition velocity, $(\hat{T}_x, \hat{T}_y)$, are shown in Figures 22 and 23, and the corresponding estimated errors $(\tilde{T}_x, \tilde{T}_y)$ are presented in Figures 24 and 25, respectively. It can be seen that the proposed CVEA still obtains stable and accurate estimates of the composition velocity in the presence of time-varying velocities of a ground-based moving target and background wind.

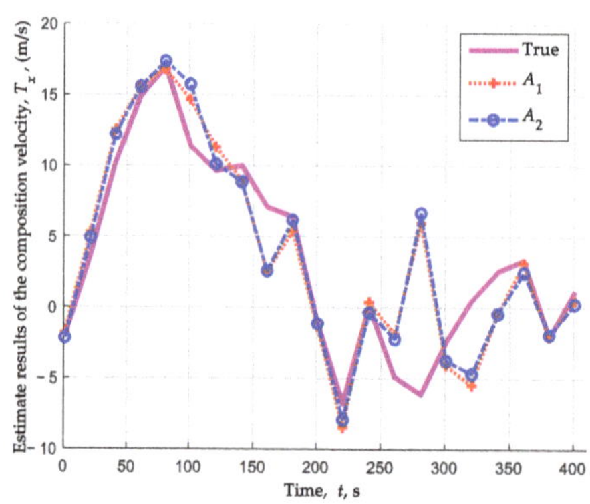

Figure 22. Composition velocity estimate results, $\hat{T}_x$ (scenario 1: group B).

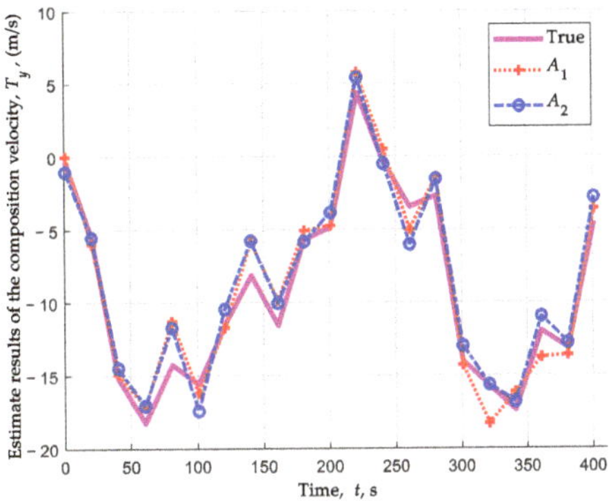

Figure 23. Composition velocity estimate results, $\hat{T}_y$ (scenario 1: group B).

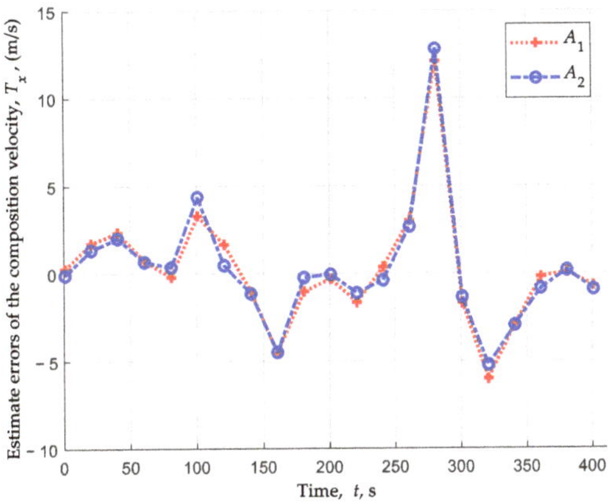

Figure 24. Composition velocity estimate errors, $\tilde{T}_x$ (scenario 1: group B).

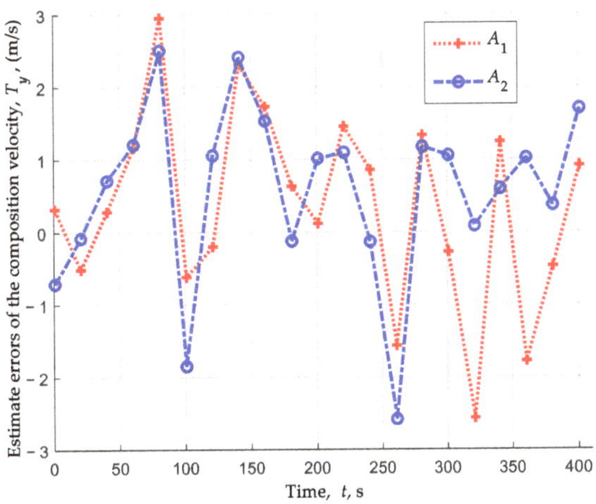

Figure 25. Composition velocity estimate errors, $\tilde{T}_y$ (scenario 1: group B).

The relative distances $r_i(i = 1, 2)$ between A_1 and the target and A_2 and the target are shown in Figure 26. It can be seen that there exist errors between the real relative distances and the desired value when the target performs greater maneuvers, for example, in the time period 300~350 s. However, the UAVs can still converge to the neighborhood of the desired orbit with small distance errors. The phase separation angle $\Delta\theta = \theta_2 - \theta_1$ is shown in Figure 27. Due to the target's maneuvers, A_2 tries to adjust its airspeed to form the desired phase separation with A_1, and thus the phase separation angle $\Delta\theta$ oscillates and converges to the neighborhood of the desired value.

Therefore, the simulation results in group B verify the performance of the proposed CSTTA approach. The CSTTA approach can guarantee that the UAVs perform successfully coordinated standoff tracking in the presence of time-varying velocities of a ground-based moving target and background wind.

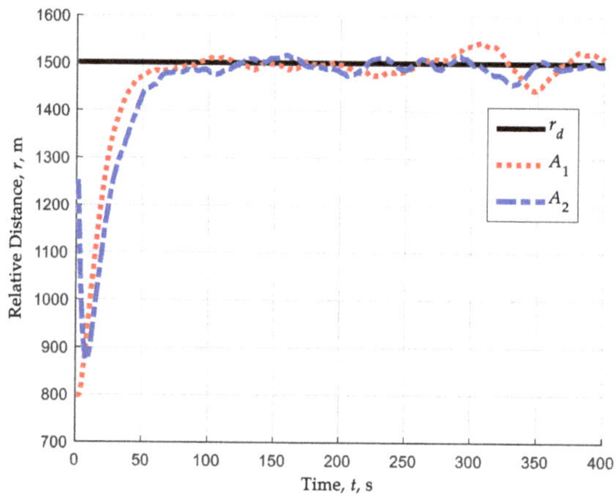

Figure 26. Distances between UAVs and target, r (scenario 1: group B).

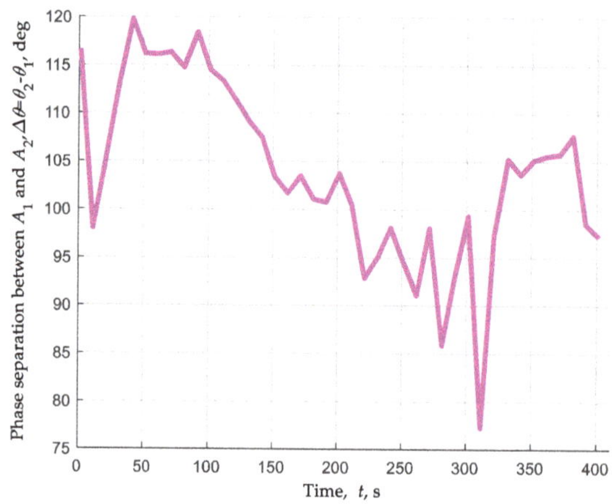

Figure 27. Phase separation between A_1 and A_2, $\Delta\theta = \theta_2 - \theta_1$ (scenario 1: group B).

7.2. Scenario 2: Tracking a Moving Target Using a Single UAV

In scenario 2, a single UAV A_1 is used to execute the standoff target tracking mission. The initial settings of A_1 are listed in Table 6, and the other simulation conditions are the same as for group B in scenario 1. In order to verify that the proposed CVEA can generate an effective and accurate composition velocity estimation to effectively improve the standoff tracking performance, the simulation results of the single drone's standoff target tracking algorithm (STTA), as shown in Equation (24) with the CVEA (denoted as "STTA+CVEA"), are compared with those of the method reported in [31].

Table 6. Initial settings of UAVs (scenario 2).

UAV A_i	Position (x_i, y_i)/m	Velocity v_i/(m/s)	Heading ψ_i/(°)
A_1	(600, 200)	100	90

Firstly, we need to define several performance indexes to evaluate the effectiveness of the CVEA approach and the method reported in [31]. Thus, the global average errors (GAEs) are introduced. It is worth noting that global means it is averaged over the N UAVs. This definition is not only appropriate for the scenario of a single UAV, but also for the scenario of multiple UAVs. At time t_k, the GAE of the relative distance regulation is defined as $e_r(t_k) = \frac{1}{N}\sum_{i=1}^{N}|r_i(t_k) - r_d|$. Similarly, the GAE of the relative course tracking is defined as $e_\chi(t_k) = \frac{1}{N}\sum_{i=1}^{N}\left|\chi_i(t_k) - \chi_i^d(t_k)\right|$; the GAE of the phase separation is defined as $e_\theta(t_k) = \frac{1}{N}\sum_{i=1}^{N}|\theta_j(t_k) - \theta_i(t_k) - \theta_d|$; the GAE of the composition velocity estimation is defined as $e_T(t_k) = \frac{1}{N}\sum_{i=1}^{N}\left(\tilde{T}_x^2 + \tilde{T}_y^2\right)^{\frac{1}{2}}$. These GAEs can be used to analyze the convergence of the designed controllers and estimator, including the heading rate controller proposed in Section 3.2, the airspeed controller proposed in Section 4.1, and the composition velocity estimator proposed in Section 5.1.

Then, in order to analyze the convergence rate of the designed controllers and estimator, the integrated time absolute error (ITAE) is defined as follows:

$$J_{ITAE}(e) = \int_0^t \tau \cdot e(\tau) d\tau \tag{77}$$

where $e(\tau)$ represents the GAEs at time τ, which includes the GAE of relative distance regulation, the GAE of relative course tracking, the GAE of phase separation, and the GAE of composition velocity estimation. It can be observed from Equation (77) that a smaller $J_{ITAE}(e)$ implies a smaller convergence error $e(\tau)$ and faster convergence rate.

Due to the process noise in the target's motion, the results of a single simulation are not sufficient to illustrate that the proposed method is effective. Hence, we run a Monte Carlo simulation 300 times for scenario 2 to further analyze the performances of the two methods using the above-defined GAEs and ITAEs.

Figure 28 shows the performance of the relative distance regulation. Figure 29 shows the performance of the relative course tracking. It can be seen that the proposed "STTA+CVEA" method has smaller GAEs and ITAEs in relative distance regulation and relative course tracking than that in [31]. This implies that the "STTA+CVEA" method has better performance than the method presented in [31] in the standoff target tracking mission of a single UAV. The reason for the above phenomenon is that in contrast to [31], in which only the radial distance of the offset is used, the "STTA+CVEA" method updates the estimation results according to the offset vector, containing more feedback information, and thus our proposed method produces a better composition velocity estimation to enhance the tracking performance of the UAV.

Figure 30 shows the performance of the composition velocity estimation. The GAE and ITAE in the composition velocity estimation produced by "STTA+CVEA" are both smaller than that in the method presented in [31]. In other words, the estimates generated in our approach can converge to the true values more quickly with smaller convergence errors.

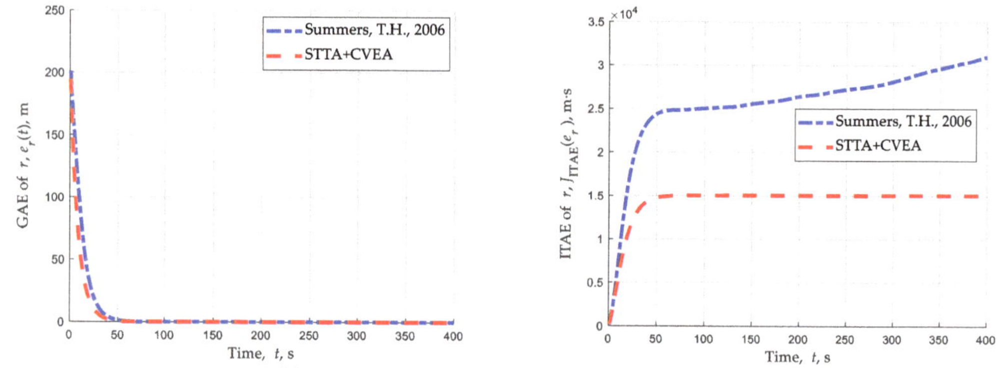

(a) GAE, $e_r(t)$ (b) ITAE, $J_{\text{ITAE}}(e_r)$

Figure 28. Performance of relative distance regulation (scenario 2). (**a**) Global average error in relative distance. (**b**) Integrated time absolute error in relative distance [31].

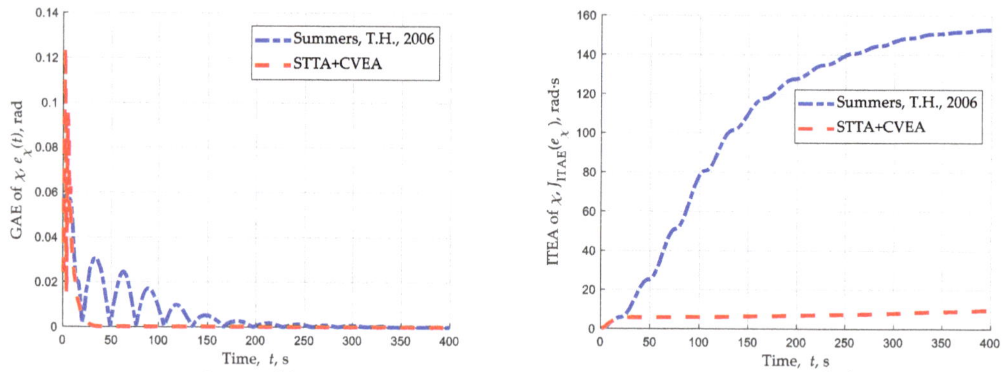

(a) GAE, $e_\chi(t)$ (b) ITAE, $J_{\text{ITAE}}(e_\chi)$

Figure 29. Performance of relative course tracking (scenario 2). (**a**) Global average error in relative course. (**b**) Integrated time absolute error in relative course [31].

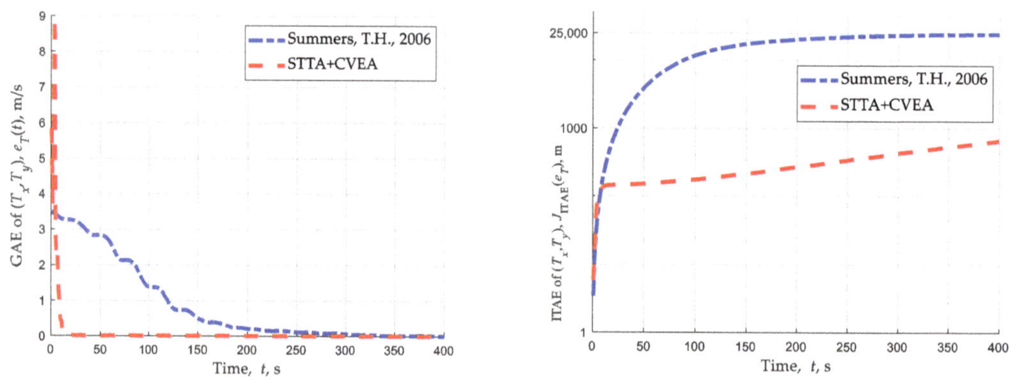

(a) GAE, $e_T(t)$ (b) ITAE, $J_{\text{ITAE}}(e_T)$

Figure 30. Performance of composition velocity estimation (scenario 2). (**a**) Global average error in composition velocity estimation. (**b**) Integrated time absolute error in composition velocity estimation [31].

Figure 31 shows a comparison of the heading rate control inputs generated using the two methods. Figure 32 shows a comparison of the relative course errors generated using the two methods. In order to show more clearly the curves of u and χ_e, Figures 31 and 32 only plot the histories of u and χ_e in the time period 0~150 s. It can be seen that the CVEA can obtain the composition velocity estimate results accurately and quickly compared to the results in [31]; the heading rate control input signal is smooth and the relative course error converges to 0 without oscillation. This confirms that the CVEA can improve the effectiveness and the robustness of the standoff target tracking algorithm (STTA) for the UAV in unknown background wind.

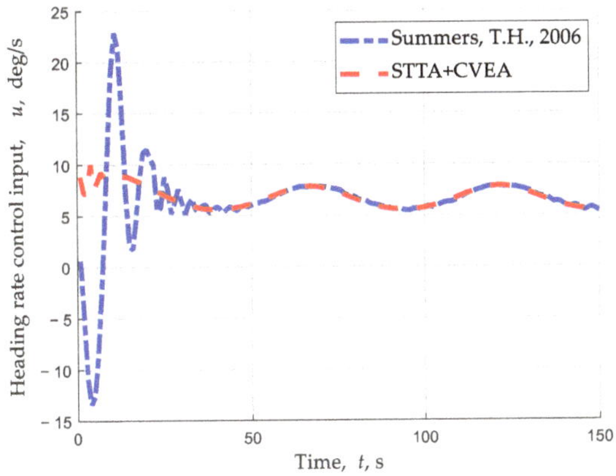

Figure 31. Heading rate control input, u (scenario 2) [31].

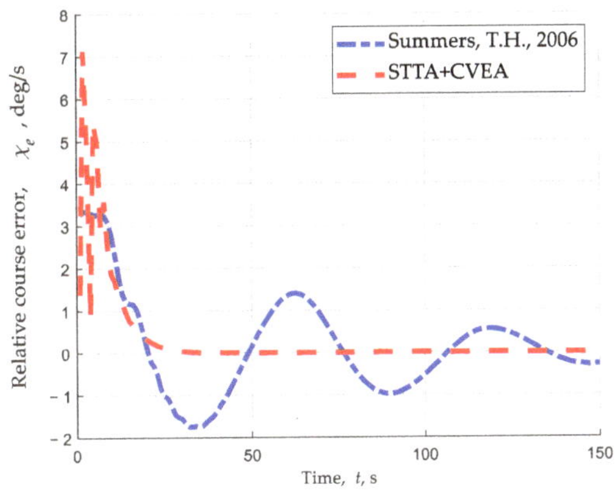

Figure 32. Relative course error, χ_e (scenario 2 [31]).

A performance comparison of "STTA+CVEA" and Ref. [31] is shown in Table 7. Compared to [31], the GAE and ITAE of the relative distance regulation are reduced by 29.44% and 43.84%, respectively, in the proposed "STTA+CVEA" method. The GAE and ITAE in the relative course tracking are reduced by 67.74% and 93.5%, respectively. The GAE and ITAE in the composition velocity estimation are reduced by 88.90% and 98.14%, respectively. Therefore, in the standoff target tracking mission using a single UAV in unknown background wind, our method has better performance than the method presented in [31].

Table 7. Performance comparison of "STTA+CVEA" and Ref. [31] (scenario 2).

Control Objective	Performance Index	STTA+CVEA	Ref. [31]	Percentage/(%)
Relative distance regulation	$e_r(t)/m$	4.2229	5.9846	29.44
	$J_{ITAE}(e_r)/(m \times s)$	1.4436×10^4	2.5706×10^4	43.84
Relative course tracking	$e_\chi(t)/rad$	0.0020	0.0062	67.74
	$J_{ITAE}(e_\chi)/(rad \times s)$	6.9507	106.9023	93.50
Composition velocity estimation	$e_T(t)/(m \times s^{-1})$	0.0954	0.8598	88.90
	$J_{ITAE}(e_T)/(m)$	313.6534	1.6874×10^4	98.14

7.3. Scenario 3: Tracking a Moving Target Using Three UAVs

In scenario 3, three UAVs are used to execute a standoff target tracking mission. Table 8 lists the initial settings of A_1, A_2, and A_3, and the other simulation conditions are the same as for group B in scenario 1. In order to verify that the proposed TPSA approach has smaller separation errors and a faster convergence rate, and thus improves the cooperative tracking performance of the UAVs, the simulation results of the TPSA are compared with the SPSA method presented in [25].

Table 8. Initial settings of UAVs (scenario 3).

UAV A_i	Position (x_i, y_i)/m	Velocity v_i/(m/s)	Heading ψ_i/(°)
A_1	(600, 200)	100	90
A_2	(−600, 300)	100	120
A_3	(0, 800)	100	−170

It is important to note that in [25] they only study the case of two collaborative UAVs tracking a ground target with background wind. Fortunately, the method in [25] is also suitable for the team of N UAVs; more details are described in [22].

In addition, a robust term is introduced to obtain disturbance rejection for wind gusts in [25]. The GAE ($e_T(t_k)$) and the corresponding ITAE ($J_{ITAE}(e_T)$) are not suitable to evaluate the effectiveness of the method used in [25]. Therefore, we mainly focus on the phase separation problem solved by implementing the proposed TPSA and the SPSA in [25].

Figure 33 shows the performance of intervehicle phase separation in the process of the cooperative standoff target tracking mission using three UAVs. Due to the discontinuity of the wrapped space separation angle leading to oscillations in the control input signal, the curve of $e_\theta(t)$ in the SPSA method shown in Figure 33a fluctuates in the convergence process. The fluctuation results in a slow convergence rate, which is confirmed by the curve of $J_{ITAE}(e_\theta)$ in the SPSA method shown in Figure 33b. In contrast to the SPSA method, the curve of $e_\theta(t)$ in our proposed TPSA approach smoothly converges to 0. In summary, comparing to the SPSA method described in [25], our proposed TPSA has a smaller phase separation error and faster convergence rate.

The comparison results for other performance indexes are shown in Table 9. Comparing to [25], the GAE and ITAE of the intervehicle phase separation are reduced by 22.51% and 4.85%, respectively, in our proposed TPSA method. From Table 9, it is verified that in the cooperative standoff target tracking mission using multiple UAVs in unknown background wind, our proposed CSTTA method has better performance than the method presented in [25].

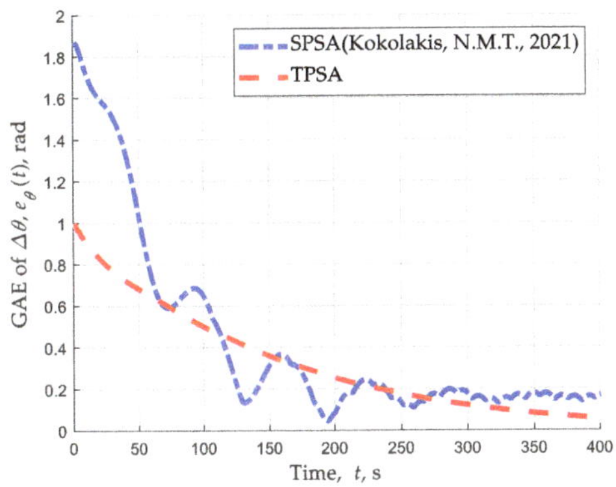

(a) GAE, $e_\theta(t)$

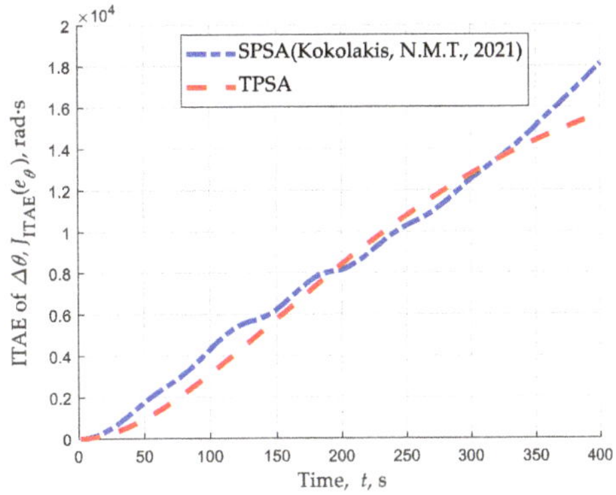

(b) ITAE, $J_{ITAE}(e_\theta)$

Figure 33. Performance of intervehicle phase separation (scenario 3). (**a**) Global average error in phase separation. (**b**) Integrated time absolute error in phase separation [25].

Table 9. Performance comparison of CSTTA and Ref. [25] (scenario 3).

Control Objective	Performance Index	CSTTA	Ref. [25]	Percentage/(%)
Relative distance regulation	$e_r(t)/m$ $J_{ITAE}(e_r)/(m \times s)$	6.5996 2.5579×10^4	10.7001 5.7246×10^4	38.32 55.32
Relative course tracking	$e_\chi(t)/rad$ $J_{ITAE}(e_\chi)/(rad \times s)$	0.0090 21.3996	0.0135 117.3008	33.33 81.76
Intervehicle phase separation	$e_\theta(t)/(rad)$ $J_{ITAE}(e_\theta)/(rad \times s)$	0.3311 8.0284×10^3	0.4273 8.4378×10^3	22.51 4.85

8. Conclusions

This paper investigates the standoff tracking of a ground-based moving target using multiple fixed-wing UAVs in unknown background wind. The main contribution of this paper is to develop a cooperative standoff target tracking algorithm (CSTTA), which considers the control input constraints of the fixed-wing UAV, the target's motion, and an unknown wind. The following conclusions can be obtained.

(i) A fundamental cooperative standoff target tracking problem includes three control objectives: relative distance regulation, relative course convergence, and intervehicle phase separation. In addition, in the case of unknown background wind, it is essential to enhance the wind resistance capacity of the UAV.

(ii) A heading rate control law based on LGVF is introduced to regulate the position of a UAV on a circle around the target with a constant standoff distance. It is proved that the proposed heading rate controller can achieve standoff target tracking for a single UAV in the condition of an arbitrary initial position and heading. Due to the heading rate input being constrained, the predefined standoff distance between the UAV and the target has an allowable lower bound. The minimum allowable standoff distance is formulated in this paper.

(iii) A new temporal phase separation algorithm (TPSA) is proposed to achieve the desired temporal phase separation in a cooperative standoff target tracking mission. The TPSA approach takes into account the minimum and maximum airspeed constraints, and can avoid the discontinuity of wrapped space phase angles. The results of comparison simulations show that the TPSA has a smaller convergence error and faster convergence rate than the previously reported space phase separation method.

(iv) The offset between the actual vehicle trajectory and the desired LGVF orbit can be utilized to estimate the composition velocity of the target's motion and background wind. The results of comparison simulations show that the proposed composition velocity estimation algorithm (CVEA) can effectively estimate the composition velocity, and thus enhance the tracking performance of the UAVs in the presence of wind and a moving target.

There are many potential directions for future consideration. Firstly, the current models and algorithms will be extended to three-dimensional coordinates. Secondly, the target location algorithm will be designed to explore the possibility of using an onboard observation sensor such as a camera to facilitate tracking. Thirdly, more complex mission environments will be considered, and the methods of terrain obstacle and intervehicle collision avoidance will be introduced to enhance aircraft safety. The proposed algorithms will be implemented on real UAVs in future works. Finally, the communication constraints, such as limited communication range and communication delays, will be considered.

Author Contributions: Conceptualization, Z.L.; data curation, Z.L. and L.X.; formal analysis, L.X.; funding acquisition, Z.Z.; investigation, Z.L.; methodology, Z.L.; project administration, Z.Z.; resources, L.X.; software, Z.L.; supervision, Z.Z.; validation, L.X.; visualization, Z.L.; writing—original draft, Z.L.; writing—review and editing, L.X. All authors have read and agreed to the published version of the manuscript.

Funding: This work received funding from the Scientific Research Program of the Education Department of Hubei Provincial Government, under Grant No. D20212902, and the Program of National undergraduate innovation and entrepreneurship, under Grant No. 202210514008.

Data Availability Statement: The data presented in this study are available on request from the corresponding author. The data are not publicly available due to privacy restrictions.

Conflicts of Interest: The authors declare no conflict of interest.

Appendix A. Proof of Lemma 1

Proof: Construct the following function

$$F(\psi) = \frac{v_r}{\lambda_u(\psi)} \tag{A1}$$

Substituting the relative speed v_r given by Equation (6) and $\lambda_u(\psi)$ given by Equation (9) into Equation (A1) yields

$$F(\psi) = \frac{\left(v_s^2 + T^2 - 2v_sT\xi\right)^{\frac{3}{2}}}{v_s^2 - v_sT\xi} \tag{A2}$$

where $T = \sqrt{T_x^2 + T_y^2}$, $\eta = arctan\left(\frac{T_x}{T_y}\right) \in [-\pi, \pi]$, and $\xi = sin(\psi + \eta) \in [-1, 1]$. Differentiating Equation (A2) with respect to ξ, one obtains

$$F_\xi(\psi) = \frac{v_sT\left(v_s^2 + T^2 - 2v_sT\xi\right)^{\frac{1}{2}}\left(T^2 + v_sT\xi - 2v_s^2\right)}{\left(v_s^2 - v_sT\xi\right)^2} \tag{A3}$$

when $\xi \in [-1, 1]$, then $F_\xi(\psi) < 0$. This means that $F(\psi)$ decreases monotonically with increasing ξ. Therefore, when $\xi = -1$, $F(\psi)$ reaches a maximum, i.e., $F(\psi) \leq \frac{(v_s+T)^2}{v_s}$. Thus, the proof for Lemma 1 is completed. □

Appendix B. Proof of Lemma 2

Proof: Firstly, two cases are discussed in the proof of Conclusion ①.

(a) When $\chi_e = 0$, then for all $k_1 > 0$, $\dot{\chi}_d = \frac{4v_rr_d^3}{(r^2+r_d^2)^2} \geq -\lambda_u(\psi)\omega_{max} + k_1sin\chi_e$ is always true.

(b) If $\chi_e \in (0, \pi)$, then we construct the following function

$$F(\chi_e, r) = \frac{1}{sin\chi_e}\left(\dot{\chi}_d + \frac{4v_r}{r_d}\right) \tag{A4}$$

Partially differentiating Equation (A4) with respect to χ_e, one obtains

$$\frac{\partial F(\chi_e, r)}{\partial \chi_e} = \frac{-4v_r}{r_d sin^2\chi_e}\left[\frac{r_d^4}{\left(r^2+r_d^2\right)^2} + cos\chi_e\right] \tag{A5}$$

It can be observed from Equation (A5) that when $cos\chi_e^* = -\frac{r_d^4}{\left(r^2+r_d^2\right)^2} < 0$ it implies $\chi_e^* \in \left(\frac{\pi}{2}, \pi\right)$, then $F(\chi_e, r)$ reaches the maximum $F(\chi_e^*, r)$, which is expressed as follows:

$$F(\chi_e^*, r) = \frac{v_r}{r}\left(\frac{\zeta}{sin\chi_e^*}\right) + 4\frac{v_r}{r_d}\left(\frac{1}{sin\chi_e^*}\right) \tag{A6}$$

where

$$\zeta = sin\phi_{GF}(1 + cos\phi_{GF})cos\chi_e^* + \left(cos\phi_{GF} - sin^2\phi_{GF}\right)sin\chi_e^* \tag{A7}$$

For the first term in Equation (A6), one can deduce

$$\frac{\zeta}{sin\chi_e^*} \geq \frac{r_d^4 - r^4 - 4r^2r_d^2}{\left(r^2+r_d^2\right)^2} \tag{A8}$$

For the second term in Equation (A6), one can deduce

$$\begin{aligned}\left(\tfrac{4v_r}{r_d}\right)\left(\tfrac{1}{sin\chi_e^*}\right) \geq \left(\tfrac{4v_r}{r_d}\right)sin\chi_e^* \geq \left(\tfrac{4v_r}{r_d}\right)\left(\frac{\left[\left(r^2+r_d^2\right)^2 - r_d^4\right]^2}{\left(r^2+r_d^2\right)^4}\right)^{\frac{1}{2}} \\ = \left(\tfrac{4v_r}{r_d}\right)\left(\frac{r^4+2r^2r_d^2}{\left(r^2+r_d^2\right)^2}\right)\end{aligned} \tag{A9}$$

According to Equation (A6), one derives

$$\begin{aligned}F(\chi_e, r) \geq F(\chi_e^*, r) &\geq \tfrac{v_r}{r}\lambda_2 + \left(\tfrac{4v_r}{r_d}\right)sin\chi_e^* \\ &\geq \tfrac{v_r}{r}\left(\frac{r_d^4 - r^4 - 4r^2r_d^2}{\left(r^2+r_d^2\right)^2}\right) + \left(\tfrac{4v_r}{r_d}\right)\left(\frac{r^4+2r^2r_d^2}{\left(r^2+r_d^2\right)^2}\right) \\ &\overset{\varepsilon=\frac{r}{r_d}}{\Rightarrow} \tfrac{v_r}{r_d}\left(1 + \frac{1-\varepsilon-4\varepsilon^2+6\varepsilon^3-\varepsilon^4+3\varepsilon^5}{\varepsilon\left(\varepsilon^2+1\right)^2}\right) > \tfrac{v_r}{r_d}\end{aligned} \tag{A10}$$

Therefore, let $k_1 = \frac{v_r}{r_d}$, we obtain $F(\chi_e, r) \geq k_1$, and thus

$$\frac{1}{sin\chi_e}\left(\dot{\chi}_d + \frac{4v_r}{r_d}\right) \geq k_1 \Rightarrow \dot{\chi}_d + \frac{4v_r}{r_d} \geq k_1 sin\chi_e \tag{A11}$$

According to the conclusion of Lemma 1 and Equation (26), it can be obtained that

$$\lambda_u(\psi)\omega_{max} \geq \frac{4v_r}{r_d} \tag{A12}$$

Substituting Equation (A12) into Equation (A11) yields

$$\dot{\chi}_d + \lambda_u(\psi)\omega_{max} \geq \dot{\chi}_d + \frac{4v_r}{r_d} \geq k_1 sin\chi_e \Rightarrow \dot{\chi}_d \geq -\lambda_u(\psi)\omega_{max} + k_1 sin\chi_e \tag{A13}$$

The proof of Conclusion ① in Lemma 2 is completed. □

Secondly, we prove Conclusion ② of Lemma 2 based on Conclusion ①. To render the proof process simpler, the following function is introduced

$$G(\chi_e) = \begin{cases} \dot{\chi}_d - \lambda_u(\psi)\omega_{max} - k_1 sin\chi_e, \chi_e \in [-\pi, 0) \\ \dot{\chi}_d + \lambda_u(\psi)\omega_{max} - k_1 sin\chi_e, \chi_e \in [0, \pi) \end{cases} \tag{A14}$$

The function $G(\chi_e)$ has this property: if $\chi_e \in [-\pi, 0)$, then $G(\chi_e) = -G(\chi_e + \pi)$. Thus, Conclusion ① of Lemma 2 can be summarized as $G(\chi_e) = \dot{\chi}_d + \lambda_u(\psi)\omega_{max} - k_1 sin\chi_e \geq 0$ when $\chi_e \in [0, \pi)$. Therefore, when $\chi_e \in [-\pi, 0)$ it implies $\chi_e + \pi \in [0, \pi)$, and then $G(\chi_e) = \dot{\chi}_d - \lambda_u(\psi)\omega_{max} - k_1 sin\chi_e = -G(\chi_e + \pi) \leq 0$. That means Conclusion ② of Lemma 2 is true. Thus, the proof for Lemma 2 is completed. □

Appendix C. Proof of Lemma 3

Proof: The proposed heading rate control law given by Equation (24) can be rewritten as follows

$$\dot{\chi} = \begin{cases} \Omega, |\Omega| \leq \lambda_u(\psi)\omega_{max} \\ sgn(\Omega)\lambda_u(\psi)\omega_{max}, |\Omega| \geq \lambda_u(\psi)\omega_{max} \end{cases} ; \Omega = -k\lambda_u(\psi)\chi_e + \dot{\chi}_d \tag{A15}$$

Next, we prove Lemma 3 based on Lemma 2. The following trigonometric inequalities are used in the proof.

$$\begin{cases} \chi_e \geq sin\chi_e, \chi_e \in [0, \pi) \\ \chi_e \leq sin\chi_e, \chi_e \in [-\pi, 0) \end{cases} \tag{A16}$$

(a) When $\chi_e \in [-\pi, 0)$, there are three cases.

- If $\Omega > \lambda_u(\psi)\omega_{max}$, then $\dot{\chi}_e = \lambda_u(\psi)\omega_{max} - \dot{\chi}_d \geq -k_1 sin\chi_e \geq -k_2 sin\chi_e > 0$;
- else if $\Omega < -\lambda_u(\psi)\omega_{max}$, then $\dot{\chi}_e = -\lambda_u(\psi)\omega_{max} - \dot{\chi}_d \geq -k\lambda_u(\psi)\chi_e \geq -k_2\chi_e \geq -k_2 sin\chi_e > 0$;
- else $|\Omega| \leq \lambda_u(\psi)\omega_{max}$, then $\dot{\chi}_e = -k\lambda_u(\psi)\chi_e \geq -k_2\chi_e \geq -k_2 sin\chi_e > 0$.

In summary, when $\chi_e \in [-\pi, 0)$, $\dot{\chi}_e \geq -k_2 sin\chi_e > 0$. This means that Conclusion ① of Lemma 3 is proved.

(b) When $\chi_e \in [0, \pi)$, there are three cases.

- If $\Omega > \lambda_u(\psi)\omega_{max}$, then $\dot{\chi}_e = \lambda_u(\psi)\omega_{max} - \dot{\chi}_d \leq -k\lambda_u(\psi)\chi_e \leq -k_2 sin\chi_e < 0$;
- else if $\Omega < -\lambda_u(\psi)\omega_{max}$, then $\dot{\chi}_e = -\lambda_u(\psi)\omega_{max} - \dot{\chi}_d \leq -k_1 sin\chi_e \leq -k_2 sin\chi_e < 0$;
- else $|\Omega| \leq \lambda_u(\psi)\omega_{max}$, $\dot{\chi}_e = -k\lambda_u(\psi)\chi_e \leq -k_2 sin\chi_e < 0$.

In summary, when $\chi_e \in [0, \pi)$, $\dot{\chi}_e \leq -k_2 sin\chi_e < 0$. This means that Conclusion ② of Lemma 3 is proved. Thus, the proof for Lemma 3 is completed. □

Appendix D. Proof of Lemma 4

Proof: According to Equation (9), we can obtain $\dot{r} = v_r cos(\chi - \theta) = v_r cos(\phi_{GF} + \chi_e) \leq v_r$. It can be observed from Equation (6) that $v_r \leq v_s + T$, which implies $\dot{r} \leq v_r \leq v_s + T$. Thus, it can be concluded that

$$r \leq r_0 + (v_s + T)t_c \tag{A17}$$

where r_0 presents the initial relative distance between the UAV and the target at time t_0. t_c presents the time of the relative course error χ_e converging to 0.

(a) If $|\chi_{e0}| \in [0, \frac{\pi}{2}]$, the convergence time t_{c1} of process ($\chi_e \to 0$) can be obtained:

$$t_{c1} = \frac{1}{k_2} ln \left| \frac{tan(\frac{\chi_{e0}}{2})}{tan(\frac{\alpha_0}{2} - \frac{\pi}{4})} \right| \tag{A18}$$

Substituting (A18) into (A17), the upper bound of r is obtained as follows:

$$r \le r_0 + (v_s + T)\frac{1}{k_2}ln\left|\frac{tan(\frac{\chi_{e0}}{2})}{tan(\frac{\alpha_0}{2} - \frac{\pi}{4})}\right| \quad \text{(A19)}$$

(b) If $|\chi_{e0}| \in \left(\frac{\pi}{2}, \pi\right]$, the relative course error $|\chi_e|$ firstly converges to $\pi/2$ in time t_{c2}, and then converges from $\pi/2$ to 0 in time t_{c3}. According to Equation (32), one has

$$t_{c2} = \frac{1}{k_2}ln\left|tan\left(\frac{\chi_{e0}}{2}\right)\right|; t_{c3} = \frac{1}{k_2}ln\left|\frac{1}{tan(\frac{\alpha_0}{2} - \frac{\pi}{4})}\right| \quad \text{(A20)}$$

Thus, the upper bound of r is defined as

$$r \le r_0 + (v_s + T)\cdot(t_{c2} + t_{c3}) = r_0 + (v_s + T)\frac{1}{k_2}ln\left|\frac{tan(\frac{\chi_{e0}}{2})}{tan(\frac{\alpha_0}{2} - \frac{\pi}{4})}\right| \quad \text{(A21)}$$

In summary, the relative distance r between the drone and the target is bounded, and Lemma 4 is proved. □

References

1. Guo, B.; Vu, D.; Xu, L.; Xue, M.; Li, J. Ground moving target indication via multichannel airborne SAR. *IEEE Trans. Geosci. Remote Sens.* **2011**, *49*, 3753–3764. [CrossRef]
2. Liu, Z.; Fu, X.; Gao, X. Co-Optimization of communication and sensing for multiple unmanned aerial vehicles in cooperative target tracking. *Appl. Sci.* **2018**, *8*, 899. [CrossRef]
3. Li, Z.; Chen, X.; Zhao, Z. Design of standoff cooperative target-tracking guidance laws for autonomous unmanned aerial vehicles. *Math. Probl. Eng.* **2021**, *2021*, 6682160. [CrossRef]
4. Zhoufan, X.; Ruixuan, W.; Xiaolin, Z.; Shulei, W. Coordinated standoff target tracking guidance method for UAVs. *IEEE Access* **2018**, *6*, 59853–59859.
5. Hyondong, O.; Seungkeun, K.; Hyo-sang, S.; Antonios, T. Coordinated standoff tracking of moving target groups using multiple UAVs. *IEEE Trans. Aerosp. Electron. Syst.* **2015**, *51*, 1501–1514. [CrossRef]
6. Rysdyk, R.; Lum, C.; Vagners, J. Autonomous orbit coordination for two unmanned aerial vehicles. In Proceedings of the AIAA Guidance, Navigation, and Control Conference and Exhibit, San Francisco, CA, USA, 15–18 August 2005.
7. Ducard, G.; Kulling, K.C.; Geering, H.P. A simple and adaptive on-line path planning system for a UAV. In Proceedings of the 15th Mediterranean Conference on Control and Automation, Athens, Greece, 27–29 June 2007; IEEE: Piscataway, NJ, USA, 2007; pp. 1–6.
8. Park, S.; Deyst, J.; How, J.P. Performance and Lyapunov stability of a nonlinear path-following guidance method. *J. Guid. Control Dyn.* **2007**, *30*, 1718–1728. [CrossRef]
9. Park, S. Guidance law for standoff tracking of a moving object. *J. Guid. Control Dyn.* **2017**, *40*, 2948–2955. [CrossRef]
10. Frew, E.W.; Lawrence, D.A.; Steve, M. Coordinated standoff tracking of moving targets using Lyapunov guidance vector fields. *J. Guid. Control Dyn.* **2008**, *31*, 290–306. [CrossRef]
11. Chen, H.; Chang, K.; Agate, C.S. UAV Path Planning with tangent-plus-Lyapunov vector field guidance and obstacle avoidance. *IEEE Trans. Aerosp. Electron. Syst.* **2013**, *49*, 840–856. [CrossRef]
12. Pothen, A.A.; Ratnoo, A. Curvature-constrained Lyapunov vector field for standoff target tracking. *J. Guid. Control Dyn.* **2017**, *40*, 2729–2736. [CrossRef]
13. Lim, S.; Kim, Y.; Lee, D.; Bang, H. Standoff target tracking using a vector field for multiple unmanned aircrafts. *J. Intell. Robot. Syst.* **2013**, *69*, 347–360. [CrossRef]
14. Sun, S.; Wang, H.; Liu, J.; He, Y. Fast Lyapunov vector field guidance for standoff target tracking based on offline search. *IEEE Access* **2019**, *7*, 124797–124808. [CrossRef]
15. Bishopa, A.N.; Fidanb, B.; Andersonb, B.D.O.; Dogancayc, K.; Pathirana, P.N. Optimality analysis of sensor-target localization geometries. *Automatica* **2010**, *46*, 479–492. [CrossRef]
16. Kim, S.; Oh, H.; Tsourdos, A. Nonlinear model predictive coordinated standoff tracking of a moving ground vehicle. *J. Guid. Control Dyn.* **2013**, *36*, 557–566. [CrossRef]
17. Sun, S.; Liu, Y.; Guo, S.; Gang, L. Observation-driven multiple UAV coordinated standoff target tracking based on model predictive control. *Tsinghua Sci. Technol.* **2022**, *27*, 948–963. [CrossRef]
18. Zhang, M.; Liu, H.H.T. Cooperative Tracking a Moving Target Using Multiple Fixed-wing UAVs. *J. Intell. Robot. Syst.* **2015**, *81*, 505–529. [CrossRef]

19. He, S.; Wang, J.; Lin, D. Unknown ground moving target tracking using multiple unmanned aerial vehicles. *Proc. Inst. Mech. Eng. Part G J. Aerosp. Eng.* **2019**, *233*, 1021–1032. [CrossRef]
20. Yoon, S.; Park, S.; Kim, Y. Circular motion guidance law for coordinated standoff tracking of a moving target. *IEEE Trans. Aerosp. Electron. Syst.* **2013**, *49*, 2440–2462. [CrossRef]
21. Yao, P.; Wang, H.L.; Su, Z.K. Cooperative path planning with applications to target tracking and obstacle avoidance for multi-UAVs. *Aerosp. Sci. Technol.* **2016**, *54*, 10–22. [CrossRef]
22. Kokolakis, N.M.T.; Koussoulas, N.T. Coordinated standoff tracking of a ground moving target and the phase separation problem. In Proceedings of the 2018 International Conference on Unmanned Aircraft Systems (ICUAS), Dallas, TX, USA, 12–15 June 2018; IEEE: Piscataway, NJ, USA, 2018; pp. 473–482.
23. Song, Z.; Li, H.; Chen, C.; Zhou, X.; Xu, F. Coordinated standoff tracking of moving targets using differential geometry. *J. Zhejiang Univ. Sci. C* **2014**, *15*, 284–292. [CrossRef]
24. Che, F.; Niu, Y.; Li, J.; Wu, L. Cooperative standoff tracking of moving targets using modified Lyapunov vector field guidance. *Appl. Sci.* **2020**, *10*, 3709. [CrossRef]
25. Kokolakis, N.M.T.; Koussoulas, N.T. Robust standoff target tracking with finite-time phase separation under unknown wind. *J. Guid. Control Dyn.* **2021**, *44*, 1183–1198. [CrossRef]
26. Harinarayana, T.; Hota, S.; Kushwaha, R. Vector field guidance for standoff target tracking. *Proc. Inst. Mech. Eng. Part G J. Aerosp. Eng.* **2022**, *236*, 2963–2973. [CrossRef]
27. Rysdyk, R. Unmanned aerial vehicle path following for target observation in wind. *J. Guid. Control Dyn.* **2006**, *29*, 1092–1100. [CrossRef]
28. Sun, S.; Dong, K.; Guo, C.; Tan, D. A wind estimation based on unscented Kalman filter for standoff target tracking using a fixed-wing UAV. *Int. J. Aeronaut. Space Sci.* **2020**, *22*, 366–375. [CrossRef]
29. Beal, T.R. Digital simulation of atmospheric turbulence for Dryden and von Karman models. *J. Guid. Control Dyn.* **1993**, *16*, 132–138. [CrossRef]
30. Gawronski, W. Modeling wind-gust disturbances for the analysis of antenna pointing accuracy. *IEEE Antennas Propag. Mag.* **2004**, *46*, 50–58. [CrossRef]
31. Summers, T.H.; Akella, M.R.; Mears, M.J. Coordinated standoff tracking of moving targets: Control laws and information architectures. *J. Guid. Control Dyn.* **2006**, *32*, 56–69. [CrossRef]

drones

MDPI

Article

Multiple-Target Matching Algorithm for SAR and Visible Light Image Data Captured by Multiple Unmanned Aerial Vehicles

Hang Zhang [1,2,*], Jiangbin Zheng [1] and Chuang Song [2]

1 School of Software, Northwestern Polytechnical University, Xi'an 710072, China; zhengjb@nwpu.edu.cn
2 Science and Technology on Complex System Control and Intelligent Agent Cooperation Laboratory, Beijing 100074, China; zhh0993@163.com
* Correspondence: zhanghang90@mail.nwpu.edu.cn; Tel.: +86-010-88538953

Abstract: Unmanned aerial vehicle (UAV) technology has witnessed widespread utilization in target surveillance activities. However, cooperative multiple UAVs for the identification of multiple targets poses a significant challenge due to the susceptibility of individual UAVs to false positive (FP) and false negative (FN) target detections. Specifically, the primary challenge addressed in this study stems from the weak discriminability of features in Synthetic Aperture Radar (SAR) imaging targets, leading to a high false alarm rate in SAR target detection. Additionally, the uncontrollable false alarm rate during electro-optical proximity detection results in an elevated false alarm rate as well. Consequently, a cumulative error propagation problem arises when SAR and electro-optical observations of the same target from different perspectives occur at different times. This paper delves into the target association problem within the realm of collaborative detection involving multiple unmanned aerial vehicles. We first propose an improved triplet loss function to effectively assess the similarity of targets detected by multiple UAVs, mitigating false positives and negatives. Then, a consistent discrimination algorithm is described for targets in multi-perspective scenarios using distributed computing. We established a multi-UAV multi-target detection database to alleviate training and validation issues for algorithms in this complex scenario. Our proposed method demonstrates a superior correlation performance compared to state-of-the-art networks.

Keywords: unmanned aerial vehicle; multi-target recognition; multi-objective matching; target tracking; target drones

Citation: Zhang, H.; Zheng, J.; Song, C. Multiple-Target Matching Algorithm for SAR and Visible Light Image Data Captured by Multiple Unmanned Aerial Vehicles. *Drones* **2024**, *8*, 83. https://doi.org/10.3390/drones8030083

Academic Editor: Carlos Tavares Calafate

Received: 8 January 2024
Revised: 20 February 2024
Accepted: 20 February 2024
Published: 27 February 2024

1. Introduction

The development of deep learning technology has greatly improved the accuracy of multi-target recognition algorithms employed in a wide variety of fields, including public surveillance. Multi-target matching for multiple unmanned aerial vehicles (UAVs) has important applications in military, production, multimedia and other fields. However, surveillance algorithms in multi-target associations for multiple unmanned aerial vehicles (UAVs) are still subject to a number of deficiencies and still challenging. There still exist technical challenges, especially in the collaborative perception tasks between drone swarms and external sensors such as satellites. This is because it involves the mutual coordination of the two processes of multi-target detection and matching (as shown in Figure 1).

The process of multi-target matching for multiple UAVs can be considered a process of association or correlation, whereby the same targets in the images captured by a surveilling group of UAVs within a given environment are properly associated for all image data captured by the UAVs in the group. Most research focused on the multi-target correlation problem formulates the decision-making strategy of UAVs as a regional monitoring problem, in which the applied correlation objective function usually requires UAVs to reduce their target tracking error by covering all targets over a large monitoring area. It is generally possible for UAVs to track all targets over a large coverage area when the number of UAVs is less than the number of targets [1–5]. Moreover, the guidance target function applied in multi-target tracking usually seeks to minimize the uncertainty of the current observation of a target, which is represented according to information entropy. However, it is inevitable that some targets will not be observable by all UAVs at a given time [6–10]. Therefore, the possibility that some targets will not be detected by all UAVs in the area must be considered when constructing the correlation objective function. One way to address this possibility is to add a multiplier to the correlation objective function representing the detection probability. In addition, the multi-target association task involving multiple UAVs still suffers from a number of problems, including a reliable means of determining the occurrence of FP and FN target detections and a method for improving the consistency of multi-target association for multiple UAVs [11].

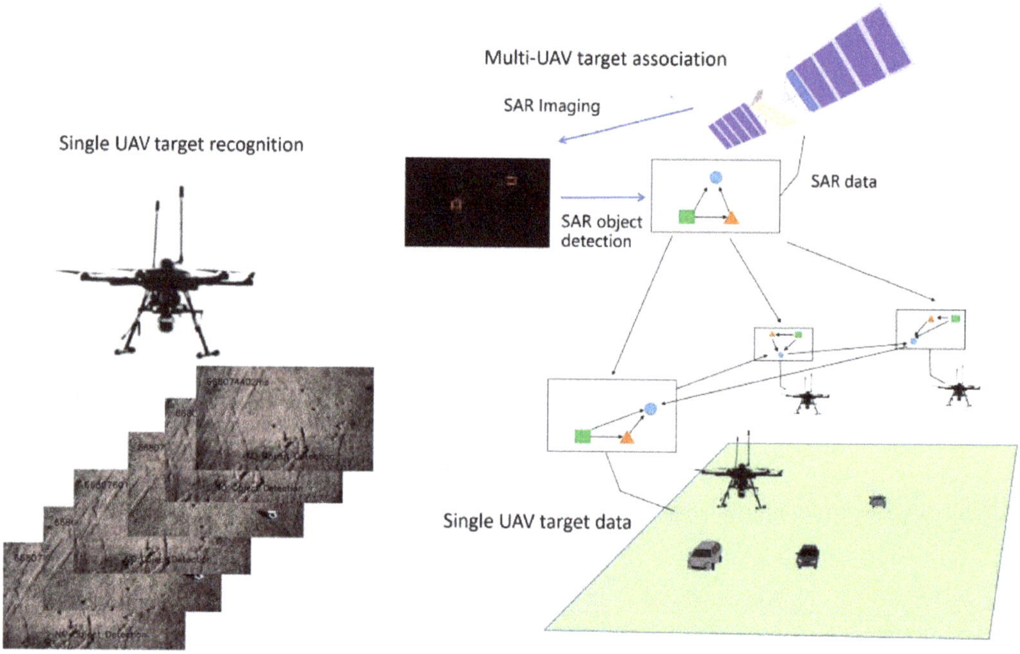

Figure 1. From single- to multi-UAV missions.

This paper investigates the target association problem in the context of collaborative detection involving multiple unmanned aerial vehicles (UAVs). The challenge of this problem lies in the weak discriminability of features for Synthetic Aperture Radar (SAR) imaging targets, leading to a high false alarm rate in SAR target detection. Compared to optical sensors, SAR sensors have a lower imaging quality. However, due to their ability to cover large areas, they often become a crucial means for detecting and searching for specific targets in designated regions of interest. Simultaneously, the uncontrollable false alarm rate during electro-optical proximity detection results in a high false alarm rate as well. Consequently, the cumulative error propagation problem arises when SAR and electro-optical observations of the same target from different perspectives occur at different times. This paper proposes a neural network-based approach to address the problems of multi-target association and tracking.

The main contributions include the following:

a. An improved triplet loss function was constructed to effectively assess the similarity of targets detected by multiple UAVs.

b. A consistency discrimination algorithm is proposed for targets from multiple perspectives based on distributed computing. On UAVs equipped with optical sensors, the algorithm utilizes optical image features and the relative relationships between targets to achieve consistency discrimination in scenarios with a high false alarm rate. On UAVs equipped with SAR sensors, the algorithm employs SAR-detected local situational information and optical image detection for consistency judgment, effectively achieving consistency judgment from a global perspective.
c. A multi-UAV multi-target detection database is established, and an open-source core code was developed, addressing the training and validation issues for algorithms in this scenario.

2. Related Work

2.1. Object Detection Algorithms

Object detection algorithms can be divided into single-stage and two-stage algorithms. Single-stage object detection algorithms are mainly divided into two types: anchor based and anchor free. Typical anchor-based algorithms include you only look once (YOLO) and single-shot detector (SSD). YOLO divides the input image into cells, and a bounding box prediction is performed for each cell [12–15]. Similarly, SSD predicts the confidence and offset of a set of anchor targets of different sizes using a Feature Pyramid Network (FPN) structure [16]. However, such anchor-based object detection algorithms require a reasonable anchor hyperparameter, which is often not suitable. Therefore, anchor-free object detection algorithms have been designed to avoid this issue by detecting objects through the prediction of key points instead of bounding boxes. Representative anchor-free models include CornerNet, CenterNet, FCOS, NanoDet, ExtremeNet, and TTFNet. In contrast to single-stage object detection algorithms, two-stage algorithms apply a region proposal network (RPN) for foreground–background classification [17–22]. The features extracted from the region of interest (ROI) proposed in the RPN are passed to a classification head to determine the class label, and to a regression head to determine the boundary box position [16]. A representative two-stage object detection algorithm employs a region-based convolutional neural network (R-CNN) [23]. This algorithm first selects regions that may contain targets through a candidate region generation method [24]. Then, a CNN is used to extract feature representations for each region. These features are input into a classifier to determine whether the target is included in an image, and input as well into a regression head to locate the target position accurately. Finally, the target detection results are obtained through post-processing steps. The fast-RCNN object detection algorithm improves the speed of the standard R-CNN algorithm by applying a CNN to extract the features of the entire image directly, and then applying an ROI pooling layer to obtain the features corresponding to the proposed region of the image [25]. However, the real-time performance is still not ideal. In contrast, ThunderNet has achieved a two-stage object detection algorithm with real-time performance using an efficient RPN and a small backbone network [26]. However, not all algorithms can directly adapt to multi-target recognition tasks in UAV scenarios. Considering this, we combined existing modules suitable for UAV scenarios to obtain a multi-target recognition model suitable for this scenario.

2.2. Data Association for Multi-Target and Multi-Camera Tracking

Multi-target and multi-camera tracking (MTMCT) technology can be mainly divided into two categories: correlation clustering and sliding time window schemes. Correlation clustering calculates the correlation of features between different detected targets, and then produces clusters of these correlations for data association. For example, correlations in target appearance and motion have been combined to improve the data association effect. The sliding time window technique performs data association within a relatively small range associated with a window of time, while utilizing a priori knowledge that the probability of the continuous occurrence of targets within the time window considered is higher [27]. First, the bounding boxes detected over the time window are connected to form tracklets [28,29]. Then, these tracklets are associated into target trajectories for each individual camera using a shorter sliding window. Finally, a longer sliding window is applied to match these target trajectories across different cameras. For UAV scenarios, SAR data related to the target can generally be obtained. This study used SAR data and image data to jointly address the target association task and propose a more accurate UAV target association model [30].

3. Proposed Algorithm

The overall multi-target association scheme proposed in this paper is illustrated in Figure 2 for two UAVs denoted as UAV 1 and UAV 2. In detail, the scheme is mainly composed of five components, including image enhancement modules and target recognition networks for the image data captured by individual UAVs, an association feature extraction network, an algorithm for eliminating false pedestrian detections and correcting missed pedestrian detections for the image data captured by individual UAVs, and an algorithm facilitating the multi-target matching of multiple UAVs.

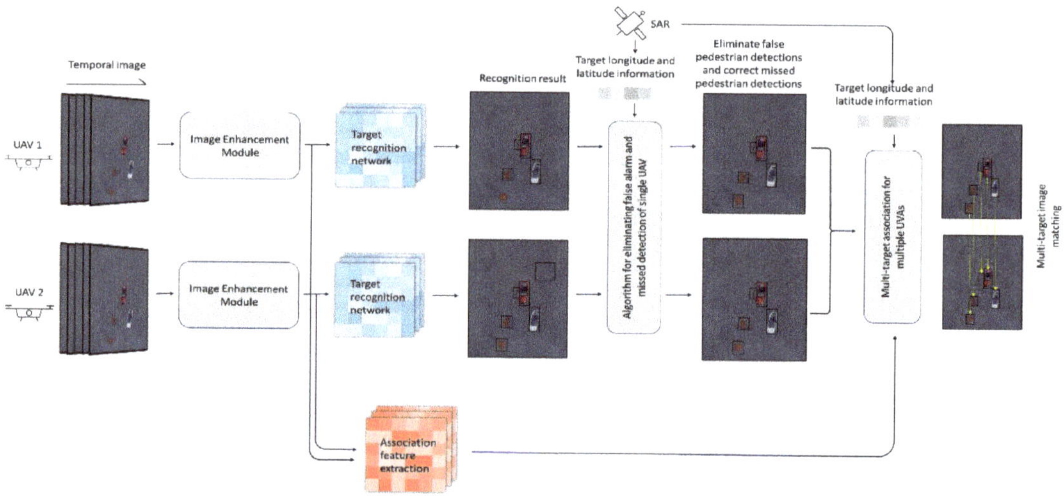

Figure 2. Overall pipeline.

3.1. Image Enhancement

The image enhancement module employed in the present work based on the MIRNet model is illustrated in Figure 3. As can be seen, the main body of the module consists of dual attention units (DAUs) composed of p multi-scale residual blocks (MRBs), where each MRB is mainly composed of a kernel selection feature fusion module, a dual attention unit, and a residual adjustment module. For a given image, the network first extracts its low-level feature, which is then applied to obtain deep features through N recursive residual groups (RRGs), where a cyclic residual group consists of two convolutional layers and P multi-scale residual blocks (MRB). The feature is convolved once to obtain the residual image, and the original image I is combined with R to obtain the final enhanced image. The network structure also allows information exchange across parallel streams via selective kernel feature fusion (SKFF), which helps to consolidate high-resolution features using low-resolution features, and vice versa.

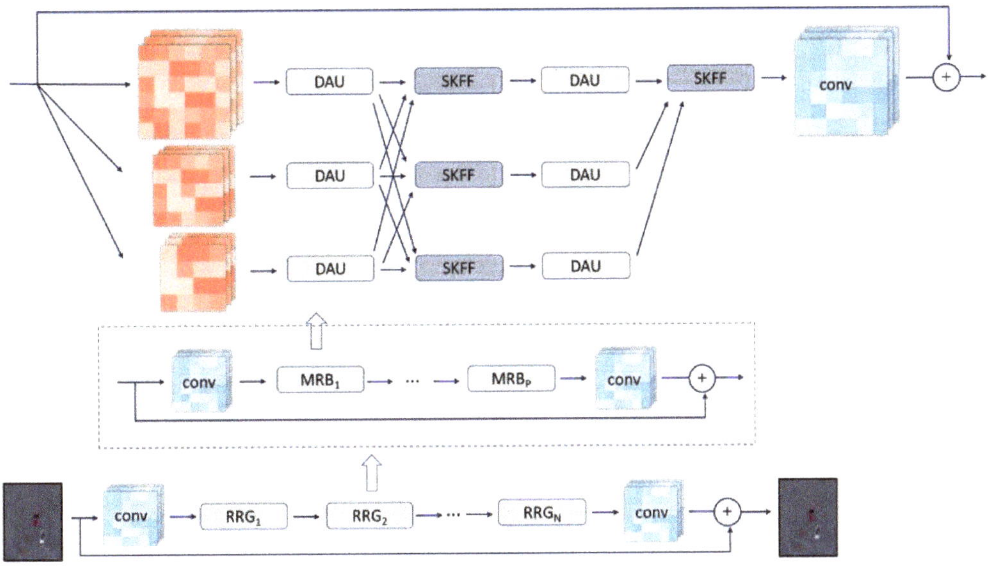

Figure 3. MIRNet model employed in the image enhancement module.

3.2. Target Recognition

The overall target recognition network architecture is illustrated in Figure 4, and it is composed of a backbone network, neck network, and head network. The backbone network is applied for extracting features from the images. Here, ResNet-50 and Darknet-53 CNN-based architectures are typically employed for feature extraction. The neck network resides between the backbone network and the head network, and it is applied for extracting features with more complex targets or strong semantic information. The head network is mainly used to make predictions regarding the type and location of targets using the features extracted from the backbone and neck networks.

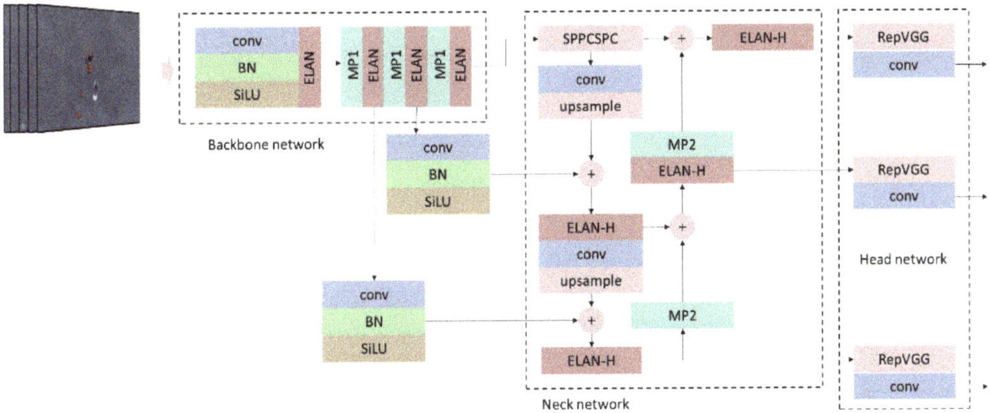

Figure 4. Target recognition network architecture.

3.2.1. Backbone Network

The present work applied the new CSP-Darknet-53 architecture as a backbone network for feature extraction [31]. The standard CONV-BN-SiLU network unit is composed of a convolution (CONV) operation with convolution kernel sizes that were mostly 3×3 and 1×1, the batch normalization (BN) process, and a sigmoid weighted linear unit (SiLU) activation function. The CONV operation can reduce the number of model parameters and improve the training speed to some extent. The BN process can avoid overfitting the network model during training, prevent gradient explosion and gradient disappearance, and accelerate the convergence speed. The SiLU function is computed as $\text{SiLU}(x) = x \cdot \text{sigmoid}(x)$. Unlike the linear rectifier unit (ReLU) function and other commonly used activation functions, the SiLU function is not monotonically increasing. For $x \approx -1.28$, the global minimum value of SiLU(x) is about -0.28. The most positive feature of the SiLU activation function is that it is self-stable. The global minimum obtained during the training process under a zero-gradient condition acts as a flexible limit in the obtained network weights, and thereby inhibits the learning of a large number of weights. As can be seen in Figure 4, the CONV-BN-SiLU units are used in conjunction with efficient layer aggregation networks (ELANs) and maximum pooling (MP) layers, and the outputs are submitted to the neck network.

3.2.2. Neck Network

While the sizes of input images have no effect on the parameters of the convolution layers, they do affect the parameters of the fully connected layers because these layers must be connected to all pixels in an input image. Therefore, the fixed length constraint was applied to the fully connected layers. Then, a spatial pyramid pooling (SPP) network, CSPnet, and a Conv-BN-LeakyReLU (SPPCSPC) structure were connected to the pyramid pooling layer of the backbone network. In this way, the network can accept input images of arbitrary sizes and generate a fixed-size output. The upsampling method applied in the neck network was nearest neighbor interpolation.

The present work applied an improved SPP network, denoted as the SPP fast (SPPF) network, which serializes the input through multiple 5×5 MP layers, where the calculation results obtained from two 5×5 MP layers in series are equivalent to those obtained from a single 9×9 MP layer. Similarly, the calculation results obtained from three 5×5 MP layers in series are equivalent to those obtained from a single 13×13 MP layer. Accordingly, the SPPF network involves fewer parameters, and can therefore provide greatly improved training and prediction speeds. The feature information is output to the head network after being remapped by the ELAN-Higher (ELAN-H) modules, which are an extension of the ELAN module with enhanced network learning capabilities.

3.2.3. Head Network

As can be seen, the outputs of the neck network are input into the RepVGG (REP) block, which is a VGG convolutional architecture applied in the head network. After passing through CONV operations, the head network outputs predictions based on the features extracted from the backbone and neck networks. Some recognition results are illustrated in Figure 2 as well.

3.3. *Association Feature Extraction*

The input images processed by the image enhancement module are also input into the association feature extraction network composed of a CNN architecture, which included standard components, such as a deep neural network for convolution calculation, MP downsampling layers, and BN processes. The loss functions applied during network training included triple loss, center loss, and category (instance) loss functions. However, the number of categories considered during testing is usually different from that considered during training because the feature vectors obtained prior to the last fully connected layer are used for prediction, and equivalent instances are determined by comparing the similarity of the corresponding feature vectors obtained for different images.

3.4. Matching Based on Individual UAVs

The different perspectives of different UAVs ensure that the number of actual targets detected by the different UAVs traversing a given area can also be different. Therefore, the present work matched the boundary boxes between the different frames of a target detected by a single UAV over time in accordance with the IoU evaluation metric. These matching results are then combined with geographical relationships to associate the target matched by an individual UAV over time with a global target. This local process for each individual UAV is summarized in Algorithm 1, and it is described in detail as follows.

We first build a global array M_i to store all information N_i^t for a given target retained by the i-th UAV during its traversal of the monitoring area, which includes the geographical location of the target, its prediction category, category confidence, frame coordinates, number of consecutive frames, and the number of times the target is matched over some number of consecutive frames (i.e., the number of matches). The first frame in a set of consecutive frames is denoted as frame 1, and the number of matches for this target is initially set as 0. In addition, the information pertaining to a given target detected by the i-th UAV in each frame $t(t = 1, 2, 3, \ldots)$ is stored in another array N_i^t. When $t = 1$, $M_i = N_i^1$. For $t > 1$, each frame is evaluated on whether the target border information stored in N_i^t is the same as that in M_i based on the IoU calculated as follows:

$$IoU = \frac{A \cap B}{A \cup B}, \tag{1}$$

where A and B respectively correspond to the collection of border coordinates of the targets in N_i^t and M_i. IoU stands for Intersection over Union. It is a key metric for assessing the reliability of object detection. A refers to the ground truth box, and B refers to the predicted box. The IoU value ranges from 0 to 1, where a higher IoU indicates greater reliability in object detection.

Target matching is based on the value of a threshold α according to the condition of $IoU > \alpha$, where α was set herein as 0.6. Under a matching condition, the information of a given target in M_i is updated by taking the target in M_i with the largest calculated IoU value, adding 1 to the number of matches for the target, and then comparing the category confidence of the target in M_i with the category confidence of the target in N_i^t. If the category confidence of the target in N_i^t is greater, the category, category confidence, border coordinates, and geographic location information of the target in M_i is updated. Otherwise, the information pertaining to this target in M_i is not updated. In contrast, under a condition of not matching when the $IoU \leq \alpha$, the target information in N_i^t is added to M_i as an element.

This process is conducted for each target in M_i. If the number of matches N_{match} is greater than or equal to three in five consecutive frames, then the target is considered to be a real target, and its information is retained in M_i. This process results in the collection of N real targets in M_i. For each real target, we calculate the vector between this target and the remaining n targets in M_i as the relative position relationship, and a total of n relative position relationships are obtained.

These relative position relationship results are compared with the corresponding points in the SAR data obtained for the area traversed by the UAVs. More specifically, we calculate the cosine similarity between these relative position relationships and the relative position relationships corresponding to the points in the SAR data. The cosine similarity S_{cos} between two vectors a and b is calculated as follows:

$$S_{cos} = \frac{a \cdot b}{||a|| \, ||b||} \tag{2}$$

We take the n targets with the largest cosine similarity and divide it by the number of remaining targets not including the target to be queried (i.e., n), which represents the cosine similarity between the current target and the target to be matched. Finally, the cosine similarity of all targets is obtained, and these values are employed to determine the occurrence of FP target detections.

To this end, if the cosine similarity of the relative position relationship between a target to be matched in the UAV image data and the corresponding target in the SAR data is less than a threshold value β, which was set as 0.8 in the present work, the target observed in the SAR data is considered a false detection. Otherwise, we find the target in the SAR data with the largest cosine similarity, take it as the final matching result of the target to be associated in the UAV data, and write its number and category into the final result of UAV target association (for example, A_2 is the second instance of object category A in the detection). If the number of matches is less than three in five consecutive frames, the target is considered a false target, and its pertinent information is removed from M_i. If the number of matches is less than three but the number of consecutive frames of the target is less than five, we increment the consecutive frames by 1, and then use the same threshold and similarity calculation method in this new frame to determine whether the current target is retained.

Applying the above process to the targets detected by each UAV ensures that all legitimate target information is retained, and illegitimate information is avoided to the greatest extent possible.

Algorithm 1 Multi-target matching process for individual UAVs that eliminates false pedestrian detections and corrects missed pedestrian detections

```
Input: multi-target recognition images, and features extracted from the target recognition network and corresponding SAR data
Apply input to define a global array M_i with all pertinent multi-target information
While not the last video frame captured by the i-th UAV
Define array N_i^t with information pertaining to a given target detected by the i-th UAV in each frame t (t = 1, 2, 3, ...)
Calculate the IoU value of the target border information stored in N_i^t and the target border information stored in M_i separately according to Equation (1)
If IoU > α
    Take the largest calculated IoU value corresponding to the target in M_i
    Increment N_match for the corresponding target in M_i by 1
  If the confidence of the target classification in M_i is less than or equal to the corresponding confidence in N_i^t
        Update the target information in M_i with the information in N_i^t
Else if IoU ≤ α
        Add target information in N_i^t to M_i as an element
For each target in M_i
If N_match ≥ 3 in 5 consecutive frames
    Set this target as the subject of query
    Calculate the relative position relationships between this target and the remain-ing n targets in M_i
    Compare these relative positions with the corresponding relative positions of the points in the SAR data based on the S_cos defined in Equation (2)
    If S_cos < β
        The target in SAR data is a false detection
    Else
        Find the target in the SAR data with the largest S_cos and use it as the final matching result of the target to be associated with the UAV image data
Else
    Remove the target from M_i
```

3.5. Matching Based on Multiple UAVs

The different perspectives of UAVs ensure that the targets detected by different UAVs are subject to relative position relationships in space. These relative position relationships were employed in the present work to associate the targets detected by different UAVs with a single global target based on the features extracted from the association feature network and corresponding SAR data.

In terms of the system illustrated in Figure 2, the present discussion assumes that the monitoring system includes only UAV 1 and UAV 2, where the total number of targets detected by UAV 1 is n_1, and the total number of targets detected by UAV 2 is n_2. The proposed multi-target association algorithm for multiple UAVs is summarized in Algorithm 2. In detail, the possible values of n_1 and n_2 have some logical bearing on the target association process, which can be defined according to the following four conditions:

(1) When $n_1 + n_1 \leq 1$, the relative position relationships cannot be matched because no relative position relationships can be obtained, and UAV target data cannot be associated with the targets in the SAR data.

(2) When $n_1 = 1$ and $n_2 \geq 1$, we first calculate the correlation similarity S between the apparent features of all targets detected by UAV 1 and UAV 2. If $S > \alpha$ for the pair of targets detected by UAV 1 and UAV 2 with the largest value of S, the targets are assumed to match, and the category of the target is set according to the category with the greatest confidence among the two detection results. Then, the relative position relationships of the targets detected by UAV 2 are compared with the corresponding relative position relationships in the SAR data according to the following method, where it is assumed that n2 = 2 as an example.

We set the single target obtained by UAV 1 as the target to be queried and set its geographical location as (x_1, y_1). Similarly, we set the geographical locations of the two targets obtained by UAV 2 as (x_2, y_2) and (x_3, y_3). The relative position relationships can therefore be expressed as (x_2-x_1, y_2-y_1) and (x_3-x_1, y_3-y_1). The cosine similarity between these two relative positions and the relative position corresponding to the point of the SAR data is calculated, and it is denoted herein as S. We take the two targets with the largest cosine similarity and divide it by the number of remaining targets except the target to be queried (assumed to be 2 here), that is, the cosine similarity of the target corresponding to the target to be matched, and finally obtain the cosine similarity of the target. A false detection condition is again assessed by comparing the cosine similarity of the relative position relationship between the UAV target to be matched and the corresponding SAR target data with the threshold β. If the cosine similarity is less than or equal to β, the target in the SAR data is assumed to be a false detection. Otherwise, the matching is deemed successful, and the target in the SAR data with the largest cosine similarity is taken as the final matching result with the target to be associated in the UAV data. The corresponding number and position in the SAR data and the category obtained from the matching between the UAV and the UAV are written into the final UAV target association result.

(3) When $n_1 > 1$ and $n_2 = 1$, the condition is similar to that of the second case. Again, S is calculated for all targets detected by UAV 1 and UAV 2. If $S > \alpha$ for the pair of targets detected by UAV 1 and UAV 2 with the largest value of S, the targets are assumed to match, and the category of the target is set according to the category with the greatest confidence among the two detection results. However, under this condition, we set the single target obtained by UAV 2 as the target to be queried when establishing the relative position relationships and applying the cosine similarity to determine the matching condition, as discussed for condition (2) above.

(4) When $n_1 > 1$ and $n_2 > 1$, we first calculate the association similarity between the apparent features of all targets detected by UAV 1 and UAV 2, which yields a correlation similarity matrix A_{n_1,n_2} of dimensions $n_1 \times n_2$. If the largest element in A_{n_1,n_2} (i.e., $A_{i,j} = \max\{A_{i,*}\}$) is greater than a threshold γ (set as 0.8 herein), the two targets are assumed to match, and the category of the target is set according to the category with the greatest confidence among the two detection results.

However, a condition of $A_{i,j} \leq \gamma$ requires additional processing. First, if $A_{i,j} < \delta$ (set as 0.5 herein), the matching of the two targets is assumed to have failed, and the coordinate points are discarded. Otherwise, if $\delta \geq A_{i,j} < \gamma$, additional verification is needed in combination with the relative position relationships between the UAV targets, where a single target obtained by either UAV 1 or UAV 2 is selected as the target to be queried depending on whether $n_1 \geq n_2$ (select the UAV 1 target) or $n_1 < n_2$ (select the UAV 2 target). We then calculate the cosine similarity between the relative position vector obtained between the selected target and the other targets of the two UAVs, and two targets are assumed to match when the largest cosine similarity is greater than a threshold ε (set as 0.9 herein).

Algorithm 2 Multi-target association algorithm for multiple UAVs

Input: multi-target association features obtained from the association feature extraction network and corresponding SAR data
Set n_1 = number of targets detected by UAV 1
Set n_2 = number of targets detected by UAV 2
If $n_1 \leq 1$ and $n_2 \leq 1$
No positional relationship matching possible
If $n_1 = 1$ and $n_2 > 1$
Set $S = \text{maximum}\{S_{\cos}$ between the apparent features of all targets detected by UAV 1 and UAV 2$\}$
If $S > \alpha$, matching successful
    Set the target category according to the category with the greatest confidence among the two detection results
    Match the relative positions of UAV 2 targets with the corresponding targets in the SAR data one by one
Else
    No action taken
If $n_1 > 1$ and $n_2 = 1$
$S = \text{maximum}\{S_{\cos}$ between the apparent features of all targets detected by UAV 1 and UAV 2$\}$
If $S > \alpha$, matching successful
    Set the target category according to the category with the greatest confidence among the two detection results
    Match the relative positions of UAV 1 targets with the corresponding targets in the SAR data one by one
Else
    No action taken
If $n_1 > 1$ and $n_2 > 1$
Set $\mathbf{A}_{n_1,n_2} = S_{\cos}$ between the apparent features of all targets detected by UAV 1 and UAV 2
$A_{i,j} = \max\{A_{i,*}\}$
If $A_{i,j} > \gamma$
    Set the target category according to the category with the greatest confidence among the two detection results
If $A_{i,j} < \delta$, matching has failed
    Discard all target information
If $\delta < A_{i,j} < \gamma$
    Verification based on the value of $S_{\cos}$ calculated for the relative position vector obtained between the selected target and the other targets of the two UAVs
If $\max\{S_{\cos}\} > \varepsilon$
    Two targets are matched. Set the target category according to the category with the greatest confidence among the two detection results

4. Experiments

4.1. Training the Model

During the training phase, the model's loss functions include Box Loss, Objectness Loss, and Classification Loss. The curves depicting the changes in training loss and evaluation metrics during the training process are shown in Figure 5. It can be observed that with an increase in the number of training iterations, various losses on the training set exhibit a decreasing trend, gradually converging. The precision and recall on the validation set show upward trends with the number of training iterations, ultimately reaching stability.

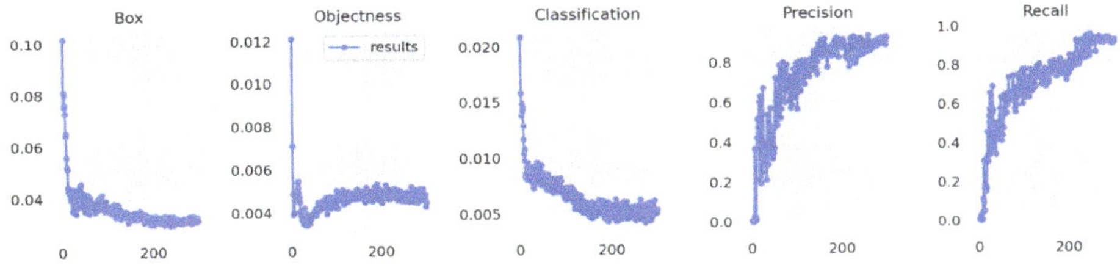

Figure 5. Model's loss functions.

4.2. Datasets

Few public datasets are available specifically for the multi-target association task involving multiple UAVs with vehicle targets. Therefore, we constructed an appropriate dataset for this task and made it public at http://103.30.41.98/ (accessed on 31 January 2024) The dataset displays results of target imaging by UAVs from various flight altitudes and perspectives. The spaceborne Synthetic Aperture Radar (SAR) images along with the corresponding object detection results are shown in Figure 6. The dataset is composed of 400 multi-target images in the training dataset and 200 multi-target images in the testing dataset. Each image in the dataset includes three to five targets. The images were modeled and rendered using the Blender 3D modeling tool to ensure that their characteristics are similar to what would be obtained under actual working conditions. We manually added FP and FN target detections to the training and testing datasets to facilitate a quantitative analysis of the target matching performance. An example of these conditions is illustrated in Figure 7. Here, four cases are considered with different FP and FN target detection rates for UAV 1 and UAV 2, which are presented in Table 1. The false positive detection rate in the table is based on a manually set random detection box, which can be generated via large-scale random generation employing large-range fluctuations in detection box parameters and small-scale random generation employing small-range fluctuations in detection box parameters according to the parameters listed in Table 2, where the settings of the fluctuation ranges are defined in Table 3. Modifications were made in the saved document in the target detection section to obtain the random detection box. The corresponding SAR data are also publicly available at https://github.com/TimeToLive404/sarsata.git (accessed on 31 January 2024).

Figure 6. SAR images and the corresponding object detection results.

Table 1. Cases of false positive (FP) and false negative (FN) detection rates considered.

Case	UAV 1		UAV 2	
	FP Rate	**FN Rate**	**FP Rate**	**FN Rate**
1	24.00	0.00	25.17	0.00
2	24.00	16.67	25.17	16.67
3	33.50	0.00	31.33	0.00
4	33.50	16.67	31.33	16.67

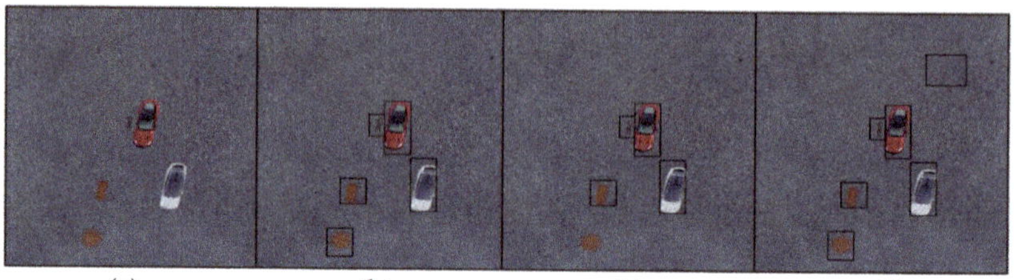

Figure 7. Illustrations of associated target data: (**a**) original multi-objective image data; (**b**) correct multi-target recognition data; (**c**) incorrect multi-target recognition data with a single FN detection; (**d**) incorrect multi-target recognition data with a single FP detection. (The suqre means the target has been detected).

Table 2. Random distribution of false positive detection frames in the four cases.

Case	UAV 1		UAV 2	
	Large-Range Random	**Small-Range Random**	**Large-Range Random**	**Small-Range Random**
1	30.54	69.46	33.77	66.23
2	30.54	69.46	33.77	16.67
3	50.24	49.76	30.54	49.76
4	50.24	49.76	50.24	49.76

Table 3. Random fluctuation range of false positive detection frames.

	Category	**Abscissa Fluctuation of Center Point**	**Longitudinal Coordinate Fluctuation of Center Point**	**Width Fluctuation**	**Height Fluctuation**
Large fluctuation	0–4	10–300	10–100	5–40	5–40
Small fluctuation	3	20–23	100–105	35–40	35–40

Considering the process of target recognition by UAVs at different altitudes and perspectives, variations in sensor errors for target detection and localization are observed, with both large- and small-scale distributions. We separately established models for the distribution of target errors, incorporating factors such as target type, the pixel center position of the target in the image, and the length and width of the target in the image pixels. By constructing this model for sensor-based target detection and localization, we can accurately describe the high-dynamic mapping relationship between UAV clusters and targets.

4.3. Experimental Conditions

All experiments were conducted on a GeForce RTX 3090 graphic card using C++ and python. The association performance of the proposed algorithm was compared with those obtained using currently available advanced MobileNetV3 and ShuffleNetV2 network based on the accuracy rate calculated as $P = \frac{TP}{TP+FP}$ and the recall rate calculated as $R = \frac{TP}{TP+FN}$, where TP is the true positive rate defining the proportion of correctly identified targets [32,33]. We applied the mean average precision (mAP) based on the calculated P and R values, and rank-1, rank-5, and rank-10 as accuracy evaluation metrics. The hyperparameters of the three experimental models were set as follows: using cross entropy loss, the Adam optimizer, a learning rate of 3.5×10^{-4}, a weight attenuation of 5.0×10^{-4}, and 100 training epochs.

4.4. Experimental Results

A display of heatmaps for target features can illustrate that the feature extractor obtains different points of interest under different perspectives. After visualizing the features of a cluster of aircraft from multiple angles (Table 4), this paper can demonstrate this viewpoint.

Table 4. Visualizing the features of a cluster of aircraft from multiple angles (The color represents the confidence level in the image where the target is located, and the darker the color, the higher the detection probability).

Bus	Box Truck	Pickup Truck	Van

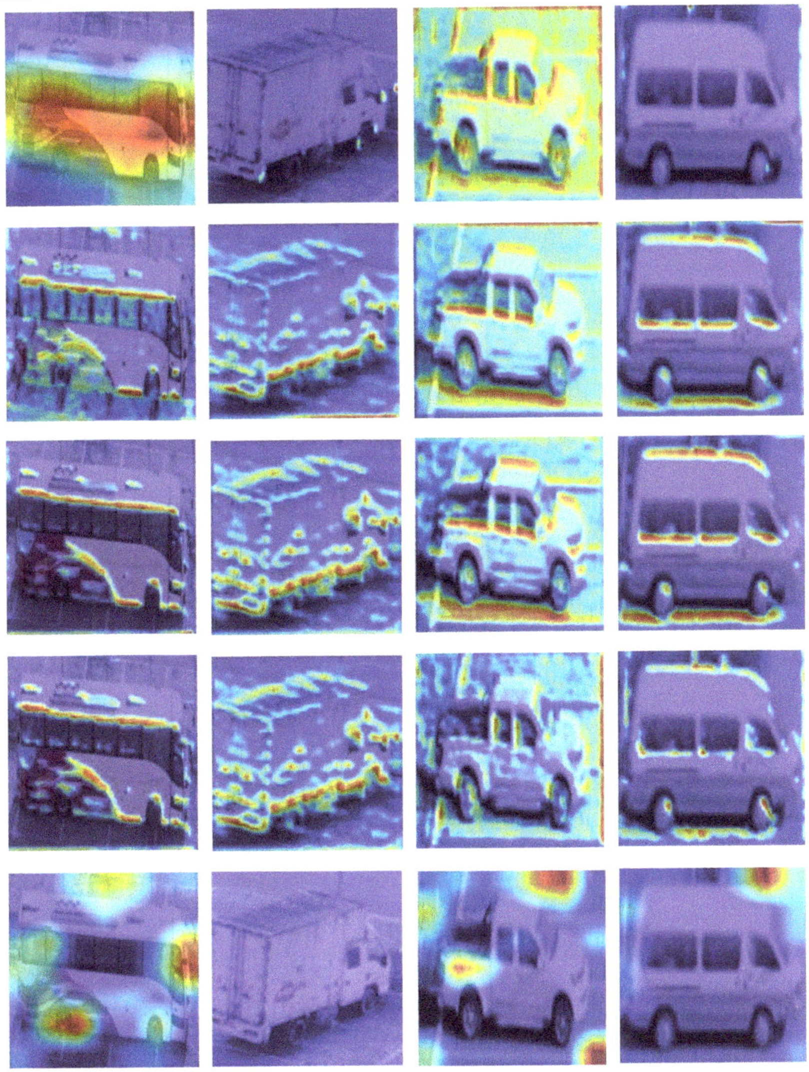

Table 4. *Cont.*

Bus	Box Truck	Pickup Truck	Van

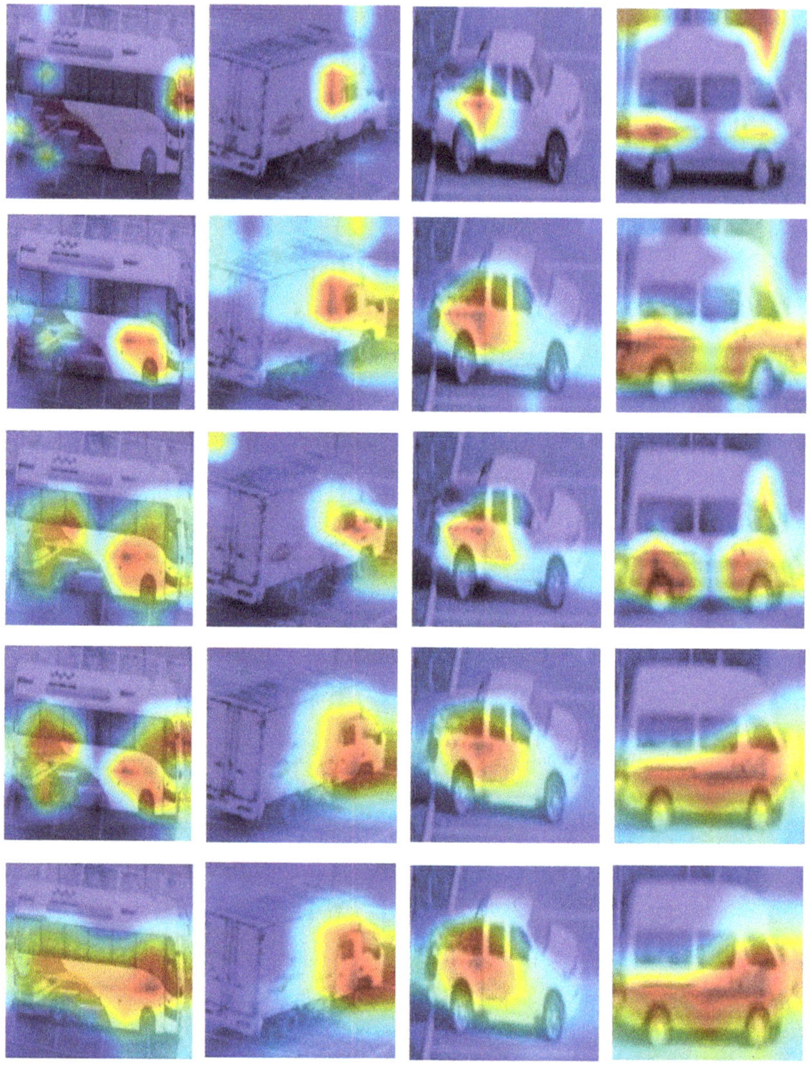

The performance metric results obtained by the three models considered for the testing dataset are listed in Table 5. As can be seen, the proposed algorithm exhibited a substantially improved matching performance over the two existing networks considered. Specifically, the mAP value obtained by the proposed algorithm was 57% and 96% greater than those obtained by the MobileNetV3 and ShuffleNetV2 networks, respectively. Moreover, the rank-1 accuracy obtained by the proposed algorithm was 3.5 times greater than those obtained by the existing networks, while the rank-5 and rank-10 accuracy values of the proposed network were 1.5 to 2 times greater than those obtained by the existing networks. Overall, the FP and FN rates obtained by the proposed algorithm were 16.67% and 0%, respectively. In fact, it is difficult to filter out the manually added recognition noise of which the position and size parameters fluctuate in a small range. In practice, this situation occurs only when the accuracy of the target recognition algorithm itself is not high. Therefore, we can conclude that the algorithm proposed herein effectively solves the multi-target association task for multiple UAVs. Finally, we note that the computation times required by all three algorithms were of the same order of magnitude, commensurate with the real-time computational performance. Accordingly, the algorithm proposed in this paper also offers a good real-time target association performance.

Table 5. Comparison of test results.

Model	mAP	Rank-1	Rank-5	Rank-10
Proposed	0.384	0.609	0.74	0.87
MobileNetV3	0.245	0.174	0.348	0.566
ShuffleNetV2	0.196	0.174	0.304	0.435

5. Conclusions

This paper addressed the issues associated with the high rate of FP and FN target detections obtained during the multi-target association of image data captured by multiple UAVs by proposing a high-performance multi-target matching algorithm. Two different networks and corresponding algorithms were established for (1) extracting the features of targets observed in the image data of individual UAVs over time and for (2) extracting features pertaining to associations between the targets extracted by multiple UAVs in space. The first process reduces the occurrence of FP and FN target detections, which greatly facilitates the subsequent association process. The proposed algorithm was demonstrated to provide a substantially improved association performance for vehicle targets compared with those obtained by the existing MobileNetV3 and ShuffleNetV2 networks in conjunction with a specially developed publicly available dataset comprising three to five targets in 400 multi-target images and 200 multi-target images in the training and testing datasets, respectively. The high association performance and computational performance of the proposed algorithm demonstrate its effectiveness and practicability for coordinating multiple UAVs in the identification of multiple targets.

Author Contributions: Data curation, H.Z.; Formal Analysis, H.Z.; Writing—Original Draft, H.Z.; Conceptualization, J.Z.; Resources, J.Z.; Investigation, H.Z., J.Z., and C.S.; Supervision, J.Z. and H.Z.; Project Administration, C.S.; Writing—Review and Editing, H.Z. and C.S. All authors have read and agreed to the published version of the manuscript.

Funding: This research received no external funding.

Data Availability Statement: 3rd Party Data Restrictions apply to the availability of these data. Data were obtained from Northwestern Polytechnical University and are available from the authors with the permission of Northwestern Polytechnical University.

Acknowledgments: We would like to thank the School of Software, Northwestern Polytechnical University and Science and Technology on Complex System Control and Intelligent Agent Cooperation Laboratory. All authors have read and agreed to the published version of the manuscript.

Conflicts of Interest: The authors declare no conflicts of interest.

References

1. Grocholsky, B.; Keller, J.; Kumar, V.; Pappas, G. Cooperative air and ground surveillance. *IEEE Robot. Autom. Mag.* **2006**, *13*, 16–25. [CrossRef]
2. Sinha, A.; Kirubarajan, T.; Bar-Shalom, Y. Autonomous surveillance by multiple cooperative UAVs. *Signal Data Process. Small Targets* **2005**, *2005*, 616–627.
3. Capitan, J.; Merino, L.; Ollero, A. Cooperative decision-making under uncertainties for multi-target surveillance with multiples UAVs. *J. Intell. Robot. Syst.* **2016**, *84*, 371–386. [CrossRef]
4. Oh, H.; Kim, S.; Shin, H.; Tsourdos, A. Coordinated standoff tracking of moving target groups using multiple UAVs. *IEEE Trans. Aerosp. Electron. Syst.* **2015**, *51*, 1501–1514. [CrossRef]

5. Ragi, S.; Chong, E.K.P. Decentralized guidance control of UAVs with explicit optimization of communication. *J. Intell. Robot Syst.* **2014**, *73*, 811–822. [CrossRef]
6. Jilkov, V.P.; Rong Li, X.; DelBalzo, D. Best combination of multiple objectives for UAV search & track path optimization. In Proceedings of the 2007 10th International Conference on Information Fusion, Québec, QC, Canada, 9–12 July 2007; pp. 1–8.
7. Pitre, R.R.; Li, X.R.; Delbalzo, R. UAV route planning for joint search and track missions—An information-value approach. *IEEE Trans. Aerosp. Electron. Syst.* **2012**, *48*, 2551–2565. [CrossRef]
8. Ousingsawat, J.; Campbell, M.E. Optimal cooperative reconnaissance using multiple vehicles. *J. Guid. Control. Dyn.* **2007**, *30*, 122–132. [CrossRef]
9. Hoffmann, G.; Waslander, S.; Tomlin, C. Distributed cooperative search using information-theoretic costs for particle filters, with quadrotor applications. In Proceedings of the AIAA Guidance, Navigation, and Control Conference and Exhibit, Keystone, CO, USA, 21–24 August 2006; p. 6576.
10. Hoffmann, G.M.; Tomlin, C.J. Mobile sensor network control using mutual information methods and particle filters. *IEEE Trans. Autom. Control* **2010**, *55*, 32–47. [CrossRef]
11. Sinha, A.; Kirubarajan, T.; Bar-Shalom, Y. Autonomous ground target tracking by multiple cooperative UAVs. In Proceedings of the 2005 IEEE Aerospace Conference, Big Sky, MT, USA, 5–12 March 2005; pp. 1–9.
12. Redmon, J.; Divvala, S.; Girshick, R.; Farhadi, A. You only look once: Unified, real-time object detection. In Proceedings of the 2016 IEEE Conference on Computer Vision and Pattern Recognition (CVPR), Las Vegas, NV, USA, 27–30 June 2016; pp. 779–788.
13. Redmon, J.; Farhadi, A. YOLO9000: Better, faster, stronger. In Proceedings of the 2017 IEEE Conference on Computer Vision and Pattern Recognition (CVPR), Honolulu, HI, USA, 21–26 July 2017; pp. 6517–6525.
14. Redmon, J.; Farhadi, A. YOLOv3: An incremental improvement. *arXiv* **2018**, arXiv:1804.02767. [CrossRef]
15. Liu, W.; Anguelov, D.; Erhan, D.; Szegedy, C.; Reed, S.; Fu, C.-Y.; Berg, A.C. SSD: Single shot MultiBox detector. In *Computer Vision—ECCV 2016*; Springer: Cham, Switzerland, 2016; pp. 21–37.
16. Lin, T.-Y.; Dollár, P.; Girshick, R.; He, K.; Hariharan, B.; Belongie, S. Feature pyramid networks for object detection. In Proceedings of the 2017 IEEE Conference on Computer Vision and Pattern Recognition (CVPR), Honolulu, HI, USA, 21–26 July 2017; pp. 936–944.
17. Law, H.; Deng, J. CornerNet: Detecting objects as paired keypoints. *Int. J. Comput. Vis.* **2020**, *128*, 642–656. [CrossRef]
18. Zhou, X.; Wang, D.; Krähenbühl, P. Objects as points. *arXiv* **2019**, arXiv:1904.07850. [CrossRef]
19. Tian, Z.; Shen, C.; Chen, H.; He, T. FCOS: Fully convolutional one-stage object detection. In Proceedings of the 2019 IEEE/CVF International Conference on Computer Vision (ICCV), Seoul, Republic of Korea, 27 October–2 November 2019; pp. 9626–9635.
20. Liu, F.; Li, Y. NanoDet ship detection method based on visual saliency in SAR remote sensing images. *J. Radars* **2021**, *10*, 885–894. [CrossRef]
21. Zhou, X.; Zhuo, J.; Krähenbühl, P. Bottom-up object detection by grouping extreme and center points. In Proceedings of the 2019 IEEE/CVF Conference on Computer Vision and Pattern Recognition (CVPR), Long Beach, CA, USA, 15–20 June 2019; pp. 850–859.
22. Liu, Z.; Zheng, T.; Xu, G.; Yang, Z.; Liu, H.; Cai, D. Training-time-friendly network for real-time object detection. In Proceedings of the AAAI Conference on Artificial Intelligence, New York, NY, USA, 7–12 February 2020; Volume 34, pp. 11685–11692. [CrossRef]
23. Arani, E.; Gowda, S.; Mukherjee, R.; Magdy, O.; Kathiresan, S.; Zonooz, B. A comprehensive study of real-time object detection networks across multiple domains: A survey. *arXiv* **2023**, arXiv:2208.10895. [CrossRef]
24. Girshick, R.; Donahue, J.; Darrell, T.; Malik, J. Rich feature hierarchies for accurate object detection and semantic segmentation. In Proceedings of the 2014 IEEE Conference on Computer Vision and Pattern Recognition, Columbus, OH, USA, 23–28 June 2014; pp. 580–587.
25. Wang, X.; Shrivastava, A.; Gupta, A. A-fast-RCNN: Hard positive generation via adversary for object detection. In Proceedings of the 2017 IEEE Conference on Computer Vision and Pattern Recognition (CVPR), Honolulu, HI, USA, 21–26 July 2017; pp. 2606–2615.
26. Qin, Z.; Li, Z.; Zhang, Z.; Bao, Y.; Yu, G.; Peng, Y.; Sun, J. ThunderNet: Towards real-time generic object detection on mobile devices. In Proceedings of the 2019 IEEE/CVF International Conference on Computer Vision (ICCV), Seoul, Republic of Korea, 27 October–2 November 2019; pp. 6717–6726.
27. Ristani, E.; Tomasi, C. Features for Multi-target Multi-camera Tracking and Re-identification. In Proceedings of the 2018 IEEE/CVF Conference on Computer Vision and Pattern Recognition, Salt Lake City, UT, USA, 18–23 June 2018; pp. 6036–6046.
28. Ristani, E.; Solera, F.; Zou, R.; Cucchiara, R.; Tomasi, C. Performance measures and a data set for multi-target, multi-camera tracking. In *Computer Vision—ECCV 2016 Workshops*; Springer: Cham, Switzerland, 2016; pp. 17–35.
29. Tesfaye, Y.T.; Zemene, E.; Prati, A.; Pelillo, M.; Shah, M. Multi-target tracking in multiple non-overlapping cameras using fast-constrained dominant sets. *Int. J. Comput. Vis.* **2019**, *127*, 1303–1320. [CrossRef]
30. Hou, Y.; Zheng, L.; Wang, Z.; Wang, S. Locality aware appearance metric for multi-target multi-camera tracking. *arXiv* **2019**, arXiv:1911.12037. [CrossRef]
31. Bochkovskiy, A.; Wang, C.-Y.; Liao, H.-Y.M. YOLOv4: Optimal speed and accuracy of object detection. *arXiv* **2020**, arXiv:2004.10934. [CrossRef]

32. Howard, A.; Sandler, M.; Chen, B.; Wang, W.; Chen, L.-C.; Tan, M.; Chu, G.; Vasudevan, V.; Zhu, Y.; Pang, R.; et al. Searching for MobileNetV3. In Proceedings of the 2019 IEEE/CVF International Conference on Computer Vision (ICCV), Seoul, Republic of Korea, 27 October–2 November 2019; pp. 1314–1324.
33. Ma, N.; Zhang, X.; Zheng, H.-T.; Sun, J. ShuffleNet v2: Practical guidelines for efficient CNN architecture design. In *Computer Vision—ECCV 2018*; Springer: Cham, Switzerland, 2018; pp. 122–138.

MDPI AG
Grosspeteranlage 5
4052 Basel
Switzerland
Tel.: +41 61 683 77 34

Drones Editorial Office
E-mail: drones@mdpi.com
www.mdpi.com/journal/drones

www.ingramcontent.com/pod-product-compliance
Lightning Source LLC
Chambersburg PA
CBHW061848210926
51850CB00006B/1002
* 9 7 8 3 7 2 5 8 2 2 2 3 2 *